Grapes

CROP PRODUCTION SCIENCE IN HORTICULTURE SERIES

This series examines economically important horticultural crops selected from the major production systems in temperate, subtropical and tropical climatic areas. Systems represented range from open field and plantation sites to protected plastic and glass houses, growing rooms and laboratories. Emphasis is placed on the scientific principles underlying crop production practices rather than on providing empirical recipes for uncritical acceptance. Scientific understanding provides the key to both reasoned choice of practice and the solution of future problems.

Students and staff at universities and colleges throughout the world involved in courses in horticulture, as well as in agriculture, plant science, food science and applied biology at degree, diploma or certificate level will welcome this series as a succinct and readable source of information. The books will also be invaluable to progressive growers, advisers and end-product users requiring an authoritative, but brief, scientific introduction to particular crops or systems. Keen gardeners wishing to understand the scientific basis of recommended practices will also find the series very useful.

The authors are all internationally renowned experts with extensive experience of their subjects. Each volume follows a common format covering all aspects of production, from background physiology and breeding, to propagation and planting, through husbandry and crop protection, to harvesting, handling and storage. Selective references are included to direct the reader to further information on specific topics.

Titles Available:

1. **Ornamental Bulbs, Corms and Tubers** A.R. Rees
2. **Citrus** F.S. Davies and L.G. Albrigo
3. **Onions and Other Vegetable Alliums** J.L. Brewster
4. **Ornamental Bedding Plants** A.M. Armitage
5. **Bananas and Plantains** J.C. Robinson
6. **Cucurbits** R.W. Robinson and D.S. Decker-Walters
7. **Tropical Fruits** H.Y. Nakasone and R.E. Paull
8. **Coffee, Cocoa and Tea** K.C. Willson
9. **Lettuce, Endive and Chicory** E.J. Ryder
10. **Carrots and Related Vegetable** *Umbelliferae* V.E. Rubatzky, C.F. Quiros and P.W. Simon
11. **Strawberries** J.F. Hancock
12. **Peppers: Vegetable and Spice Capsicums** P.W. Bosland and E.J. Votava
13. **Tomatoes** E. Heuvelink
14. **Vegetable Brassicas and Related Crucifers** G. Dixon
15. **Onions and Other Vegetable Alliums, 2nd Edition** J.L. Brewster
16. **Grapes** G.L. Creasy and L.L. Creasy
17. **Tropical Root and Tuber Crops: Cassava, Sweet Potato, Yams and Aroids** V. Lebot

18. **Olives** I. Therios
19. **Bananas and Plantains, 2nd Edition** J.C. Robinson and V. Galán Saúco
20. **Tropical Fruits, 2nd Edition, Volume 1** R.E. Paull and O. Duarte
21. **Blueberries** J. Retamales and J.F. Hancock
22. **Peppers: Vegetable and Spice Capsicums, 2nd Edition** P.W. Bosland and E.J. Votava
23. **Raspberries** R.C. Funt
24. **Tropical Fruits, 2nd Edition, Volume 2** R.E. Paull and O. Duarte
25. **Peas and Beans** A. Biddle
26. **Blackberries and Their Hybrids** H.K. Hall and R.C. Funt
27. **Tomatoes, 2nd Edition** E. Heuvelink
28. **Grapes, 2nd Edition** G.L. Creasy and L.L. Creasy
29. **Blueberries, 2nd Edition** J. Retamales and J.F. Hancock
30. **Citrus, 2nd Edition** L.G. Albrigo, L.L. Stelinski and L.W. Timmer
31. **Tropical Root and Tuber Crops: Cassava, Sweet Potato, Yams and Aroids, 2nd Edition** V. Lebot
32. **Cucurbits, 2nd Edition** Todd C. Wehner, Rachel Naegele, James Myers, Kevin Crosby and Narinder P. S. Dhillon
33. **Carrots and Related Apiaceae Crops** Emmanuel Geoffriau and Philip W. Simon
34. **Strawberries** J.F. Hancock
35. **Sweet Cherries** Lynn E. Long, Gregory A. Lang and Clive Kaiser
36. **Cut Flowers and Foliages** James E. Faust and John M. Dole
37. **Mushrooms:** ***Agaricus bisporus*** Youssef Najib Sassine
38. **Date Palm** Jameel M. Al-Khayri, Shri Mohan Jain, Dennis V. Johnson and Robert R. Krueger
39. **Peach** George Manganaris, Guglielmo Costa and Carlos H. Crisosto
40. **Vegetable Brassicas and Related Crucifers** Geoffrey R. Dixon and Rachel Wells
41. **Tropical Fruits, 3rd Edition, Volume 1** R.E. Paull and O. Duarte
42. **Grapes, 3rd Edition** G.L. Creasy and L.L. Creasy

Grapes

3rd edition

Glen L. Creasy
Viticulturist
Sabrosia Winegrowing Services
Montpellier, France

and

Leroy L. Creasy†

CABI

CABI is a trading name of CAB International

CABI
Nosworthy Way
Wallingford
Oxfordshire OX10 8DE
UK

Tel: +44 (0)1491 832111
E-mail: info@cabi.org
Website: www.cabi.org

CABI
200 Portland Street
Boston
MA 02114
USA

Tel: +1 (617)682-9015
E-mail: cabi-nao@cabi.org

A catalogue record for this book is available from the British Library, London, UK.

ISBN-13: 9781800627048 (paperback)
9781800627055 (ePDF)
9781800627062 (ePub)

DOI: 10.1079/9781800627062.0000

Commissioning Editor: Rebecca Stubbs
Editorial Assistant: Emma McCann
Production Editor: James Bishop

Typeset by Straive, Pondicherry, India
Printed and bound in the UK by CPI Group (UK) Ltd, Croydon, CR0 4YY

Contents

Foreword

I was honoured to be invited to write this Foreword to a fine textbook on viticulture. My association with the father and son authors goes back to my days as a graduate student at Cornell University in the early 1970s. My friend Bob Pool and I had morning coffee in Le Creasy's lab. Since then I have watched and admired the progress of his son Glen in his studies of viticulture.

This Crop Production Series of economically important horticultural crops is substantial in its scope, and it is appropriate that the grapevine, the world's largest fruit crop is included. It is also noteworthy that the book is being revised so soon after the first publication.

Creasy and Creasy is the 'go to' reference on my bookshelf when I am searching for thoughts on current production systems. This is as designed, that the book should provide sufficient scientific background in subjects like physiology to explain modern methods. Given the authors' experience, there is good awareness of viticultural practices in different climate zones.

The range of subjects covered is comprehensive, from the early origins of grapevine cultivation to the management of modern problems like grapevine trunk disease.

I trust readers will find as much benefit and enjoyment with this updated volume as I found with the first.

Dr Richard Smart, 'the flying vine-doctor'
Cornwall, UK
www.smartvit.com.au

Preface

Amazingly, it has now been 17 years since the first edition of this book went to the publishers. Also amazing is the amount of change since then! So why not let's go back to the preface from the first edition and ask how we did with our predictions? Well...

- China is dominating global grape production, but not in wine - it's table grapes that they've gone into in a big way. And even then, the amount they export is minuscule in comparison with total production. Will they enter into the global market with table grapes?
- Where grapes are being grown commercially has also expanded, but not just because of increasing knowledge and technology: climate change has made its presence known in a big way. Due to rising temperatures there are vineyards around 60° latitude in Scandinavia, and new plantings in England, Denmark, and Belgium to name just a few in the northern hemisphere. However, this is not without downsides: seasonal variation in weather have seen years with severe disease pressure, damaging spring frosts, more flooding, and some areas getting too dry and hot to support commercial grape production. The future of grape growing for those areas is not looking so good.
- Vineyard mechanization has increased due to the need to save on labour, and robotics are on the cusp of becoming significant to a large segment of the industry. What changes these could make is both exciting and concerning to think about.
- Small vineyards have been successful, particularly with the "buy local" movement, and for wine producers tourism is increasingly important to add economic value to them and the entire region they're in.

There is still an oversupply of wine grapes in the world, which is not good, but on the other hand, table grape consumption is on the rise. In the next two decades we may see this shift continue, as consumers move away from wine as a lifestyle beverage.

The menacing cloud hanging over all our heads, climate change, is no longer approaching, but here and spitting rain on us right now. This is happening faster than we had hoped, and the question is still out on how bad the storm will get before it's done. As well, we found out that pandemics weren't just a scary story told by scientists, they're a real thing and can wreak global havoc with our lives, business and trade. Odds are, it won't be the last one we experience, so our operations need to be strong and resilient.

Challenges are nothing new to viticulturists: the specifics may change, but still, the best way to meet them is with determination and knowledge - you, dear reader, can supply the former, and hopefully, this book can help with the latter.

This book is dedicated to the memory of my parents, Leroy L Creasy and Min T Creasy, both of whom have passed since the last edition.

April 2025

Preface to the Second Editon

Nine years on, and here we are, with the second edition of this book. In nine years, how much has changed in the world of grapes and their culture? Has our understanding increased such that our jobs in growing grapes has gotten easier? Have there been technological advances that have changed the way we manage the vines? Well, (inevitably) it seems yes and no.

Yes, we understand the vine that little bit better, and this has allowed us to manage them to perform in the environment they need to grow in. Yes, we understand more about how rootstocks perform. Yes, we know more around how grapes ripen and develop quality parameters like aroma and colour. Yes, there are new fungicides, herbicides and disease-resistant cultivars. Yes, modern equipment is better, faster and, hopefully, less expensive. The large touch- screen smartphone was still a curiosity in 2009, and now many of us carry one in our pockets, each with amazing amounts of computing power, high- resolution cameras, GPS and other sensors, along with the ability to access the internet at high speed. And the commercial use of robots in the vineyards is tantalizingly close . . . so yes, knowledge of the grapevine and our ability to manage it is very different now to what it was nine years ago.

Despite all this, is our job easier? That's a harder question to answer. In the same time frame, the market demands better quality yet cheaper products. We are growing grapes in more challenging locations. New diseases have been described or spread to new areas. Pesticides that used to work are no longer effective and there are fewer replacements in the pipeline. And we are all a bit older and more tired!

So, the challenges remain, but that is as it should be. These encourage us to never stop learning, never stop asking new questions and never stop experimenting with ways to meet those challenges.

In the first edition we said that the book is a snapshot in time and this is still true. However, we do think that this is a good platform from which to start, or help you along, a journey of learning about the grape and its culture.

We were honoured by the reception the first edition received, and that it was felt worthy of revision. We hope you find this edition of value in increasing your knowledge, and enjoyment, of the grape and its products.

November 2017

Preface to the First Editon

This book targets advanced plant science students with specific interest in viticulture and those who are producing grapes for the first time. We think it will be of interest to enthusiasts of the vine and its products who want to learn a bit more about how grapes are grown and a lot about the whys behind the methods. There are answers to many questions in the following chapters but, as always seems to be the case, the more you learn the more you discover that there is always more to learn!

Grapes have so many uses and are so unique that no fruit can challenge their superiority. The production of grapes continues to grow: China now pro- duces more table grapes than the next nine major producing countries combined. It seems inevitable that China will also dominate wine grape production at some point, which makes life in the industry terribly interesting and exciting!

In fact, methods of grape production and how the fruit is utilized are changing rapidly. For example, the advent of year-round supplies of table grapes, facilitated through economic means of shipping between the northern and southern hemispheres, has bolstered demand for the product, leading to more innovation and investment in the industry and improvement in our ability to grow and market them. Grapes are being produced in many places where it was not considered possible before, largely due to advances in understanding of how the vine works, as well as to improvements in technology and management skills.

Labour is a traditional limiting factor in grape production, but progress is being made to further mechanize production and harvesting. Mechanical harvesters have become the standard for wine and juice grapes and have even changed the way vines are trained because hand harvesting benefits from large clusters in well-defined areas, while for mechanical harvesters it doesn't matter. Improvements in methods of pest control, with new application technologies and improved chemicals incurring reduced environmental impact, also help the bottom line.

Bigger machines and more complex equipment would suggest a trend to larger producers and the decline of small producers. This has not been the case in the wine industry, at least, as the number of small wineries has increased greatly over the past 25 years, and has become the basis for significant tourism industries. This helps to avoid wine becoming a commodity, and therefore engenders curiosity and enthusiasm by producers and consumers alike.

Many small vineyards are being started with great enthusiasm by people with limited experience in grape growing. We hope this book, with its mix of theory and science, will be useful to them when they are pioneering their own path into the world of the vine.

The grapevine is a plastic and adaptable plant. It will grow in the most unlikely of places, and can be shaped into a myriad of different forms in our quest to extract the highest quality product from these. The fruit of this vine is made into many different products, of varying tastes and aromas. We like grapes and all the things that are made from them. Le retired early from academic life to grow grapes, and knows every day that it was the right thing to do. Glen became interested in grapes at about the same time and now lives, breathes and, importantly, drinks, around the subject area.

LLC: I would like to thank the many individuals who have supported my research on grapes and encouraged me to continue this research to the present. Special thanks go to Bill Wagner and his excellent staff at Wagner Vineyards for making available his vines for research. Special thanks to Bruno and Marcello Ceretto of Alba, Italy for introducing me to the best wine and food in this world.

GLC: I wish to acknowledge the large number of people who have contributed to my experiences with all things grape – growers, students, scientists, teachers, winemakers and all the rest that have guided me to where I am today. With reference to this work, I thank my wife Kirsten and kids Bella and Maddy for putting up with the process of writing a book, Le for the chance to write with you and my mom, Min Creasy, for handling the index. I feel truly lucky to have so many supportive people around me!

A few parting words. A book is frozen in time. It is written and endures (or not). Grape growing and viticulture, however, change daily. A book is intended to bring readers up to a specific moment in time and to stimulate them to continue seeking more information. Principles never change but their applications change greatly, so may the quest for knowledge never stop!

June 2008

History, Uses and Production 1

Abstract: The grapevine is a popular and highly adaptable plant whose fruit is used for many different products. It evolved in many temperate areas of the globe, but the primary species of interest, *Vitis vinifera*, originated in southern Caucasia. There are about 7.1 million hectares under cultivation, producing 80 million tonnes of fruit that is used widely for fresh consumption, drying, as juice or jam or fermented into wine.

The grapevine (*Vitis* spp.) is cultivated all over the world and the grape itself is used for a myriad of products, many of which are well known to all of us. Viticulture is one of the major horticultural industries of the world, with the area of grapevines cultivated exceeding 7.1 million ha (OIV, 2025a). Most grapes are grown for the production of wine but, when first discovered, its appeal as fresh eating fruit was probably what attracted the first hunter-gatherers. Today the fruit is used in a wide variety of products, ranging through fresh fruit, preserves, juice, wine, raisins and extracts.

The grapevine is a vigorously growing plant and in some places is considered an invasive weed (Uva *et al.*, 1997). Fortunately, the fact that the grape is used for a great many products means that it can be considered more of a crop plant than a weed. That the grapevine is a climbing plant lends it an unusual plasticity of form. The viticulturist can manipulate it in many ways and change the manner in which it is trained, almost yearly if desired. Few perennial plants have this kind of flexibility, which forms part of its fascination as a food crop.

Geographical Origins of Grapevine Species

Grapevines have evolved in several different areas of the world, leading to a great many different species developing. The origin of cultivation of the *V. vinifera* grape, now planted throughout the world, is probably in southern Caucasia, now occupied by north-west Turkey, northern Iraq, Azerbaijan and Georgia (Mullins *et al.*, 1992). At first, grapes (*V. vinifera* spp. *sativa*) were probably

DOI: 10.1079/9781800627062.0001

gathered from the wild, with the vines growing up into the trees. The association of grapes with oak, now used in the winemaking process in the form of barrels in which wine is aged, may have begun with the vine using oak trees as support, since *Saccharomyces cerevisiae* (or the winemaking yeast) strains have been isolated from oak trees (Sniegowski *et al.*, 2002). The people living in these areas in ancient times discovered the utility of *V. vinifera* grapes and took the vine with them on their trading routes, to Palestine, Syria, Egypt, Mesopotamia and then to the Mediterranean. The Greeks, Phoenicians and Romans (Buxó, 2008) took to the vine readily and spread it, and methods for its cultivation, throughout Europe and as far north as Britain.

From Europe, *V. vinifera* was taken to North America, Peru and Chile, with the Dutch ensuring it travelled with them to South Africa (1616). The English packed grapevines on the First Fleet to Australia (1788) and on travels to New Zealand.

Species of grapes native to North America are numerous, having originated in many different types of environments, from moist to dry (see Fig. 1.1). The *V. labrusca* grape, native to north-eastern USA, is widely used for juice production in many US states and in South America. Other species are not so commonly used for grape production, including many native to Asia, but they do make an important contribution to the production of grapes, as will be discussed later.

Family, Genus, Species and Related Plants

Botanically, the grapevine is a liana, a climbing vine. As such, it does not invest heavily in something as solid as a tree trunk for support, rather having developed

Fig. 1.1. *Vitis riparia* grapes growing wild in the north-east of the United States. The berries are densely coloured and strong of flavour. Content author's own.

to take advantage of trunks already occurring in forests and bushlands. In the wild, the grapevine starts as an understorey plant, growing rapidly and upward, clinging to other plants to eventually reach the top of their canopies (see Fig. 1.2). Once there, it fruits heavily, producing dark-coloured berries that birds, in particular, eat, thus disseminating the seed. These fruits are not very appetizing to humans, being strong in flavour and astringency, high in acidity and relatively

Fig. 1.2. *Vitis riparia* smothering a tree in an upstate New York winter. Content author's own.

low in sugars. In some cases, wild grapes can be palatable, such as *V. amurensis*, the Amur grape, which has its origin in north-eastern Asia. This species has been useful in breeding programmes due to its cold hardiness and resistance to some diseases, but it has also been used to make still table wines.

The seedling, when it sprouts, shows off two rudimentary leaves, thus indicating that it is a dicotyledon, like most broadleaved plants (and unlike monocots, e.g. maize and other grasses). The fruit is a true berry (botanically speaking), containing the seeds within, and thus the grapevine is classified as an angiosperm.

The grapevines belong to the family Vitaceae, which are mostly woody, tree-climbing vines, though a few have a shrubby growth habit. They are characterized by tendrils and inflorescences opposite the leaves. There are 12 genera within the family including *Vitis*, *Ampelocissus*, *Clematicissus*, *Parthenocissus* (Virginia creeper), *Ampelopsis* and *Cissus* (kangaroo vine).

The genus *Vitis* is the part of Vitaceae that the grapevine industry is most interested in and it consists of two subgenera, *Vitis* and *Muscadinia*. One peculiarity of the genus is that the flower petals separate from the bottom of the flowers, not from the top as in most other plants (see Fig. 1.3), hence it is referred to as a calyptra (ancient Greek for 'veil'), or cap.

The subgenera are distinct because they have different chromosome numbers (38 for *Vitis* and 40 for *Muscadinia*) and morphological features.

Fig. 1.3. Grape cluster showing individual florets, some of which have their fused petals (calyptra) separating from the basal part of the flower (top and left). Content author's own.

Muscadinia (*V. rotundifolia*), in comparison with *Vitis*, have differences in seed shape, simple as opposed to branched tendrils, smooth bark, continuous pith inside canes (i.e. there is no interruption at the node position), fewer berries per cluster and berries that tend to abscise from the rachis (Williams 1923; Bailey, 1933; Einset and Pratt, 1975). There are three named species in this group, the most important of which is *V. rotundifolia* – discussed further in Chapter 2, V. *munsoniana* and V. *popenoei*.

Because of their different chromosome number, plants in this subgenus will not naturally interbreed with *Vitis* species. However, with persistence, some crosses were successful (Detjen, 1919) and subsequently used by other breeders, such as Alain Bouquet in France (Bouquet, 1986), whose disease-resistant vines are now beginning to be planted.

Vitis includes many species, including *V. vinifera*, the most widely planted grape species in the world, which is used primarily for wine, table consumption, juice and raisin production.

Another well-known species is *V. labrusca*, native to North America. Its advantages over *V. vinifera* are that it is much more tolerant of pests and diseases, but for many, the flavour of the fruit, described as foxy, is an acquired taste. Hence, though they are made into wine, the grapes are mostly used for production of juice and preserves, though there is also a significant, but regional, industry producing them for fresh consumption.

Regardless of how the fruit tastes, many of the non-*V. vinifera* species have been vitally important to the commercial development of *V. vinifera* cultivars, in finding a solution to the problem of phylloxera (an insect that attacks grapevine roots) and other soil-related pests and conditions (further discussed in Chapters 5 and 10).

Natural Growth Conditions

The grape is a hardy perennial plant, meaning that it can grow and survive in areas where the temperature goes well below freezing in the winter season: under the right conditions, some grape species can survive temperatures as low as –40°C (Pierquet *et al.*, 1977) and there are active breeding programmes working to develop more cold-hardy cultivars for both table and wine purposes (such as the University of Minnesota's 'Frontenac'). However, *V. vinifera* is much more tender, and generally cannot withstand temperatures below –15°C without suffering damage (Clore *et al.*, 1974). The above-ground parts of the vine develop a hardy outer covering called a periderm as the growing season ends, which is an indicator that the vine tissues are developing resistance to environmental extremes.

The grapevine evolved in temperate climates, so it grows when conditions are warm enough, but stops when temperatures fall below about 10°C. Unlike tree fruits, e.g. apple or pear, which set a terminal bud as winter approaches, grapevines will continue to grow as long as conditions permit.

The vine is highly adaptable to different environments which, in part, is why it is found growing in many and varied climates. Typically, at least in terms of commercial production, it is cultivated between the 10°C and 20°C annual global isotherms, although grapes are now grown in many areas outside these boundaries due to identification of suitable mesoclimates and increased knowledge about their cultivation and climate change. There would be few areas in the world that would be too hot for grapevines to grow, assuming that other plants can also survive there. However, obtaining a reliable commercial crop from those vines may be the most challenging aspect, especially with climate change increasing the risk of damaging weather events.

Historical Cultivation

It did not take long for people to recognize the advantages of cultivating crops rather than collecting from the wild. The Egyptians were using grapes from approximately 3000 BC, and pictures showing vines growing on structures date back to around 1500 BC (Singer *et al.*, 1954, cited in Janick, 2002). The Chinese had probably started cultivating *V. vinifera* vines by 2000 BC (Huang, 2000) and native cultivars (based on *V. amurensis*, for example) before that. The plastic nature of the vine lends itself well to manipulation and doing so with some form of trellis exploits the vine's tendency to fruit heavily. Early trellising systems appeared to be forms of arbours, rather than the more common hedge or overhead trellises seen today. However, there has been little commercial adoption of trellis systems that have been introduced over the past 100 years. Be that as it may, structures for supporting vines can and do take many different forms and are discussed in more detail in Chapter 6.

Uses

Just as the grapevine is highly adaptable to where and how it is grown, its fruit is also highly adaptable to different uses. Somewhat unique, the grape attains a high concentration of sugar when ripe, and also (depending on cultivar) pectin, as well as a wide range of aromatic compounds. These factors, in concentration with the presence of relatively high levels of acids (particularly tartaric acid), mean that the fruit is amenable to many different end uses. Almost 45% of global grape production is used for wine and related fermented products, with table grapes increasing in importance (at the expense of wine) to 45% of the total in 2023. Raisins make up 6.5%, with juice and other uses making up a bit less than 4% (OIV, 2024).

Fermented grape products

Evidence that humans were fermenting grapes with the specific purpose of making an alcoholic beverage can be traced back to around 7000 BC in China

(McGovern *et al.*, 2004), in the Near East around 6000 BC (McGovern *et al.*, 2017). There is an association between grapes and various types of yeasts (usually living on the surface of the berry) (Parish and Carroll, 1985; Martini *et al.*, 1996; Cavalieri *et al.*, 2003), so it is likely, at least initially, that fruit which had been picked and stored may have started fermenting naturally. This will have produced an alcoholic mixture that some found enjoyable, and thus people wanted to be able to repeat the process. This was possibly the beginning of winemaking.

Much of the world production of grapes ends up as wine, and it is made into a bewildering array of types and price points. Wine can be thought of as a naturally made storage form of the fruit as it retains characteristics of the grape and, protected from oxidation, can remain palatable for many years. The production of wine from grapes can be simple – with few additions postharvest – to complex, with addition of a variety of substances designed to modify its appearance, aroma and taste.

Wines can be further manipulated in the form of sweetening and the addition of additional alcohol, which produces fortified wine products such as port and sherry, which were even more stable and travelled better on long sea journeys. With distillation, wine can be transformed into products such as brandy, grappa and marc, which often carry the characteristics of the grape variety from which they were made. However, the skill of the grape grower, in producing the starting material for the beverage, is still paramount in the production of a quality wine and subsequent products, such as those made through distillation.

Table grapes

The first use of grapes by man was probably as fresh fruit, and this continues to be an important industry. Table grapes have their own important attributes, such as looking attractive, having good eating properties (e.g. flavour and berry firmness), large and consistently sized berries, a sturdy rachis and strong attachment of the berry to it. Ideally, grapes for fresh consumption should also be resistant to both injury caused by handling and postharvest diseases (discussed in more detail in Chapters 10 and 11).

Major table grape cultivars include 'Sultana' (also known as 'Sultanina', 'Thompson Seedless', 'Oval Kishmish'; Winkler *et al.*, 1974; Fig. 1.4), 'Almeria' ('Ohanez'), 'Cardinal', 'Dattier' ('Waltham Cross'), 'Emperor', 'Malaga', 'Perlette', 'Ribier' ('Alphonse Lavallée'), 'Rish Baba', 'Tokay' and 'Ruby Seedless'. New cultivars are always under development, and there is a tendency for them to be sold by their colour rather than by a cultivar name so as to create more flexibility in growing and marketing. Seedless French hybrid cultivars developed in New York (see Fig. 1.4), Arkansas and other places have more winter hardiness and shorter growing season requirements and therefore can be planted in colder climates than the major cultivars planted in California and most of Europe.

Fig. 1.4. Left, examples of 'Sultana', also known as 'Thompson Seedless'. This cluster has not been grown for commercial table grape production and so the berries are smaller than those found on clusters in a shop. Middle: 'Einset Seedless', a French-American hybrid grape. Right: 'Red Globe', a seeded cultivar. Content author's own.

Curiously, 'Zinfandel' (also known as 'Primitivo' or 'Kratošija'), a cultivar now well known as a wine grape in California, was first planted on the eastern coast of the USA as a table grape, before being taken west and used as a wine grape (Sullivan, 2003).

The table grape industry has changed several times in the last 100 years. In North America the development of the iced railroad car signalled the end of the eastern North American table grape industry, as table grapes from the western states were then available in eastern markets. Improved storage and transport methods also greatly contributed to the change, but developing markets have made a huge impact in some countries.

Globally, there has been strong and stable growth in table grape production for the past 30 years, a 2.6-fold increase since 1995. Of note is China, which grew by a massive 18 times to over 13.5 million tonnes, 42% of global production (https://www.oiv.int/what-we-do/data-discovery-report?oiv). Peru has also made a large jump, though starting from a smaller base, going from 27000 tonnes in 1995 to 770000 tonnes in 2023.

The popularity of seeded grapes, usually capable of being stored for longer periods of time, has declined in favour of fresh seedless grapes (Alston *et al.*, 1997). More information about the postharvest handling of table grapes is presented in Chapter 11.

Raisins

Raisins, or dried grapes, may have also occurred in nature as berries that had not been eaten by birds or other animals and left to dry on the vine to create a very sweet and decay-resistant product. They have been eaten probably since grapes were first found to be edible, but evidence exists to suggest that they have been created on purpose since Egyptian times (Janick, 2002). Today, they are an important facet of grape production but there are relatively few cultivars used for their production, with 'Sultana' being the most important. Grapes can be dried out in the open (see Fig. 1.5) or in forced-air heaters, the latter usually reserved for the more premium market fruit due to the extra cost.

Fig. 1.5. Raisins drying out in between the vineyard rows (photograph courtesy Matthew Fidelibus, used with permission).

Turkey, the USA and China lead the world production of raisins, contributing more than 60% of the total but with significant amounts made in Iran, South Africa, Uzbekistan and Chile (USDA, 2025c). Unlike table grapes, the increase in world-wide production of raisins in the last 30 years has been slight, with a recent downwards trend. The production of raisins is highly dependent on the weather, as the vast majority of grapes are dried in the sun. Because of this vulnerability, production of raisins is decreasing in areas where rain is more likely at the end of the season (risk increasing with climate change), and also because of fluctuations in the need for other products made from 'Sultana' grapes (e.g. for fresh market, juice or bulk wine). See Chapter 11 for further information about the production of raisins.

Grape juice

The juice of grapes is appealing to the human palate, through a combination of the sweetness, acidity and flavours (Morris, 1989). Juice can be consumed immediately or further processed (e.g. pasteurized) to create a longer-lasting product. Grapes are also used in the production of jellies and other preserves, and juice is widely used in the bottled drinks industry as a natural sweetener as well as on its own (Winkler *et al.*, 1974; Olien, 1990). However, the amount of world production used for juice and these other products is very small compared with that used for wine or table grapes.

Some cultivars, such as 'Sultana' and 'French Colombard' (another white grape), produce a neutral-flavoured juice well suited to concentration and then for use as a sweetener, whereas other juices, such as that from *V. labrusca* grapes, have a distinctive flavour all their own. Red juices can be made from heat-extracted grapes of 'Grenache'/'Garnacha' (a *V. vinifera* grape), 'Concord' (*V. labrusca*) or other cultivars. In general, however, juice from *V. vinifera* grapes does not process well and has less flavour and acidity than the *V. labrusca* grape juices, and so can be less palatable for consumers (Morris, 1989). Postharvest processing of juices is discussed in Chapter 11.

A relatively recent use of grapes is the production of health-related products, as grapes are a natural source of readily extractable phenolics that have antioxidant properties (Teissedre *et al.*, 1996; Nuttall *et al.*, 1998; Yilmaz and Toledo, 2004). In many cases, the phenolic compounds are extracted out of the material left over from primary processing of the grapes, e.g. pomace. These are quite active and can be sold as nutritional supplements or used to bolster the antioxidant capacity of foodstuffs (Bonilla *et al.*, 1999).

Grape-Producing Countries

Many of the world's countries produce at least some grapes. Table 1.1 shows production by country from 2017 to 2022. The top ten producers are responsible for approximately 70% of world production, and it is notable that mainland China, which was ranked fifth in 2005 is now the top producer by a significant margin. Note also the variation in production from vintage to vintage, usually caused by weather events.

The amount of land given over to grape production from 2017 to 2022 and the percentage change during that time are shown in Table 1.2. Declines in area are the norm for the past number of years, particularly in Chile and the USA. Though Italy and France posted some increases from 2017, in more recent years they have lost total area. There is a definite trend for the net removal of vineyards, globally.

Table 1.1. Estimated average annual grape production in millions of tonnes by country (from OIV, 2023).

Country	2017	2018	2019	2020	2021	2022
China	13.1	13.7	14.2	14.3	14.9	15.6
Italy	6.9	8.6	7.9	8.2	8.1	8.1
France	5.0	6.3	5.5	5.9	5.1	6.2
USA	6.7	6.9	6.2	5.4	6.1	6.0
Spain	5.0	6.9	5.4	6.5	5.3	6.0
Turkey	4.2	3.9	4.1	4.2	3.7	4.2
India	2.9	2.9	3.2	3.4	3.5	3.4
Chile	2.4	2.8	2.7	2.4	2.9	2.7
Australia	2.2	1.9	1.9	1.7	2.2	2.1
Argentina	2.1	2.7	2.6	2.1	2.4	2.1
World	73.4	80.6	77.5	77.4	77.6	80.1

Table 1.2. Estimated area of land cultivated to grapes by country in thousands of hectares. The percentage change in area since 2017 is also indicated (OIV, 2024).

Country	2017	2018	2019	2020	2021	2022 prov	2023 prelim	2017
Spain	968	972	966	961	963	955	945	−2.4%
France	788	792	794	799	795	795	792	0.5%
China	760	779	781	766	754	758	756	−0.5%
Italy	699	705	714	719	722	718	720	3.0%
Turkey	448	448	436	431	419	413	410	−8.5%
USA	434	408	407	402	393	392	392	−9.7%
Argentina	222	218	215	215	211	207	205	−7.7%
Chile	207	208	210	207	182	182	172	−16.9%
Portugal	194	192	195	195	194	193	182	−6.2%
Romania	191	191	191	190	189	188	187	−2.1%
Other countries	2418	2428	2446	2485	2433	2435	2444	1.1%
World total	7329	7341	7355	7370	7255	7237	7202	−1.7%

Despite the net loss in vineyard area, some of the most significant vineyard developments have come from countries with centuries of growing tradition, such as Spain and Kazakhstan, for instance, where the high-labour input and low-yielding vineyards are being replaced with modern cultivars and management practices (Smart, 1996). The level of investment in China remains high, making it well placed to have a large impact on world table grape production. If its domestic market becomes saturated, the rest of the world can expect competition on a significant scale soon after. Wine production in China, having peaked in 2012, has dropped precipitously since, to just one-fifth the volume in 2024 (OIV 2025a).

In South America, Chile is the country that seems to have the greatest export presence, with established international markets for wine and table grapes. They have always been strong in table grape export, currently 75% of their production (USDA https://apps.fas.usda.gov/psdonline/app/index.html#/app/downloads), but their wine industry, represented by Viños de Chile A.G. since 2007, has increased the amount of exported wine to where it is 67% of their total production (OIV 2023). Peru's rapidly increasing area devoted to table grapes drives their exports, with 76% of their production going to their international neighbours.

South Africa is another country that has a strong focus on international customers, with 75% of table grapes and 43% of wine exported (OIV 2023; USDA https://apps.fas.usda.gov/psdonline/app/index.html#/app/downloads).

Overall, wine producers realize that in order to be competitive in the shrinking world marketplace for wine, they must produce better-quality products at lower cost. As the highest profit margins lie in the more expensive bottles of wine, for example, there will be thinner margins on the basic table wine and more competition to sell fewer bottles of premium product. Grapes make up a significant portion of the cost of a bottle of wine, so the pressure filters down to the growers as well.

For growers of grapes for the fresh market, the trend is for increased consumption but also greater international competition. China is by far the largest producer of grapes for fresh consumption, well ahead of India, Turkey and Egypt (https://www.oiv.int/what-we-do/data-discovery-report?oiv), but the vast majority of the fruit is currently consumed domestically, and forecasts are for increasing demand for more imported grape products as well. China will be a country to watch, as it has already had a huge impact on the world apple juice market (Huang and Gale, 2006), and predictions are that it could have a similar impact on the grape (and wine) market. However, at this point, despite producing over 40% of the world's table grapes, only 3.0% were exported in 2022 (USDA, 2025b).

Cultivars, Anatomy and Improvement

2

Abstract: There are approximately 24,000 named cultivars, but because there are often more than one name for the same grape, the number of distinct cultivars is more on the order of 4,000. Of these, 33 make up around 50% of the planted area, but there is increasing interest in lesser-known indigenous cultivars that could have characteristics better suited to present and future growing conditions. Clones are a subset of cultivars, closely related genetically, but usually having some distinguishable characteristics. Knowledge of vine anatomy is essential for proper management of vines, as well as grapevine-specific terminology. Conventional breeding of new cultivars is possible, but a lengthy process, but newer technology, like CRISPR, may bypass this process to allow introduction of desirable traits without resorting to genetic modification. The methods of determining grape quality, or suitability for purpose, vary with end use, but technology to provide objective measurements is advancing.

Grapes are one of the world's most-planted horticultural crops, and there is a wide range of cultivars in use, many of them well suited to a particular area or end use. In addition to listing the most used cultivars, this chapter will discuss clones, or selections within a cultivar, which are another important variable in the growing of quality grapes. After a review of the parts of the vine there will be a brief discussion about vine breeding and improvement.

Main Cultivars for Various Uses

Though there are many different kinds and names of grapes, there are relatively few that make up the bulk of those produced. The terms cultivar and variety are used interchangeably in this industry to refer to particular types of grapes. There is something in the order of 24,000 named cultivars (Viala and Vermorel, 1909), but because there is often more than one name for the same grape, the number of different and distinguishable cultivars is in the order of 4000 (OIV, 2013). For example, 'Syrah' is also known as 'Shiraz' and 'Sultana' is known as 'Sultanina','Thompson Seedless', 'Oval Kishmish', 'Kouforrogo', or 'Tchekirdeksiz'. For the latter example, it was said that the grapevine was

DOI: 10.1079/9781800627062.0002

originally brought to California named as 'Lady De Coverly' by a William Thompson, but growers referred to the cultivar as 'Thompson Seedless' and the name stuck (Christensen and Peacock, 2000a). The Organisation Internationale de la Vigne et du Vin (OIV) list approximately 250 cultivars as being significant to the wine industry (OIV, 2000), and a 2017 report states that 33 cultivars are responsible for 50% of the global vine area, and 13 cover more than 33% (OIV, 2017). Hence, there are many cultivars that are planted in only very small amounts.

Due to this complexity in naming, in addition to printed resources (Robinson *et al.*, 2012) there are internet databases that attempt to maintain clarity in the relationships between species, cultivars and their common names (for example, see the OIV Vine Varieties Database https://www.oiv.int/what-we-do/viticulture-database-report?oiv= and the Foundation Plant Services Grape Registry, https://fps.ucdavis.edu/fgrvarieties.cfm).

The entire genome of *V. vinifera* was completely sequenced many years ago (Velasco *et al.*, 2007), the second food plant to have reached this milestone, after rice. The genomic sequences of a range of *Vitis* species and clones have been published since, with the data available online (e.g. Gramene Grapevine, https://vitis.gramene.org/). This is an invaluable resource for grapevine researchers and breeders, which will ultimately help the industry reduce chemical inputs, increase resilience to climate change, and provide a better quality product for the consumer.

The addition of such tools as DNA fingerprinting is helping to elucidate the genetic relationship between cultivars. For example, Goto-Yamamoto *et al.* (2006) have demonstrated the genetic similarity of a number of grape cultivars through the use of microsatellite DNA markers. Figure 2.1 is a diagrammatic

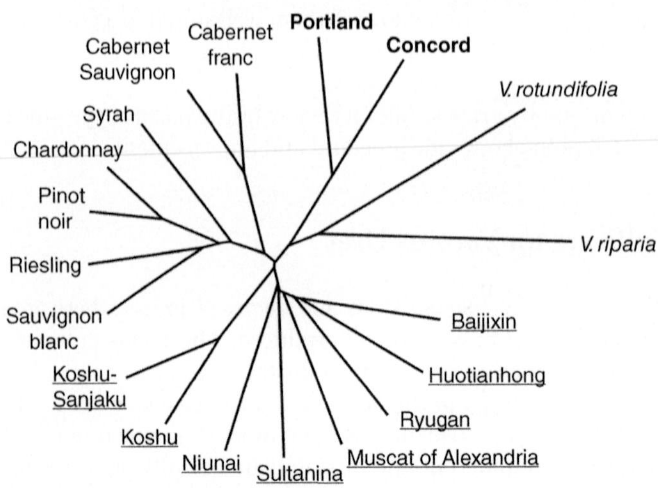

Fig. 2.1. A diagram representing the genetic similarity of grape species and cultivars through the length of the lines between them (Used with permission, from Goto-Yamamoto *et al.* © *2006 American Society for Enology and Viticulture. AJEV 57:105-108*).

representation of these relationships, where the length of the branches represents the genetic similarity (e.g. 'Chardonnay' and 'Pinot noir' are quite closely related, but *V. rotundifolia* and *V. riparia* are not).

The main cultivars used for grape production on a worldwide basis are changing rapidly. The latest data available place Kyoho, grown as a table grape, as the world's most-planted cultivar (OIV, 2017), having supplanted Cabernet Sauvignon at the top position as a result of plantings in China. Wine cultivars such as 'Airén', which was grown in large areas of Spain partly due to the use of low vine density in those very dry areas, have been pulled out and replaced by more profitable irrigated vineyards planted to cultivars such as 'Cabernet Sauvignon', 'Merlot', 'Syrah'/'Shiraz' and 'Chardonnay'. Similar trends are being seen in many other traditional grape-growing areas – a result of the internationalization of the wine market.

Within individual grape-growing regions, particularly in places that have been producing grapes for many years, there are local cultivars that are not much planted outside those areas. Particularly with the preference for labelling wines by cultivar instead of by region of origin (e.g. 'Pinot noir' instead of Burgundy), the average consumer has come to recognize a small number of cultivar names. This puts pressure on grape growers to plant those cultivars, resulting in traditional cultivars being pulled out: a potential loss in terms of diversity of plant material and different wines from which to choose. This has alarmed some in the industry, but there is renewed interest in some of these varieties because they may be better suited to the warmer and drier conditions arriving with climate change.

Even though there are selections of cultivars in table, raisin and juice production, they appear not to be as numerous as in the wine industry. This could be due to the added layer of complexity in postharvest processing of wine grapes, as well as to the product being more carefully analysed by the consumer.

Still further cultivars are not used for their fruit but in breeding programmes or as rootstocks, e.g. *V. champini, V. berlandieri, V. rupestris, V. riparia*, etc. Though these are not as well known, they are vitally important to the long-term health of the grape industry, as they represent a reservoir of diversity that can be used to develop disease resistance or cultivars better suited to a particular region or use.

The crossing of *V. vinifera* and other species of grapevine was not widespread until the 19th century, when North American vine diseases and phylloxera spread throughout Europe. Vine breeders recognized that the best way to combat these problems was to breed new cultivars and rootstocks that were resistant. To do this, they crossed species of North American vines that had co-evolved with the pests and had resistance to them, with the *V. vinifera* grapes that produced the fruit they needed for making wine. And so were developed the French hybrids, which could produce fruit without succumbing completely to the disease and root pests. Specialized rootstocks were also developed by crossing a wide range of *Vitis* species, for example, to develop vines with tolerance to high-pH soils (rootstocks are discussed further in Chapter 5).

Muscadines, the close cousin to the *Vitis* grapes, are grown primarily in the south-eastern USA, where they support a small but persistent wine, table grape, juice and preserves market. Well-known cultivars include 'Scuppernong', 'Carlos', 'Magnolia' and 'Fry', the latter being the most widely planted (Olien, 1990). Famously, 'Fry' was the cultivar that was most used in Virginia Dare, the best-selling wine in the USA before Prohibition set in (Adams, 1985; Olien, 1990).

With each category of use, there are cultivars of grapes that have been specially selected as they perform best for those uses. In table grape and raisin production the market consists of more and more grapes without seeds (Alston *et al.*, 1997), whereas in wine grapes it is desirable to have seeds, which contribute flavour and mouthfeel characteristics to wine, as well as aiding in juice extraction during processing. Table 2.1 shows examples of cultivars used for different end products. 'Sultana' and other neutral-flavoured grapes are used for multiple purposes, which brings some advantages in terms of finding a market for them.

Clones

The *Oxford Companion to Wine* defines a clone as a 'population of vines all derived by vegetative propagation from cuttings or buds from a single "mother vine" by deliberate clonal selection' (Smart and Vouillamoz, 2023). In practical terms, a clone is a selection of a cultivar that has some distinguishing characteristic that someone noticed and thought was significant enough to warrant separate propagation of the vine.

Examples of characteristics that may be used to select a clone are vine growth (vigour), disease resistance, leaf or shoot appearance, cluster shape, fruitfulness, fruit composition, etc. So far, clones have been selected on the basis of morphological differences. It is already possible to distinguish clones using

Table 2.1. Examples of grape cultivars used for various end products.

End use	Cultivars
Fresh consumption: seeded	'Kyoho', 'Muscat of Alexandria', 'Red Globe', 'Isabella', 'Emperor', 'Concord', 'Tokay', 'Noble'
Fresh consumption: seedless	'Sultana', 'Red Globe', 'Crimson Seedless', 'Perlette', 'Sugraone'
Raisin: seeded	'Muscat of Alexandria'
Raisin: seedless	'Black Corinth', 'Sultana', 'Monukka', 'Fiesta'
Juice/preserves	'Sultana', 'Grenache', 'Concord', 'Niagara'
Wine	'Airén', 'Grenache', 'Cabernet Sauvignon', 'Marechal Foch'
Fortified wines	'Palomino', 'Pedro Ximénez', 'Touriga National'
Distillation	'Muscat of Alexandria', 'Gouais blanc', 'Ugni blanc'

DNA screening (Cervera *et al.*, 1998; Scott *et al.*, 2000), which could eventually be used to help select desirable new clones as well as identify unknown clones already planted. In addition, techniques to identify clone based on radiation reflected from the underside of leaves have been found to have strong potential (Fernandes *et al.*, 2015), with the added bonus of not requiring destructive sampling. Work in this area, using various spectral techniques, is still ongoing.

Since the historical appearance of clones is due to natural variation in the population of vines, they are often the result of a single gene mutation – some part of the DNA not being expressed, being expressed differently or perhaps being over-expressed. These point mutations are relatively common, and many desirable clones (and even cultivars) have arisen by chance through this pathway. There are colour sports of many cultivars: white and pink grapes appear to be a mutation of red grapes – there having been no white grapes before this random change. Two chance mutations that affected the ability of the plant to turn on synthesis of red pigments (anthocyanins) was all it took (however unlikely the occurrences were) to create white-fruited grapevines (Walker *et al.*, 2007). These changes are so obvious and have implications for what the fruit can be used for, so they end up being designated as new cultivars, but the same process can result in more subtle changes, which might be identified as a new clone.

Most cultivars of grape have different clones available for planting, the number of which depends on the natural genetic stability of the cultivar and how carefully someone is looking for new sports of it. Genetic mutants of a vine appear spontaneously in the vineyard, and in some cases, these display improved growth or fruiting properties and can be propagated asexually to plant new areas with the new clone (or, in some cases, cultivar). ‘Pinot noir’, for example, expresses a considerable amount of variability in comparison with other cultivars (Konrad *et al.*, 2003). Over the centuries, it is thought that ‘Pinot noir’, ‘Pinot Meunier’, ‘Pinot gris’ and ‘Pinot blanc’ all have a common ancestry, and it is still relatively easy to spot variants in the vineyard (Fig. 2.2).

Note that, in some countries, clonal names may differ, i.e. the plant material may be the same but the name is different, as happens with cultivars. In addition, in some cases, due to natural variation or mix-ups in labelling, the name given to clones in some areas may not be correct. This highlights the importance of being able to reliably distinguish clones, a task that is not yet fully completed through the use of ampelography (discussed later in this chapter), as clones growing in different areas may exhibit slightly different growth habits, which can make consistent typing of a vine difficult. Fortunately, DNA analysis is close being able to rapidly distinguish clones (Urra *et al.*, 2023), but robust paper trail is still the best means of knowing which ones you have.

In a practical sense, clones have value because they are all the same variety, but not quite. This can result in differences that make them more suited to specific management systems (e.g. an upright clone is better suited to a trellising system that relies on shoots all growing upward), seasonal weather

Fig. 2.2. Top, 'Pinot gris' clusters with a non-coloured mutation. Bottom, mutation on a 'Pinot blanc' cluster that has resulted in production of anthocyanins in the skins (note the berry that is half green and half red...). Content author's own.

(e.g. flowering occurs at slightly different times, or fruit set percentage is different under similar conditions), disease pressure (e.g. some clones may have more open clusters that result in less disease) and qualities of the end product (e.g. flavour, colour). In general, it is good practice to have a few different clones in a planting to spread some of the risk by having slightly greater genetic diversity, while still being able to produce a varietal product.

Anatomy and Physiology

The grapevine structure is very similar to that of many other woody perennials. It has a root system that serves to anchor the plant in the ground, but also gathers and sends water and nutrients to support plant growth and acts as a carbohydrate reservoir for carrying over energy from season to season. There is also a trunk, which serves structural, carbohydrate storage and conduit roles, and the branches (called canes or cordons) that support the shoots, fruit and leaves.

Roots

The root system is made up of the larger arms and branches, down to root tips and root hairs. The latter organ is the workhorse of the root system, where the vast majority of nutrient and water uptake occurs (Pratt, 1974). Basic root structure is shown in Fig. 2.3.

The root system, as does the shoot system, produces plant growth regulators that can modify the growth of other parts of the vine (further discussed in Chapter 3).

The vascular tissues, mainly the xylem and phloem, are in the centre of the root and surrounded by the periderm, as found in many woody perennial roots. It is important for the xylem and phloem to remain protected from the rest of the root and the soil beyond, as this is a key control point for the entry of water and nutrients (Tanton and Crowdy, 1972).

The root tip is the most important part of the root system, in that it is the part that will allow the plant to explore new areas of soil that have the necessary nutrients needed for the vine to survive. The root cap, covering the growing point (root meristem), protects the actively dividing cells in the meristem and also contains cells with starch granules – important for indicating to the root which way is up (Wilkins, 1966).

Root hairs are the specialized organs that grow from the epidermal cells behind the tip and along for about 5 cm (zone of absorption, "A" in Fig. 2.3). As these, collectively, have a very large surface area, they are the part of the root that perform most of the water and nutrient gathering, and also interact with the mycorrhizae (see next section). As a section of root continues to grow and thicken, the root hairs die back, but are continually replaced by new ones emerging near the tip. The outer layer of cells on the older roots serve to absorb water and nutrients but, as their relative surface area is very small in comparison with the root hairs, the amount of absorption is low (Steudle and Meshcheryakov, 1996).

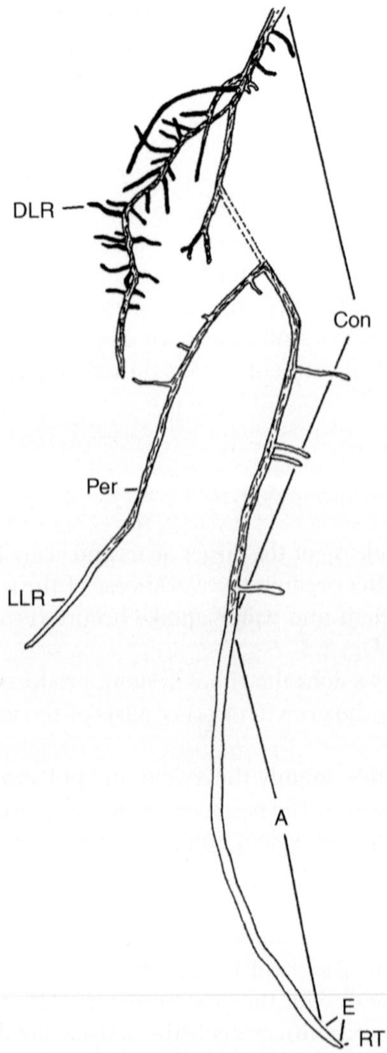

Fig. 2.3. Root of *Vitis vinifera* ('White Riesling') showing actively growing and inactive or dead portions. A, zone of absorption; Con, zone of conduction; DLR, dead lateral root; E, zone of cell elongation; LLR, living lateral root; Per, periderm; RT, root tip (Used with permission, from Pratt © 1974 American Society for Enology and Viticulture. AJEV 25:131-150).

It is worth noting that roots do not grow toward areas of higher water or nutrient availability, i.e. they cannot pinpoint rich areas and target them. Roots grow in all directions and, if an area of higher nutrient or water availability is grown into, there is a more rapid proliferation of roots there.

Roots exhibit significant turnover, with parts of the root system dying off as other parts grow (Comas *et al.*, 2000; Anderson *et al.*, 2003). Management decisions, such as severity of pruning, also affect root life, with more severe pruning reducing the average lifespan of a root (Comas *et al.*, 2000). Those parts of the root that continue to develop become more permanent branches and are a site for carbohydrate storage.

Mycorrhizae

These are a group of fungi that form beneficial relationships with most species of plants, including grapevines. In trees, shrubs and woody perennial vines, vesicular–arbuscular mycorrhizae (VAM) are the most common. These are symbiotic relationships, as both the plant and the fungi benefit. VAM are usually most associated with plants in low-phosphorus soils, as the hyphal network is more efficient at gathering this immobile nutrient than native plant roots. However, VAM associations are also known to occur in higher-phosphorus soils (Morin *et al.*, 1994).

Grapevines normally have VAM infections (Possingham and Groot Obbink, 1971; Deal *et al.*, 1972), and this has been shown to have a beneficial effect on the growth of vines (Biricolti *et al.*, 1997; Aguín *et al.*, 2004), with improved root and shoot development (see Fig. 2.4).

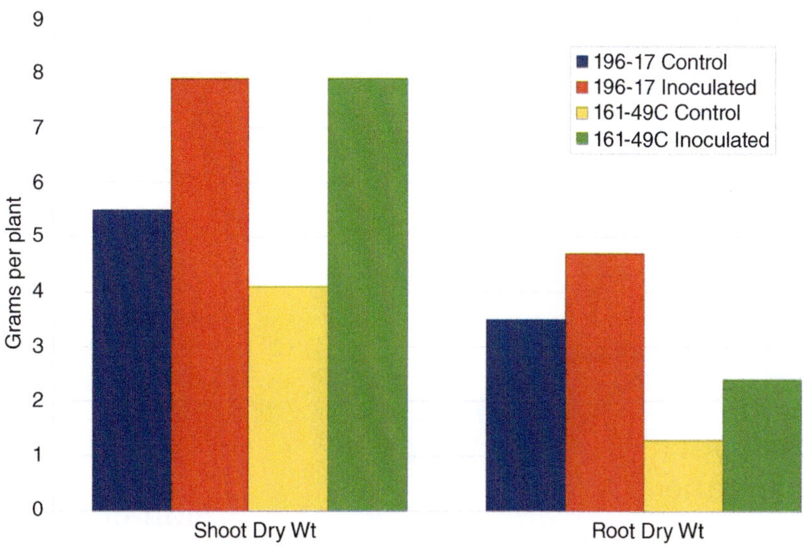

Fig. 2.4. The effect of inoculating two different rootstocks in grafted nursery vines with mycorrhizae. Both shoot dry weight and root dry weight were increased through the use of the mycorrhizal fungus *Glomus aggregatum* (reproduced with permission; redrawn from Aguín *et al.* © 2004 American Society for Enology and Viticulture. AJEV 55:108-111).

The exploration of soil is very important to the survival of the vine. Some nutrients, such as phosphorus, are immobile in the soil and thus, for the plant to obtain a supply of it, the roots must grow into areas of soil that have those nutrients available (Mullins *et al.*, 1992). This is unlike a nutrient like nitrogen, which is mobile in the soil solution and is brought to the roots by water percolating through the soil profile (see Chapter 8). Mycorrhizae expand the nutrient-gathering surface area for the vine by a significant margin (one study with the cotton plant showed that the hyphal network extended 3 cm from the root hair's surface under low P conditions, Mai *et al.*, 2019) resulting in more efficient harvest of the moisture and nutrients present.

Above the soil

If we look at a mature vine in the dormant season (Fig. 2.5), the above-ground parts consist of the trunk, which is the portion of the vine from the ground to about the fruiting wire and provides support for canopy growth as well as being a carbohydrate storage site. Also shown are new canes, which were the previous year's shoots and a non-count cane, which are shoots arising from buds buried in the bark (formerly at the base of shoots).

Along the cane are nodes, separated by internodes. At this point in the season, the nodes are where the following season's shoots will arise. Positioned

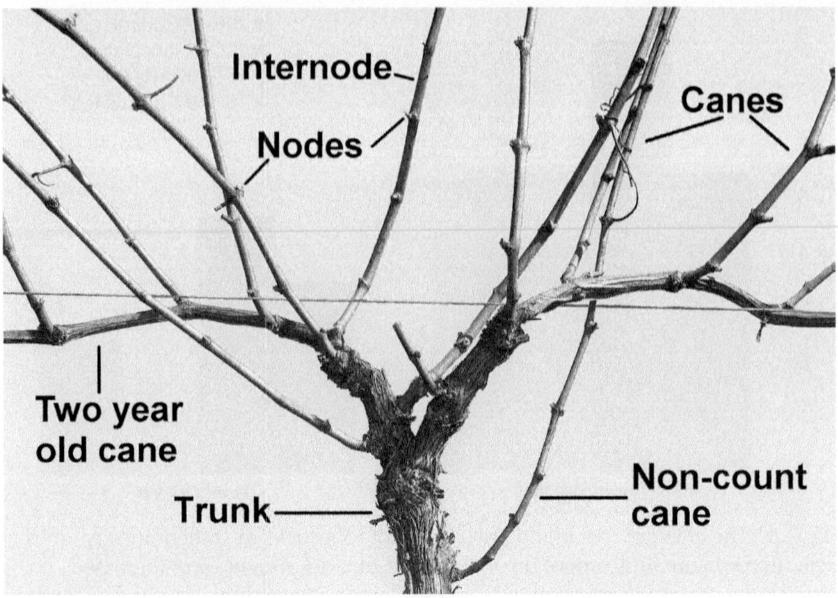

Fig. 2.5. A dormant, cane-pruned vine showing the trunk, 2-year old canes, new canes, a non-count cane, nodes and internodes. Content author's own.

at alternate sides of the cane are compound buds, so called because they contain three (the primary, secondary and tertiary) pre-formed shoots (see Fig. 2.6). Each of these will have six to nine leaf primordia and, in some cases, flower cluster primordia already formed and indeed, if these dormant buds are dissected they can be seen with a reasonably powerful stereo microscope (Antcliff and Webster, 1955).

The compound bud is designed for overwintering and is protected by tough bud scales and woolly fibres. Primary buds are less winter hardy than secondary or tertiary buds. If the primary bud is killed by cold temperatures in winter, the secondary bud will grow; if the secondary bud is killed the tertiary bud can grow. The largest clusters are found on the primary buds; in comparison, the secondary bud has inferior clusters and results in a lower yield. Tertiary buds generally do not have clusters, but following a severe winter or early-season frost damage, at least the vine is able to grow to normal size from those.

When environmental conditions become favourable, one of the shoot primordia – usually the primary one – will begin to grow and push out of the bud. Temperature is the main driver of plant growth, and shoots (and any flower clusters that may be on it) develop more rapidly as temperatures rise during the season.

The early-season shoot development is called fixed growth because it is the growth of the pre-formed nodes and leaves and results in the early appearance of leaves and flower clusters. If a shoot is examined closely, a pattern to tendril and leaf appearance emerges: every third node lacks a tendril (Mullins *et al.*, 1992), but this pattern starts only once the first flower cluster or tendril appears at the base of the shoot, which may have anything up to five or more nodes without any tendril or flower cluster appearing.

At each internode and at the base of the leaf petiole (leaf stem) there forms another compound bud and an axillary (or lateral) bud (see Fig. 2.6). The axil-

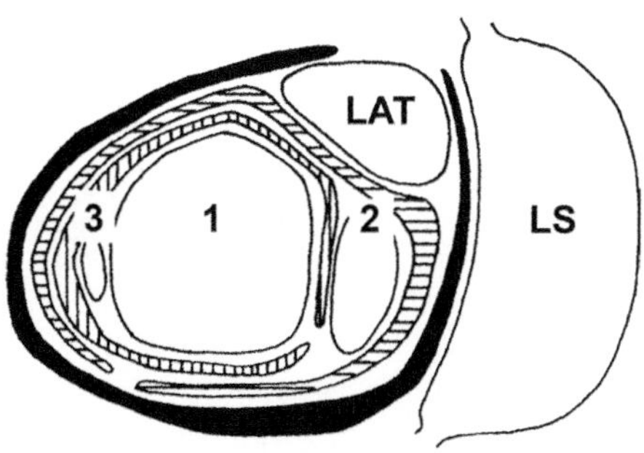

Fig. 2.6. Diagram of a transverse section through a compound bud (eye) of *Vitis labrusca* ('Concord') showing relative positions of leaf scar (LS), lateral shoot (LAT, also known as the axillary bud) and three dormant buds (1, 2 and 3) (Used with permission, from Pratt © 1974 American Society for Enology and Viticulture. AJEV 25:131-150).

lary bud will develop into a noticeable lateral shoot if conditions are favourable during the season, and the compound bud is the one that will overwinter and be the source of shoots for the following season. It is worth noting that grapevines do not form adventitious, or spontaneously formed, buds – all shoot growth originates from a previous node position. Since there can be several compressed nodes at the base of every shoot, which are not removed through normal practice, this results in many buried buds that can grow later in the vine's life (see Fig. 2.5); however, shoots will never arise from an internode area.

Further elongation of the shoot beyond the fixed growth is called the free growth, and is the result solely of the environmental conditions of the current season and the shoot tip meristem. Grapevines will grow as long as conditions (primarily temperature) permit, so their growth is called indeterminate.

The previously mentioned nodes are distinguished by being greater in diameter than the internode, much like a finger's knuckle. The internode keeps the leaves spaced apart and acts as a conduit, containing the xylem and phloem as well as having its own meristem, which causes growth in cane diameter through the season. The outer layer of meristem growth is old phloem, but as the cane grows it becomes what we see as the bark of the vine, which covers the older canes and vine trunks. A cross-section of a cane shows the relative positions of the phloem and xylem on either side of the cambial layer, as well as the way in which the inner cortical cells (pith) break down to leave the centre open (see Fig. 2.7).

Internode length influences shoot length as it is variable, probably depending on growing conditions when it is formed and shortly thereafter. Quinlan and Weaver (1970) reported that internodes were most lengthened

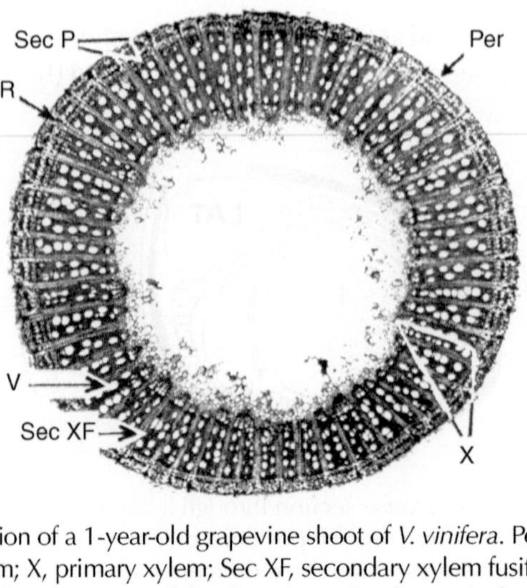

Fig. 2.7. Cross-section of a 1-year-old grapevine shoot of *V. vinifera*. Per, periderm; Sec P, secondary phloem; X, primary xylem; Sec XF, secondary xylem fusiform cell; R, ray; V, xylem vessel. Note that the centre of the shoot is open – the cortical cells have collapsed (Used with permission, from Pratt © 1974 American Society for Enology and Viticulture. AJEV 25:131-150).

when gibberellic acid (a plant growth regulator, see Chapter 3) was applied to them when they were less than 1 cm in length rather than later, demonstrating that this is a key time for determination of this parameter.

Arising from the node positions are leaf petioles, tendrils, flower clusters and, as already mentioned, they house the axillary and compound buds. At the distal (furthest from origin, as opposed to basal, closest to the origin) end of the petiole is the leaf blade, which is the main photosynthetic organ of the vine.

Photosynthesis

Photosynthesis is perhaps the most important process on Earth. In basic terms it is the harnessing of light energy to convert carbon dioxide to chemical energy, while releasing oxygen. It is the opposite of respiration, which is the process of using chemical energy by combining it with oxygen, releasing carbon dioxide. Animals survive through the process of respiration, while plants, although they respire as well, produce the chemical compounds that can be respired.

The workhorse of photosynthesis is the chloroplast, which collects light energy using (primarily) chlorophyll. Chlorophyll appears green because it absorbs most red and blue wavelengths of light. The light energy collected is used to split water molecules (H_2O), adding the hydrogen (H) to carbon dioxide (CO_2) to form carbohydrates (CHO) in the form of sugar (glucose). As a result of this process, oxygen (O_2) is released. Ribulose bis-phosphate carboxylase/oxygenase (RuBisCO) is the key enzyme in photosynthesis, with up to half of the protein in the chloroplast being just RuBisCO. Demonstrating the importance of plants in the ecosystem, RuBisCO is probably the most abundant protein on Earth (Ellis, 1979).

The ingredients that the photosynthetic process needs are light, moderate temperature, CO_2 and water. Temperature also influences respiration, where its rate increases the warmer it gets. Because the process of respiration is happening along with photosynthesis, under certain conditions there will exist a compensation point, where the amount of chemical energy photosynthesis is creating equals the amount being used by respiration. In this case, no net photosynthesis is taking place – the tissue is making energy as quickly as it is using it. If the temperature increases from there, the plant can be in a situation where it is using up stored energy, despite the fact that photosynthesis is proceeding.

In order for the chloroplast to have access to CO_2 and release the O_2 that it makes, there needs to be a transfer of gases between the inside of the leaf and the outside environment. Because the outside environment is quite hostile (in terms of relative moisture content) to the easily desiccated interior cells of the leaf, there needs to be tight regulation of the movement of gases in and out of the leaf. This is accomplished with a passage through the leaf's surface, called a stomate (pl. stomata) (see Fig. 2.8). Specialized guard cells form the pore when the cells become turgid, and the pore closes when the cells lose water. This an efficient mechanism for the vine leaf to regulate the passage of gases, including water vapour, in and out of the leaf. When

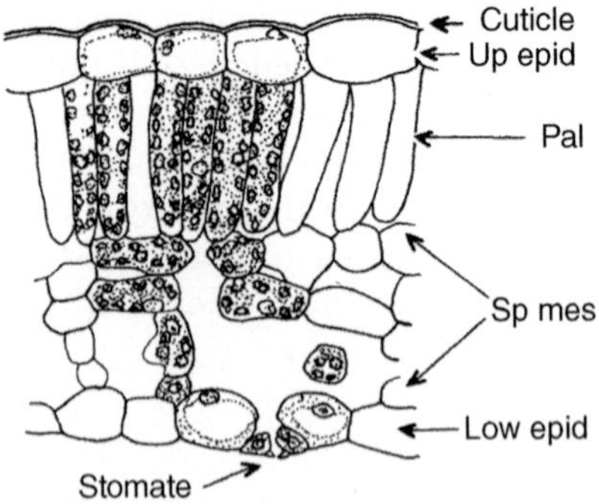

Fig. 2.8. Transverse section of a grapevine leaf (*V. vulpina* L.). The bulk of photosynthesis occurs in the palisade and spongy mesophylls. Up epid, upper epidermis; Pal, palisade layer; Sp mes, spongy mesophyll layer; Low epid, lower epidermis (reproduced with permission from Pratt © 1974 American Society for Enology and Viticulture. AJEV 25:131-150).

the vine has a good water supply, the guard cells are turgid and the stomatal pores are open; when the vine suffers water stress, the guard cells become flaccid and the pores close. A typical 'Cabernet Sauvignon' leaf may have approximately 180 pores/mm^2, but there will be more in sun-exposed than shaded leaves (Palliotti *et al.*, 2000). There are far more stomata on the abaxial (lower) side of the leaf compared with the adaxial (upper) side. Berries also have stomatal pores, though their frequency is very low in comparison (Blanke and Leyhe, 1987).

Wind can significantly affect the rate of water loss through the stomata by reducing the thickness of the boundary layer (area of still air) over the leaf. This increases the strength of the humidity gradient and thus the rate at which water exits the leaf. If the vine cannot supply enough water from the roots to match that lost through the leaves, the leaves and shoots would wilt. To prevent this, the vine shuts the stomata, preventing possibly lethal water loss, but also preventing photosynthesis and the production of carbohydrates. So as wind speed increases, more and more stomata close (Freeman *et al.*, 1982; Campbellclause, 1998). This simple response to wind is the cause of many changes to vine growth and fruit development. Making sure the vine is well supplied with water under hot and windy conditions can minimize stomatal closure, but there can be many instances where the vine roots simply cannot keep up with loss through the leaves, and the vine shuts down regardless of how much water is available in the soil.

The critical wind speed where photosynthetic efficiency starts to drop is altered according to numerous variables, but one study suggests that values above 14 km/h tend to decrease stomatal openings (Dry *et al.*, 1989). In areas with considerable wind, windbreaks will increase the number of hours during the day that the vines can retain open stomata, which results in increased photosynthesis (Freeman *et al.*, 1982) and thus more energy available for growth or fruit development.

Wind also moves away heat in the vineyard (Smart and Sinclair, 1976). Since photosynthesis is optimal at temperatures just below 30°C, if wind cools the vine canopy from 28° to 25°C, photosynthesis will not be operating as efficiently as it could be. Conversely, if wind is cooling a hot vineyard, there may be a net benefit. Light, water and temperature thus have a strong influence on the photosynthetic process, and therefore grapevine productivity.

The vine can alter itself to help maximize photosynthesis or minimize the effect of too much sunlight or heat. Leaf blades are able to change their angle relative to the petiole and follow the position of the sun through the day, to obtain the best light capture. However, this ability can also be used to tilt the blade away from the sun to reduce leaf temperature when light intensity is excessively high (Gamon and Pearcy, 1989). In this way, the vine is able to take best advantage under a range of conditions.

Continuing with the characteristics of the leaf, the blade may have noticeable hairs, usually on the abaxial side, which can influence insect and disease susceptibility as well as water loss from the vine. The hairs may physically hinder insect access to the epidermis, and they can increase the depth of the boundary layer above the stomata. On young leaves, flower clusters and shoots, there may be clear, round objects, which may look like insect eggs (see Fig. 2.9). These are a normal occurrence and have been called sap balls; their origin and purpose are unknown (Mullins *et al.*, 1992), but they tend to go more opaque as time goes on.

Flowers and berries

The grapevine flower is one of a number arranged on a panicle inflorescence (Pratt, 1971). The individual flower, or floret, can be perfect (having both male

Fig. 2.9. A sap ball, as it appears on a flower cluster. Though it may look like an egg of some kind, it is normal to see these on leaves, shoots and flower clusters. Content author's own.

and female parts), female or male, though on cultivated vines it is always perfect (which allows for self-pollination). Similarly, vines can be dioecious (having separate male and female plants) or monoecious. Many rootstock cultivars are dioecious, unlike their cultivated-for-fruit counterparts.

The fruit of grapevines is a true berry, like the blueberry or cranberry. It develops from the fertilized flower, with the bulk of the flesh being made up of ovarian tissue. *Vitis* usually have up to four seeds (Winkler and Williams, 1935), while you may find up to six in muscadine grapes (Olien, 1990). However, it is not that unusual to find more than four seeds in *V. vinifera* cultivars (Bioletti, 1938).

The flower clusters are closely related to the tendril (see discussion in Chapter 3) and usually arise from the third or higher node on a shoot. The number of flower clusters per shoot varies with cultivar, management and environmental conditions, but can range from none to five or even more.

The floret is attached to the rachis by the pedicel, and the rachis is attached to the peduncle, or stem, of the cluster. Flowers, because they open from the base rather than the tip, form a cap (called the calyptra), which pops off at flowering (anthesis) (see Fig. 1.3). As with other flowers, there are anthers, which produce the pollen, and a stigma, which receives the pollen.

The berry that develops from the fertilized flower has a waxy outer covering called the cuticle. Later in berry development this will appear as the white or greyish bloom on the skin. The cuticle helps protect the berry against water loss as well as disease invasion (Possingham *et al.*, 1967; Blanke and Leyhe, 1988). The epidermal layer of cells, or skin, of the berry also serves this function, as well as being the source of colour and some flavour compounds. In some cultivars lenticels (former sites of stomatal pores) are visible on the berry surface as well, which help exchange gas through the berry skin. Lenticels may become more visible later in the season as the berries start to senesce (see Fig. 2.10).

Seeds in the berry are high in phenolic content as well as containing the embryo that could develop into another grape plant. If the berry is pulled off the pedicel, the latter often has a portion of the berry left on it, called the brush, which is pulp and vascular tissue that has been pulled from inside the berry.

While fruit from the *V. vinifera* vines tends to be firm and have a skin that adheres well to the flesh, fruit from native American *V. labrusca* and related vines have what is known as a slipskin, where if you squeeze the berry, the pulp pops out, leaving the skin in your fingers.

Ampelography

Ampelography is the science or art of distinguishing grapevine cultivars by examining their appearance. Through the observation of leaves, shoots, shoot tips, clusters and fruit, it is possible to determine the cultivar and, in some cases, the clone of a cultivar.

Fig. 2.10. 'Chardonnay' berries near harvest, showing the brown spots (lenticels) that were formally stomatal pores in the skin. Content author's own.

Some examples of parameters that are examined in ampelography include: (i) leaf shape and angles between veins (see Fig. 2.11); (ii) the number of teeth between major leaf veins; (iii) the form and colour of the shoot tip; (iv) shoot growth habit; (v) presence and position of hairs on plant parts; (vi) cluster openness or shape; (vii) berry shape and skin thickness; and (viii) tolerance to phylloxera or botrytis, etc. (OIV, 2009). Clearly, as there is still a measure of subjectivity to identification in this manner and the potential for changes to leaf morphology in different environments, its usefulness is limited, though it is still an important method of cultivar identification.

Breeding and Genetics

In looking at the most-planted grape cultivars, many, if not all, have been in cultivation for centuries. This is not to say that new cultivars are not being produced. There are several active grape-breeding programmes around the world, usually with the goal of producing cultivars that will produce better end products in a certain region, allowing disease resistance, reduced production costs, greater yields, better quality, etc.

However, the genetics of the Vitis genus are complex enough that many thousands of seedlings must be grown for several years to reach fruiting age for only a small chance of finding a superior vine that could be launched as a new cultivar. Resources for this kind of undertaking are scarce, but luckily, there are many existing cultivars available to accommodate most needs.

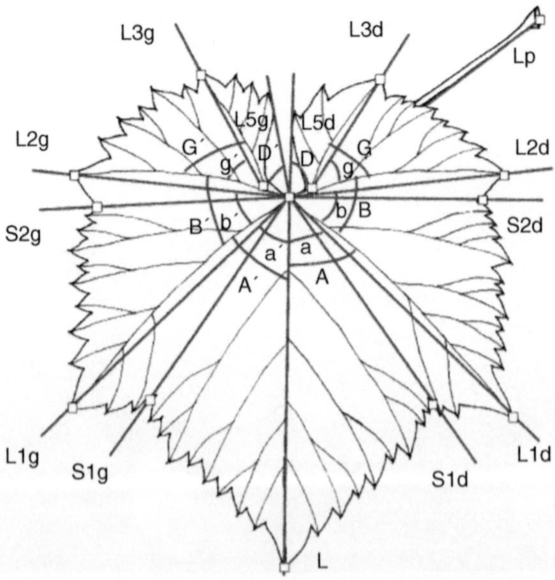

Fig. 2.11. An example of ampelographic measurements that can be made on mature grape leaves. Abbreviations correspond to parameters as set out by the OIV (Office International de la Vigne et du Vin (OIV, 2009)) (Used with permission, from Santiago *et al.* © 2005 American Society for Enology and Viticulture. AJEV 56:287-290).

Small changes to the genome may result in different clones, as discussed earlier, or they may result in what people consider new cultivars. In conventional breeding programmes, pollen from one grapevine (the male of the cross) is brushed onto the stigma of the other vine (the female parent). This results in a recombination of genes that results in a wide variety of unique individuals, some of which may be very similar (perhaps indistinguishable) to one of the parents, but most of which will be very different. So, a breeding programme requires the harvesting of many seeds associated with a particular set of parents, planting them, growing them for a number of years until they reach fruiting maturity and then evaluating their growth habits as well as the fruit they produce. Grapevines grown from seed will remain juvenile (in a vegetative state) for a time, though treatment with cytokinins has been successful in initiating flower clusters on young seedlings (Srinivasan and Mullins, 1981). There is hope that, through the manipulation of genes regulating fruiting, tools may be made that can minimize the juvenile phase and therefore increase the rate at which seedlings can be evaluated (Pillitteri *et al.*, 2004). However, and at present, the time between creating the cross and being able to examine the fruit is usually many years. To fully evaluate a vine and the fruit it produces takes over a decade due to the need to see how the vine performs in different seasons and, if wine is to be made, there are further variables such as looking at appropriate styles of wine, as well as how the wine ages in the bottle.

Molecular biology is increasingly being used in breeding programmes, in some cases to introduce desired characteristics directly into vines, such as disease resistance (Agüero *et al.*, 2005; Vidal *et al.*, 2006) and, in some cases, to aid in tracking the movement of desirable traits from parent to offspring (Grando *et al.*, 2003). Increasingly, as the debate about the use of genetically modified organisms continues, molecular biology is being used to learn more about how the plants or other organisms work (Giovannoni, 2001), which can lead to more effective management or control techniques (Purcell and Hopkins, 1996).

The development of Clustered Regularly Interspaced Short Palindromic Repeats (CRISPR) technology has led to exciting possibilities for the grape sector. CRISPR allows the modification of genes in living organisms with a specificity and ease unparalleled to earlier molecular techniques: genes can be disrupted or added to help understand physiology or create plants with new characteristics. Current research is investigating the use of CRISPR to generate Chardonnay that has better downy mildew resistance and table grapes with better botrytis resistance (Salvagnin *et al.*, 2024).

Small changes to the genome of a plant occur as a matter of course, and sometimes this is the result of an error in replicating DNA, but the importance of transposable elements (also referred to as transposons) has been realized in more recent times. These are sequences of DNA that 'jump' around the genome in a random fashion, creating mutations that result in greater variation in how plants grow. The activity of transposons varies, usually increasing with forms of stress (e.g. heat, UV radiation, water) applied to the cells, so it is thought that this is a mechanism for plants to create variation that helps them adapt to change in their environment. Changing the activity of transposons has been used to generate a population of grapevines with high rates of genetic diversity, from which new clones and cultivars of grapes or rootstocks could be produced, as well as for investigating the physiology of the vine and how it interacts with the environment (Lizamore, 2013). Research like this may also be useful for producing new vine material that will allow the grape industry to be successful in these rapidly changing times.

Breeding programmes may also look to change the ploidy level (number of sets of chromosomes the organism has) in grapes, as this usually has an effect on flower and berry size (Dermen, 1964). The development of tetraploid or higher levels of duplication grapevines from the native diploid state can be achieved through a conventional breeding programme or through the use of colchicine (Dermen, 1954). The Kyoho grape bred in Japan is a tetraploid cultivar that was developed in part for its large fruit (Okamoto, 1994), which can reach 13 g/berry and above (Wang *et al.*, 2001), compared with the typical wine grape berry weight of approximately 1–2 g.

Indicators of Quality

Although a vastly important attribute of a product, the trouble with quality is that it is not something that can be readily quantified, especially when it comes to something as complex as some grape products. Examples of basic quality parameters for table grapes are size, shape, sweetness and acidity; however, these are all relative. A small example of a large-berried table grape could still be far larger than a large-berried wine grape. Therefore, factors determining quality should be end use-specific.

It has been very difficult to quantify flavour and aroma compounds, and often we are left with human sensory analysis to determine some aspect of quality. The human perception of aroma and taste can be very sensitive (often more sensitive than many of the analytical methods currently available), but it does vary from person to person (Amerine and Roessler, 1976; Trimmer *et al.*, 2017), and being able to communicate the quality of an aroma (for example, if it smells like strawberries or raspberries), as well as the intensity of that aroma, is problematic (Noble *et al.*, 1984). As a result of this, there have been attempts to standardize the terminology used for aspects of human sensory perception (Noble *et al.*, 1987; Gawel *et al.*, 2000), so that people in different places can use the same criteria to describe these attributes. However, quantifying these attributes still requires trained sensory panels or specialized equipment (e.g. gas chromatography – olfactometry) used in concert with trained panellists.

As indicators of quality, Table 2.2 shows some examples of parameters that are important for those grapes' end use. It is evident that Brix (or percentage soluble solids, which are made up mostly of sugars in grapes) is important in many cases, but for some end uses other parameters come into play.

In some instances, objective measures have been developed for characteristics (e.g. Brix by refractometer or hydrometer, acidity by titration, moisture by weight), but in others there is no standardized method available yet (e.g. tannins relating to astringency or flavour). Because of the importance of these attributes to end use (particularly when it comes to valuing grapes according to quality) there is much research effort invested in finding ways of measuring and quantifying these aspects.

Table 2.2. Examples of grape characteristics that are important for different end uses.

End use	Fruit attributes
Wine grapes	Brix, acidity, pH, colour, tannins, flavour
Table grapes	Brix, acidity, colour, flavour, presence of bloom, Brix:acid ratio
Raisins	Brix, colour, percentage moisture, flavour, presence of bloom
Juice	Brix, acidity, colour, flavour
Preserves	Brix, acidity, pectin, solids content

3

Grapevine Growth and Fruit Development

Abstract: Grapevines are a perennial woody plant that pass through key visible phenological stages such as dormancy, budbreak, anthesis (flowering), fruit set, veraison, berry maturation, cane lignification and leaf-fall. Roots also have a period of dormancy, but begin growing before budbreak. Flower cluster initiation occurs over one year before its fruit is harvested and flower clusters are closely related to tendrils, their divergence determined by plant growth regulators. Factors such as sugar, acids, phenolic and flavour/aroma compounds change dramatically during the course of ripening. The most significant phenolics are anthocyanins, which provide fruit colour, flavan-3-ols that make up tannins, and flavonols, which are made in grape skins in response to sun exposure. Key aromas come from the monoterpene, methoxypyrazine and norisoprenoid groups. Fruit composition is altered by climate, with seasonal heat accumulation being a good indicator of potential vine performance.

This chapter reviews the growth cycle of the grapevine, along with factors that influence it. The fact that the vine is a perennial plant means that management is quite different to that for annual crops, as well as there being unique aspects to cultivating young versus established vines. An overarching concept, that of vine balance, is introduced as it has implications for decisions made from the very start of vineyard planning to how the weather affects the vine after harvest. Fruit quality parameters, which comprise a great many things but of which we can only measure a relative few, will also be mentioned, particularly with reference to grape berry maturation.

Phenology

Phenology is the study of events or growth stages of a plant or animal that recur seasonally and their relations with various climatic factors, including temperature, solar radiation and day length (Mullins *et al.*, 1992). With respect to grapevines, it can be used as a predictive tool, allowing the manager to plan vineyard operations in advance. In combination with other plants'

DOI: 10.1079/9781800627062.0003

phenological data, it can help to sequence events even though the weather patterns in the current year may be substantially different to those of previous years.

It is useful to track the growth stages of grapevines so that development can be monitored with respect to other seasons (or some other benchmark), and so that, for example, pesticide sprays can be applied in a timely fashion. Keeping track of this means determining stages and terminology, for which the modified Eichhorn-Lorenz (E-L) system performs well (Coombe 1995).

In discussing grapevine phenology, it is useful to use a starting point prior to budbreak. Before any visible signs of growth, the vine begins to come out of dormancy at about the same time as the soil temperature starts to rise (due to increased solar radiation and air temperatures). The roots reactivate, and the phloem and xylem tissues start to function again. Visibly, this leads to sap exuding from fresh cuts on the vine, as root pressure builds to fill the empty xylem vessels (Sperry *et al.*, 1987). The volume of liquid that can exude from the cut end of a cane can be high – Bennett (2002) collected 174 ml in a 24 h period.

Actual breaking (or appearance of green tissue through the bud scales) of the dormant buds has usually been said to occur when the air temperature reaches approximately 10°C (Winkler *et al.*, 1974; Williams *et al.*, 1985), though Moncur *et al.* (1989) report no physiological basis for this threshold. Indeed, their experiments on potted cuttings revealed that the base temperature for bud development in grapevines was as low as 0.4°C (depending on cultivar), which more closely matches values for other woody perennials (Anstey, 1966; Richardson *et al.*, 1975).

The progression of the budbreak process is illustrated in Fig. 3.1. Buds are at their most hardy prior to budbreak with respect to both physical and cold temperature damage. The first shoots start to grow powered by energy derived from stored carbohydrate, as there is no photosynthesis yet occurring (Winkler *et al.*, 1974; May, 1987). The pre-formed leaves expand, as do the internodes. Any flower clusters present also develop, rapidly forming individual florets (see Fig. 3.2). Shoots can grow as fast as 2–5 cm/day (1.8 cm/day, Keller *et al.*, 2005; 5 cm/day, Wolf and Warren, 1995), forming a new internode every 2–3 days (Lovisolo and Schubert, 2000). Photosynthesis occurs as soon as there is green tissue on the shoots; however, due to the high metabolic activity and use of stored carbohydrates, there is no net production of photoassimilates until several leaves have fully expanded (Hale and Weaver, 1962; Winkler *et al.*, 1974).

Unlike many other perennial woody crops, grape flowers open long after budbreak. As it takes some time for the shoot to develop leaves that are capable of supplying the carbohydrate needs of the rest of the vine, it is critical that enough stored carbohydrate is available to support the initial development of shoots, roots and flower clusters. If there is not, then it is the flower clusters that suffer the most, as they can drop off the vine due to a lack of available carbohydrate (Ollat, 1992).

When the shoots have approximately 15–17 nodes formed on them (Pratt and Coombe, 1978), the flowers begin to open and the calyptra fall from the rest of the flower (see Fig. 3.3). The flowering and fruit set process in grapes is very

Fig. 3.1. A series of photos showing early bud growth, from dormant, Stage 1 (top left); budswell, Stage 2 (top middle); wooly bud, Stage 3 (top right); green tip, Stage 4 (bottom left); rosette of first leaves visible, Stage 5 (bottom right). Stages of development are labelled according to the Modified E-L System (Coombe 1995). Content author's own.

Fig. 3.2. Young 'Pinot Meunier' shoots in spring (Modified E-L stage 10) showing, on the left, the first leaves and two flower clusters near the tip and on the right, a non-count shoot that has pushed from the secondary bud inside the compound bud. Individual florets have already differentiated by this stage. Content author's own.

Fig. 3.3. Anthesis (cap fall) in a flower cluster. This is the first day any of the florets have opened on this flower cluster. Content author's own.

weather dependent. Pollination (transfer of the pollen from the anther to the stigma) is mostly by wind, though insects may also contribute. Self-pollination, occurring as the cap comes off the ovary, is the norm (Winkler *et al.*, 1974; Kimura *et al.*, 1998), though some studies have suggested that cross-pollination results in bigger fruit and higher seed counts (Sampson *et al.*, 2001; Milne *et al.*, 2003). Airborne pollen counts have been used as a predictor for fruit set in some grape-growing areas (Cristofolini and Gottardini, 2000; Cunha *et al.*, 2003) as there is a good correlation between total pollen in the air and fruit set, but this may be due to more pollen being released when the weather is warmer and dry, which is generally conducive to fruit set.

Fertilization occurs 2–3 days after pollination, as the pollen tube must grow (a highly temperature-dependent process) down through the style and up into the micropyle in a 'J' shape (see Fig. 3.4) in order to reach the embryo. Even relatively brief spells of cool temperatures cause degeneration of embryos

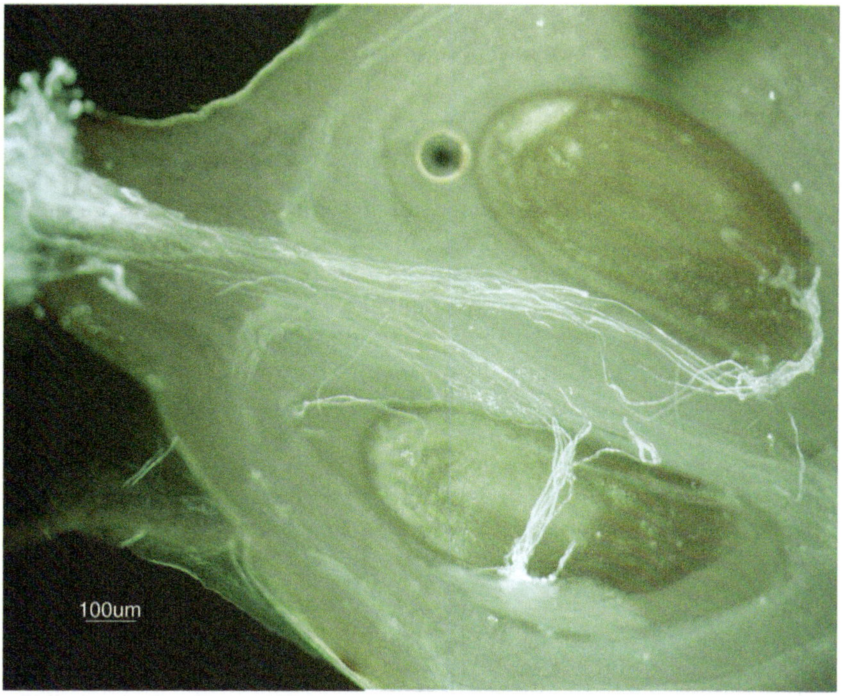

Fig. 3.4. Growth of pollen tubes (lighter paths through the style) from stigmatic surface to the micropyle, shown using a fluorescent dye (image courtesy of Mardi Longbottom, used with permission).

and decrease the chance of fruit set (Ebadi *et al.*, 1996), which goes some way to explaining why this is a problem in cool-climate grape-growing areas.

Typical success rates for fruit set may be in the 30–45% range. Flower clusters may contain from just a few to more than 1000 flowers (Woodham and Alexander, 1966), depending on conditions at flower cluster initiation and cultivar.

Flower cluster initiation occurs about 18 months before the fruit from those flowers is harvested. At about the time of flowering, the flower cluster primordia are starting to be initiated in the compound buds located in the axils of the basal leaf petioles (Winkler *et al.*, 1974). Both temperature and light are important determinants of flower cluster size and number (Buttrose, 1969). As the season progresses, the initiation of flower cluster primordia works its way up the shoot, meaning that along different parts of the shoot (next season's cane) there can be significantly different levels of fruitfulness (this has implications for methods of pruning, see Chapter 6).The amount of branching, and thus cluster size and total floret number, is determined in the current season, but is also influenced by overwintering and budbreak conditions in the following season (Woodham and Alexander, 1966; May, 2000; Petrie and Clingeleffer, 2005). Individual florets do not usually form until after budbreak the following season.

Lateral shoots arising from axillary buds may also have fruit on them, which is termed second set. Generally, these are quite far behind in terms of development compared with the primary crop and, as well as being smaller in size than the primary crop, are rarely used in production. However, in tropical climates, there would be the opportunity for this crop to be harvested. In most cases this fruit is a nuisance, as hand-pickers can mistake it for ripe fruit, and machine-harvesters may harvest these grapes, depending on its set up and other factors.

Once fruit is set on the vine, it is unlikely that the vine will lose it. Apple and other tree fruit crops have one or more times of the year when the crop can abscise naturally (Westwood, 1993); however, with grapes the fruit cannot be dropped, and the vine is pretty much committed to bringing it to maturity. As such, after fruit set it becomes the most important destination for the vine's carbohydrate supply, as carbon radiotracer studies have shown (Hale and Weaver, 1962; Quinlan and Weaver, 1970).

Berry expansion is rapid following fertilization, due to both cell division and cell expansion. The growth in terms of weight, diameter, volume or other measurements can be described as a double-sigmoid curve (see Fig. 3.5), particularly for seeded berries – for seedless ones the curve is less pronounced (Coombe, 1960; Winkler *et al.*, 1974; Friend, 2005). The first period of rapid growth is often called phase I, which is followed by phase II, a time of relatively little growth as the seed matures and begins to lignify (i.e. develop a hard outer coating). At the end of this period the berry undergoes an amazing transformation, softening, becoming translucent, beginning to colour (if a red cultivar) and increasing in size rapidly (phase III). On the inside, the berry begins to metabolize malic acid, accumulate sugar and develop characteristic flavour

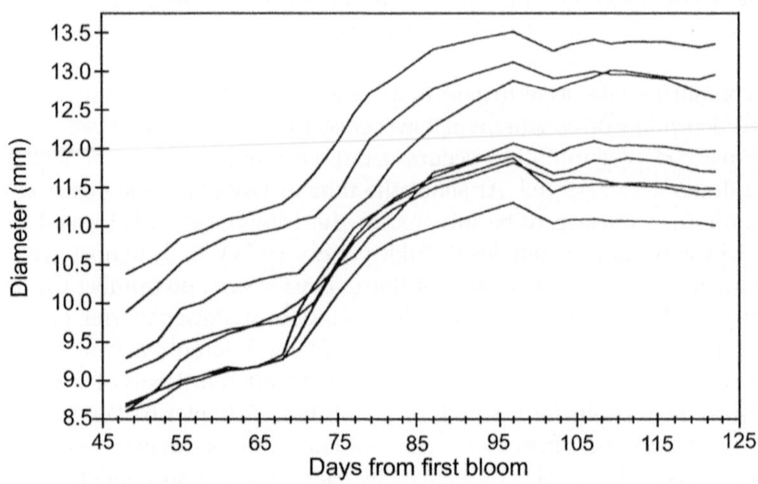

Fig. 3.5. Individual growth curves (by measurement of diameter) for eight grape berries, demonstrating the double-sigmoid shape characteristic of seeded fruit and variation in the start of the ripening phase (rapid growth) (from Creasy, 1991).

and aroma compounds. The French use the term *véraison* (usually written 'veraison' in English) to describe the shift in colour, which signals that all these other changes are taking place, too. It is not known exactly what triggers this massive and rapid change in berry physiology, but the term engustation (Coombe and McCarthy, 1997) has been proposed to describe the great upheaval of events.

Berries mature, with their contents changing through time, culminating in harvest (which, though it is listed in the modified E-L stages, is not a phenological stage of plant development as it is determined by a large range of factors unrelated to the plant itself). Following this, vine leaves may be in a state of senescence (if environmental conditions are worsening) or continue to function until temperatures drop so low as to cause the leaves to fall from the shoots. During the latter stages of berry ripening, shoots will begin developing a hard outer epidermis in preparation for overwintering. The last phenological stage is leaf-fall, usually triggered by frost, or near-frost events.

Vine (Vegetative) Development

Patterns of root growth

The exploration of soil available to a vine is extensive. Vine roots have been found to penetrate to depths of more than 10 m in favourable soils (Winkler *et al.*, 1974), and in general will make as full use as possible of the soil profile. As with any plant, many factors influence root growth, including soil structure, compaction, water availability (including precipitation and water table level), plant genotype, planting density, competing plants, etc. Most roots in a typical soil profile will be found in the top 1 m, but more sparse rooting is found to varying depths and can contribute greatly to water and (to a lesser extent) nutrient uptake.

An important point to make is that vines grown from seed have a taproot, and will explore soil to a greater depth more quickly than vines produced by cuttings (by far the more common method of propagating), which have roots that form adventitiously on the sides of the cane and thus do not grow downwards as much. Even so, different species of vines have varying tendencies to develop deeper root systems, even from cuttings (Perold, 1927), which becomes an important consideration when choosing a rootstock (see Chapter 5). Those roots that tend to explore horizontally are termed spreader roots, and those that tend to grow further down into the soil profile are termed sinker roots. Rootstocks can be classified in terms of their root angles, i.e. the degree to which roots grow away from the vertical plane (Perold, 1927). Thus, a vine with a steep angle tends to grow many sinker roots, whereas one with a shallow angle tends to grow spreader roots. In terms of determining the suitability of rootstock vines to areas with or without a high water table, this information can be very useful.

Like the above-ground parts of the vine, the roots also have a period of quiescence in the dormant season. As the soil warms in the spring, root tips begin to explore the soil, but their development is relatively slow (in terms of dry matter accumulation) compared with the growth and development of the shoots and flower clusters. After fruit set, the full canopy of the vine is able to support more root development but, as the berries ripen towards the end of the season, root growth again slows due to the shift in partitioning of carbohydrates towards the fruit. In areas where there is sufficient warmth to allow photosynthesis after harvest, there may be another flush of root growth before the vine becomes dormant, which appears to be related to the excess photosynthetic capacity on the vine after the fruit is removed (Williams, 1996; Conradie, 2005). In cooler climate areas where there is little or no time between harvest and leaf-fall this cannot, of course, happen.

Tendrils

Most, if not all, vinous plants produce tendrils as a means to support themselves on another structure. It is through this specialized lateral branch (Pratt, 1974) that the vine can invest less energy in developing a solid trunk and more into growing in length, which is in keeping with the vine's evolutionary origins.

To facilitate grasping on to a support, the tendrils grow away from light and curl around or within objects, eventually becoming lignified and woody (see Fig. 3.6). If a tendril does not latch onto something, then it will wither and fall from the shoot. Unlike shoots, tendrils have determinant growth, i.e. they grow to a finite length.

Fig. 3.6. A dormant tendril that had wrapped around a foliage wire during the growing season. Tendrils such as these can be quite woody and difficult to remove at pruning, which demonstrates their role in helping to support the vine. Content author's own.

Cane maturation

Similar to tendrils, shoots will lignify, become more woody and change colour from green to tan, brown or sometimes distinctly red as they age (see Fig. 3.7). It is this formation of a periderm that is associated with the shoot becoming more resistant to drought and cold – an adaptation that allows the plant to survive below-freezing temperatures that are found in many areas where *Vitis* species have developed. The compound buds at each node on the hardened shoot – now called a cane – also go through a similar process, with the bud scales becoming lignified and resistant to drought, temperature and mechanical damage. As shoots mature, there is a build-up of starch and/or soluble sugars within, which is associated with the vine being able to survive below-freezing temperatures (Hamman *et al.*, 1996; Chapter 4) and which provides energy for growth in the following spring.

Leaf-fall and abscission

Towards the end of the growing season environmental signals trigger the vine to accelerate its physiological preparations for the dormant season. Leaves may begin to yellow (or become red if a red-skinned grape cultivar, see Fig. 3.8) and fall from the vines. However, if the vines still have their crop, senescence

Fig. 3.7. Section of a 'Pinot noir' shoot late in the season showing the red lignified periderm on the shoot and the cluster peduncle beginning to lignify as well. Content author's own.

Fig. 3.8. Examples of teinturier type grapes. Left: 'Kolor' mid-season; right: 'Grand noir', late in the season. Not only are the flesh and juice of the berry filled with anthocyanins, but the stems and leaves also express a lot of colour. NB One of the parents of 'Kolor' is 'Pinot noir'. Content author's own.

may be delayed due to the demand for photosynthates from the fruit (Petrie *et al.*, 2000a). The ability of vines to photosynthesize after harvest, and therefore augment storage of photosynthates before the dormant period, is significant in terms of vine performance. Increased carbohydrate status in vines at the end of the season is correlated with increased winter hardiness and fruitfulness in the following season (Howell *et al.*, 1978; Bennett *et al.*, 2005). The coincidence of harvest and leaf-fall in cool-climate vineyard areas, and thus limitation on photosynthesis, may be a reason why vine yields are lower in cool climates as opposed to warmer ones (Howell, 2001).

Dormancy

The period of dormancy is integral to the survival of the vine in areas where winter temperatures fall below 0°C, although dormancy is also an important part of the vine cycle in warmer areas, too. Grapevines can be grown in tropical areas, and as shoots are not determinant they will grow indefinitely. This can create some production and management problems, however, so vines in these areas are often forced into and back out of dormancy through the use of severe pruning or chemicals such as hydrogen cyanamide (Lavee, 1973; Shulman *et al.*, 1983). It has also been proposed that cyanamide could be used to force shoots to grow even before leaves have been removed or senesce, resulting in evergreen production of grapes (Lin and Wang, 1985).

Naturally occurring dormancy is brought about by low temperatures in the late growing season as well as by shortening day length (Schnabel and Wample, 1987). Non-*V. vinifera* species, such as *V. labrusca* and *V. riparia*, seem to be more sensitive to day length than the *V. vinifera* vines (Fennell and Hoover, 1991). Once the leaves have fallen the vine is in a state of ecodormancy, which is controlled by the environment (e.g. cold temperatures) as well as in a state of endodormancy (through physiological factors within the plant). Therefore, buds will not develop immediately even if environmental conditions improve. While vines do not have a chilling requirement (i.e. needing a certain amount of time below 10°C in order to emerge from endodormancy) per se, cold temperatures do facilitate the process, decreasing the length of time needed for the plant to start growing again when environmental conditions improve (Schnabel and Wample, 1987).

The major task of pruning occurs during the dormant season (Chapter 6).

Flower Initiation, Fruit Development and Berry Maturation

The perennial nature of the grapevine is nicely illustrated through an examination of the vine's reproductive cycle. Here we will see that fruiting potential is initiated over 1 year before the crop is harvested.

The fruitfulness, or how much potential crop, a vine has can be determined well before budbreak by dissecting and examining the primordial shoots within, or by forcing the buds to grow out early. Such studies have shown than flower clusters for the next season are initiated about 2–3 weeks before the current season's flowers are opening (Lavee *et al.*, 1967; Pratt, 1979). An examination of buds at different node positions along 'Thompson Seedless' shoots in Davis, California showed that basal buds had greater fruitfulness than more distal buds at the start of June, but another sample taken in July revealed that buds at all node positions were fruitful. This led to the conclusion that the initiation of flower clusters occurred over time, and in sequence from the bottom of the shoot upwards (Winkler *et al.*, 1974).

Though this indicates when the inflorescences appear to be initiated, it does not tell us the whole story. Microscopic examination of developing buds has allowed the sequence of flower cluster initiation to be derived.

Where do tendrils and flower clusters come from?

Earlier, it was mentioned that tendrils and flower clusters are related. This is so because they are derived from the same uncommitted primordium (called an anlage) produced as another growing point off the primordial shoot's apical meristem. The primordium can become one of three things, two vegetative and one reproductive: a tendril, a shoot or a flower cluster. If it becomes a tendril it will develop, but not branch very much; if it becomes a shoot, then the main shoot will split, forming a fasciation, or bifurcated shoot (see Fig. 3.9). If it is destined to become an inflorescence it will undergo much branching, which will lead to what we eventually see as a flower cluster.

It is of note that tendrils are not found on shoots below flower cluster positions (if present), so tendrils can be thought of as the default form of the anlage, where conversion to a flower cluster happens only under certain conditions. Indeed, the distinction between an inflorescence and a tendril can become blurred, with the former having tendril-like appendages and the latter having their ends populated by groups of flowers, ending up as fruit (see Fig. 3.10).

Fig. 3.9. Left: shoot fasciation, where the shoot flattens and then splits into multiple nodes. A lateral and leaf petiole were removed before the photo was taken, but notice that there are still three leaves present. Right: a bifurcated shoot, not the development of a lateral shoot: note that the shoot arises from a position a tendril would normally be in. Content author's own.

Fig. 3.10. Tendril-like cluster of berries, the result of only a small number of flowers being initiated when the tendril/cluster was being formed. Content author's own.

For the flower cluster to appear as we would expect, it needs to undergo repeated branching. This branching can continue through the season and, by leaf-fall, much of it has been completed, laying the foundation for next season's potential crop. However, more branching can occur around the time of budbreak after the dormant season, or whenever environmental conditions allow (May, 2000).

Factors affecting branching

Because buds in the varying node positions go through the process of flower cluster initiation at different times, changing environmental factors that affect this process will also affect the potential crop for each bud in the following season.

Temperature is one such factor, as it affects all biological processes. Cool and excessively hot temperatures discourage the initiation process, though absolute values for minimum temperatures necessary for any initiation to occur vary from cultivar to cultivar (Buttrose, 1969). The optimum for maximizing potential crop is thought to be in the range of 30–35°C, but this also appears to vary with cultivar (Buttrose, 1969). In many cool climates, achieving this range of temperature during flowering, when flower cluster initiation is going on, is relatively rare, hence the chance for high crop loads is already limited by this factor.

Light is also important, as it affects the rate at which photosynthesis can occur and hence photoassimilate supply. The carbohydrate supply near bloom is an important factor affecting the number and potential size of the flower clusters being initiated (May, 1965; Sommer *et al.*, 2000), but Bennett *et al.* (2005) have also shown that vine carbohydrate status following fruit set will also affect flower cluster number and size in the following season. Hence, overly vigorous shoots are associated with fewer flower clusters because vigorous vines generally have shadier canopies, and also the growing points of a vigorous shoot are much better at drawing carbohydrates away from the developing flower clusters.

Plant growth regulators

As with many aspects of plant development, plant growth regulators play an important role in flower cluster initiation. The two most involved are gibberellic acid (GA) and cytokinins (CKs).

GA enhances the development of an uncommitted primordium that forms to the side of the shoot apex, encouraging the development of tendrils (Srinivasan and Mullins, 1980) but inhibiting the formation of inflorescences (Srinivasan and Mullins, 1980; Palma and Jackson, 1989). CKs will encourage the formation of flower clusters from the tendrils (Srinivasan and Mullins, 1980), so the net effect of applying GA to a grapevine shoot during the flower cluster initiation period will be to promote the production of tendrils rather than flower clusters. The net effect of applying CKs to the shoot would be more flower clusters.

The role that each of these plant growth regulators plays has been elucidated in part through the use of chlormequat, which has been used as a herbicide. This chemical blocks the synthesis of GA (primarily, though it may have some effect on CK production as well) (Skene, 1968; Lang, 1970).

Cytokinins are also important later on in flower development, as their presence promotes the formation of flower primordia on the inflorescence near budbreak in the following spring, and can improve fruit set during flowering as well (for a review, see Srinivasan and Mullins, 1981). Analysis of the bleeding sap from cuts on vines before budbreak show that CKs are being transported from the roots to the upper parts of the vine at this time of year (Skene and Kerridge, 1967), which demonstrates the potential importance of soil temperature early in the season, as warmer soils will have more root growth and potentially more CK production. Work by Woodham and Alexander (1966) showed that, for each of two 10°C increments (from approximately 10°C), if the root temperature was higher than that of the shoot, there was an approximate increase of 18% in flower number per inflorescence using cuttings taken from dormant vines and grown in glasshouse conditions.

Flower development and anthesis

Differentiation of flowers at the terminal positions on the inflorescences starts near the time of budburst (Srinivasan and Mullins, 1981) as soil and air temperatures begin to rise, which also means that the full cropping potential of the vine has been realized.

The pollen sacs in the anthers mature shortly before capfall (Staudt, 1999), as does the receptivity of the style itself. Thus, it is thought that the mechanical disruption caused by the movement of the cap off the ovary and style, with probable transfer of pollen from the anthers to the style (a process called pollination) as well as other factors, means that grapes are mostly self-pollinated (Lavee and Nir, 1986).

The timing of capfall has been associated with a certain number of nodes on the flower cluster's shoot, though the number seems to vary between cultivars. However, it does not seem to vary much within a cultivar or between seasons (Pratt and Coombe, 1978), which means that counting the number of nodes on shoots can be a way of predicting flowering date, as long as the number of days it takes to form a node is known. Some research suggests this is in the order of 3 days per node (Lovisolo and Schubert, 2000), though this rate depends on environmental conditions.

The duration of flowering is also highly dependent on the environment at the time. Cool, overcast weather, associated with rainfall, lengthens the flowering period, whereas warm and sunny conditions hasten it. Thus, flowering can occur over a period of a few days to longer than a month. This is thought to contribute to variation found in the fruit through the rest of the season, although there is also considerable variation found within the flower cluster itself (May, 1987; Friend *et al.*, 2003).

Capfall occurs mainly in the morning hours (Staudt, 1999), with the highest rate occurring between 7.00 and 9.00, and the final ones falling by 12.00. This is thought to be a result of changes in turgor pressure within the cells in the calyptra's abscission zone (Swanepoel and Archer, 1988). Cooler temperatures and rainfall are associated with lower rates of flower opening, helping to lengthen the flowering period (May, 1987; Friend, 2005).

Fruit set

Once viable pollen lands on a receptive stigmatic surface a germ tube emerges from the pollen grain within a short period of time, as little as 30 minutes (Staudt, 1982). It then begins to grow down through the style to the micropyle, following a 'J'-shaped path, and on to the nucellus (see Fig. 3.5). There is some thought that the size of the flower, and thus the length the pollen tube must grow, is a factor in the success of fruit set. This is because the speed at which the pollen tube grows is very much dependent on the temperature, with colder temperatures resulting in drastically lower rates of elongation (Staudt, 1982). If the growth is slowed to the point where the ovules degenerate before the pollen tube reaches them, fertilization cannot take place.

Fig. 3.11. Seed remnants within stenospermocarpic seedless table grapes. Content author's own.

Plant growth regulators are also involved in this phase of reproductive growth, as auxins are released from the pollen tube (Taylor and Hepler, 1997) as it grows, which stimulates growth of the ovarian tissues.

If the pollen tube does reach past the micropyle, the generative nucleus in the pollen unites with the synergids in the ovarian embryo sac, causing a rapid increase in metabolic activity (Kassemeyer and Staudt, 1981) that results in further development of the now fertilized embryo. The ovule becomes the seed, and the ovule wall enlarges to become the berry pericarp, or flesh.

Commonly, one or two seeds are found per berry in *V. vinifera* fruit, usually more than two on average for *V. labrusca* (Ebadi *et al.*, 1995), though there is considerable variation between cultivars (Forsline *et al.*, 1983). Berry size seems to be driven initially by the presence or absence of seeds, and then by the mass of seed within the berry rather than just the number of seed (Scienza *et al.*, 1978; Ebadi *et al.*, 1996; Trought and Tannock, 1996; Roby and Matthews, 2004; Friend, 2005). In terms of yield, it is most important to have at least one viable seed.

Seedlessness

Evolutionarily speaking, the seeds within the fruit are the most important part of the vine: the vine produces seed to propagate and spread the offspring. So how have seedless berries come about? After all, if no seeds are produced, then how can the vine broadcast its genetic material? As with the appearance of clones, random mutations can cause a vine to arise that has seedless berries. In addition, some crosses of plants can result in sterile offspring, so a seeded cultivar pollinating another seeded cultivar may result in a seedless offspring.

In most cases, seedless cultivars are not, in fact, really seedless. Many of the popular seedless table grapes are stenospermocarpic. In this case, the flowers are pollinated and the embryos fertilized, but soon after the embryo aborts; however, in the time that the fertilized embryo is developing, it produces enough plant growth regulators to encourage growth of a large berry. If a seedless grape is cut apart, often a small remnant is visible inside, which is the aborted seed (see Fig. 3.11). In some cases, there may be something that looks like a viable seed, but it is in fact the lignified shell of a seed with no embryo within. 'Einset Seedless' (developed at the NYSAES, Geneva) is one cultivar that is prone to having these crunchy seed remnants, though it appears to be seasonal in nature.

Parthenocarpy, a process that requires pollination but not fertilization to set fruit, occurs with cultivars like 'Black Corinth' (syn. 'Zante Current', Fig. 3.12). These vines have very small berries, from a lack of cell division in comparison with seeded ones (Coombe, 1973), as without the seed present, division and growth of the cells of the ovarian wall are not encouraged (Olmo, 1946).

Fig. 3.12. A 'Black Corinth' / 'Zante Current' cluster showing the small size of the parthenocarpic berries. Content author's own.

Factors influencing fruit set

There are many factors that will affect the percentage fruit set in grapevine. Among the most important are the availability of light, moderate temperatures and dry weather. The effects of temperature on pollen tube growth and root-produced plant growth regulators have already been mentioned. However, light is at the core of many vine functions as it is the source of carbohydrate energy for the plant. If conditions are overcast, the vine will not be able to produce enough photosynthates to feed the growth of shoots and all other sinks, and so flower clusters, being a relatively weak sink, will not attract a great deal of them. This correlates with how the grapevine grows in the wild, as the vine is vegetative until it has grown up the tree trunks and emerged into the higher light levels in the canopy above.

As the vine is a perennial plant, the amount of carbohydrate stored is also a factor – if the vine has sufficient reserves in its root and other permanent wood, then it is better able to supply energy to the growing parts when photosynthesis falters. This is used as evidence to support the notion that older vines (having more permanent wood than younger vines) perform better (Howell, 2001). In addition, since the early-season shoot growth is highly dependent on stored reserves, if reserves are not sufficient then shoot growth will suffer and so, too, will the leaf area available to manufacture photoassimilates.

Research has demonstrated a positive link between vine carbohydrate levels pre-budburst and percentage fruit set in a cool-climate growing region (Bennett *et al.*, 2005). This has led to the suspicion that stored carbohydrates are more important than previously thought when it comes to determining how much fruit will set at bloom. This is true in warm climates as well, as if vines are deprived of postharvest photosynthetic recovery, subsequent productivity suffers (Holzapfel *et al.*, 2006).

Water must be available to the vine near flowering to allow photosynthesis to proceed at the optimal rate, as water stress has been found to be highly detrimental to fruit set (Hardie and Considine, 1976; Jackson, 1991). However, as flowering occurs early in the season, water is usually available in the soil profile and stress must be actively avoided in only the driest of regions, or in those areas with soils of very low water-holding capacity.

One physiological disorder that is probably connected to a lack of available carbohydrate is early bunchstem necrosis (syn. inflorescence necrosis). This fruit set problem results in the loss of individual flowers, branches of flower clusters or even entire flower clusters in the weeks leading up to and including the fruit set period (Jackson and Coombe, 1988; Ibacache, 1990; Fig. 3.13). It is thought that this problem is a result of a build-up of toxic ammonium in the flower cluster tissues because of an inability of those tissues to convert ammonium to the amino acid glutamate – a process that requires carbohydrates (Jordan, 1993). Its appearance is sporadic and difficult to predict, but is associated with cool, wet springs and therefore lower carbohydrate production by vines. It can have devastating effects on yields, but the best way to prevent it would seem to be avoiding within-canopy shade, keeping the vines healthy and with a good carbohydrate reserve. A similar disorder occurs as harvest approaches - bunch stem necrosis, or waterberry (Fig 3.14) where a section of the rachis dies and the berries distal to it stop accumulating sugars and shrivel. This has been associated with shaded fruit and various nutritional disorders (perhaps similar to early bunchstem necrosis), but there is still no clear picture of the conditions needed for it to occur (Bondada and Keller, 2012).

Fig. 3.13. Early bunchstem necrosis, or inflorescence necrosis at three stages of fruit development. Left: part of a badly affected flower cluster. Middle; mid-season the necrotic flowers are still present in the cluster. Right: some clusters affected by this disorder can end up looking like this at harvest. Content author's own.

Fig. 3.14. -Late season bunch stem necrosis, or BSN, on (left) a Syrah cluster, where the necrotic section of the rachis is visible, and (right) Grenache clusters. Content author's own.

Deficiencies in certain nutrients can also reduce fruit set, chief among them being zinc and boron. Zinc is necessary for activation of many enzymes, as well as for the production of the plant growth regulator indole acetic acid (Marschner, 1986), and boron is necessary for the formation of cell walls (Thellier *et al.*, 1979), which is involved in all aspects of plant growth.

In fact, any significant stress the vine is under, be it abiotic (e.g. nutrient or water) or biotic (disease or insect pest damage), can reduce the potential of the vine to set fruit.

Berry development and maturation

Once the embryo is fertilized in grapes, the vine is committed to maturing the fruit, unlike some other fruit crops where there is fruit abscission and thus a chance to reduce what might otherwise be too high a crop. Fertilization results in immediate and rapid cell division (from 1 to 2.5 per cell on average) to move from approximately 200,000 cells at anthesis to a maximum of 600,000 by veraison (Harris *et al.*, 1968). However, the highest activity in terms of cell division occurs before flowering, where about 17 cell divisions are responsible for attaining that figure of 200,000 in the first place.

In phase 1 of growth there is no significant accumulation of sugars in the berry, as much of the photoassimilate is used for cell division and expansion. However, there is an accumulation of organic acids, primarily malic and tartaric (Hrazdina *et al.*, 1984), the latter of which is relatively unusual in fruit crops (McGovern *et al.*, 2004). The duration of phase I seems to be similar for most grape cultivars (Nakagawa and Nanjo, 1965; Coombe, 1976), while that for phase II can vary considerably depending on cultivar, management and environment (Winkler and Williams, 1935; Coombe, 1960). Many characteristics in the grape change in the period leading up to maturity.

Grape composition

The primary component of mature grapes is water, making up about 75–85% of their weight (see Table 3.1). Approximately 15–25% is in the form of sugar, a higher percentage than in many other fresh fruits. The organic acids tartaric, malic and citric make up 0.5–1.0% of the fruit, pectin about 0.25% and there is a long list of other nutritional components. If all of these are totalled they amount to over 99% of the weight of grapes but, if they were the only contributors, we would not be able tell grapes apart from many other fruits.

Fleshy fruits like grapes are made up of living cells and therefore have all the 'primary metabolites' of all plants. Primary metabolites are the proteins, amino acids, nucleic acids, fatty acids, cellulose and others found in all plant cells that facilitate respiration, photosynthesis and, to all intents and purposes, life. These primary metabolites make up a higher percentage of the total mass in leaves, buds, flowers and other non-reproductive parts compared with fruit.

Table 3.1. Raw grape composition, comparing *V. vinifera*, *V. labrusca* and *V. rotundifolia*. Values refer to amount per 100 g fresh weight (USDA, 2025a).

Component	*V. vinifera*	*V. labrusca*	*V. rotundifolia*
Water, g	80.5	81.3	84.3
Carbohydrate, g	18.1	17.1	13.9
Sugars, g	15.5	16.2	13.0
Fibre, total dietary, g	0.90	0.90	0.90
Protein, g	0.72	0.63	0.81
Potassium, g	0.19	0.19	0.20
Total lipid (fat), g	0.16	0.35	0.47
Phosphorus, mg	20.0	10.0	24.0
Calcium, mg	10.0	14.0	37.0
Magnesium, mg	7.00	5.00	14.0
Vitamin C, mg	3.20	4.00	6.50
Sodium, mg	2.00	2.00	1.00
Iron, mg	0.36	0.29	0.26
Niacin, mg	0.19	0.30	na
Vitamin E, mg	0.19	0.19	na
Vitamin B-6, mg	0.09	0.11	na
Zinc, mg	0.07	0.04	0.11
Thiamin, mg	0.07	0.09	na
Riboflavin, mg	0.07	0.06	1.50
Vitamin K, mg	14.6	14.6	na
Vitamin A, RAE, mg	3.00	5.00	3.00
Folate, DFE, mg	2.00	4.00	na

Most of the attributes that make grapes desirable to us (and to certain pest species!) accumulate or develop in phase III of berry development. One of the most important of these is sugar, in the forms of glucose and fructose.

The currency of the phloem, which is the network of conduits that moves photoassimilates from one place to another in the vine, is sucrose, which is not the same as that for photosynthesis or for ripening grapes. Photosynthesis creates glucose, which is freely interconverted with fructose. One each of these molecules is combined into sucrose prior to their journey through the phloem and then on to the sink (growing point, fruit, etc.), where it is taken out of the phloem and converted back into glucose and fructose. Therefore, very little sucrose (which is the form of sugar we are most familiar with, as it is what makes up the granulated sugar we put in much of our food) is found in ripe grapes (0.15% sucrose compared with 7% and 8% by weight glucose and fructose, respectively).

Grapes are high in carbohydrates and not a particularly good source of dietary fibre. However, they are a useful source of many minerals and vitamins B_6, C, E and K. They are also a source of antioxidant compounds through the phenolics in their skins and possibly seeds (Yilmaz and Toledo, 2004).

What is fruit maturity?

Fruit maturity and ripeness are often-used terms in the grape industry, but it is difficult making the linkage from these to the quality of the grapes. This is because quality is, by its very nature, a qualitative term, not a quantitative one – there are very few absolute measures of quality, as one person's quality may differ from another's. Today, industries talk about suitability for end use to try to get around this elusive concept of quality.

There are several measurable parameters in grapes that relate in some way to quality factors. One of these is some measure of sugar concentration, which usually is accomplished by estimating the amount of dissolved compounds in the juice. As the vast majority of dissolved compounds are sugars (glucose, fructose and sucrose), this ends up being a pretty accurate representation.

There are several different and commonly used measures. Baumé is a convenient term for winemakers as the number corresponds to the approximate alcohol percentage once the grapes have been fermented to dryness. The Brix value corresponds to the approximate percentage sugar in the solution and is an adaptation of the percentage soluble solids, which is used in many industries to measure sugar in liquids. Specific gravity is an actual measure of the density of a solution, commonly determined through the use of a hydrometer. Its value is expressed with respect to the density of water, as a solution with dissolved sugar will be heavier than water. Oechsle is a derivative of specific gravity where the value of one is subtracted from the value for specific gravity and the result multiplied by 1000. Table 3.2 shows the relationship between these different measures. Note that the specific gravity of a liquid changes with its temperature, so this is a factor that must be held constant or corrected for when comparing units.

Because many parameters associated with quality are linked with changes to sugar in the berry, Brix (or other measures related to sugar concentration) is a commonly used measure linked to meeting minimum requirements for harvest. For example, 'Concord' grapes harvested for juice production may have to reach a minimum of 15 Brix to be accepted by the processing plant. Bonuses can be associated with exceeding this target, or there may be deductions from a base pay rate if the crop comes in below the target.

As the degree of sweetness is affected by the amount of acid mixed in with the sugar, and vice versa (McBride and Johnson, 1987), the amount of acid in grapes is also often measured. This is most often performed by titrating a sample of juice with a standardized basic solution to a certain end point (usually pH 8.2).

Table 3.2. Equivalent measures of sugar concentration for four scales used in the grape industry. Note that the relationships between the measurements are altered with changes in temperature of the liquid.

Baumé	Brix	Specific gravity	Oechsle
12.5	22.5	1.094	94

For wine grapes, the titration is usually done with 0.10 normal sodium hydroxide (NaOH) and the result expressed as titratable acidity (TA):g tartaric acid (the predominant organic acid in grapes)/l solution. In some areas, data are expressed on a sulphuric acid basis and sometimes to a slightly different end point, which means that TA measurements are not always directly comparable.

Because the ratio of sugar to acid (usually Brix:TA) is important to perceived taste, it is particularly important in table grape production and has been formally recognized as being such since the 1960s (Nelson *et al.*, 1963). As long as sugar levels are high, grapes with relatively high acidity will still be acceptable to the market. At the other end, if acidity is too low in relation to sugars, the grapes can taste overly sweet and insipid (Liu *et al.*, 2007).

Related to, but not the same as TA, is the concentration of hydrogen ions (H^+) in a solution. This is expressed as pH, and is important for the biological stability of grape juice and wine as well as having an effect on the ionic forms of some molecules, such as anthocyanins, affecting their colour. pH is an important measure that is not directly related to TA as one might think; pH measures the H^+ present in solution at any given time whereas, through the process of titration, H^+ that are bound to organic acids are progressively removed and measured. So, juices with identical TA can be coupled with either low or high pH, depending on the relative amounts of organic acids and other cations present (such as potassium) that can take the place of H^+.

Sometimes it is not what makes up the fruit that is important, but other factors, such as physical damage, presence of disease or non-grape materials (leaves, stems, insects, sand, etc.). These can often be measured objectively. For example, the level of botrytis infection in grapes can be estimated by measuring the amount (or activity) of the enzyme laccase in the juice (Zouari *et al.*, 1988), as this enzyme is not found naturally in grapes but comes from the disease organism.

More recently, quantifying other aspects of quality has been attempted, such as objective measures of (i) colour through spectrophotometry; (ii) tannins through precipitation assays (Harbertson *et al.*, 2002; Herderich and Smith, 2005) or high-performance liquid chromatography (Vrhovsek *et al.*, 2001); and (iii) flavour compounds (Ortega-Heras *et al.*, 2002; Howard *et al.*, 2005). Most of these factors had previously been assessed through sensory means. This process introduces an amount of subjectivity into the picture, which is often incompatible with determination of scales such as pay rates for fruit.

However, it is worth noting that aspects that were once considered purely subjective are now able to be measured objectively (e.g. sugar concentration). It is inevitable that progress will be made in this area, though commercialization may take some time.

Content versus concentration

Distinguishing between concentration and content in berries is also of value. Many parameters are measured on a concentration basis (Brix would be an example of this) but, as the size of the berry changes considerably while ripening, it can give a skewed view of what is actually happening. For example, conventional Brix monitoring during ripening may show the solid line in Fig. 3.15, which indicates that Brix increased through ripening, but the rate of increase was less on 31 March. Usually, a decrease in Brix is associated with rainfall, but in this case there was no rain during this period. The cause of the slowing in Brix accumulation is not apparent but, if the sugar concentration per berry is calculated and plotted (dashed line), we see that the sugar content of the berries decreased at this sampling date. However, this is not possible at this stage of berry development, so another reason could be sampling error: a less ripe population of berries was taken on this date, perhaps because a different person did the berry collection. So, it is likely that there was not a dip in the Brix at that date, but rather it was the result of an artefact of sampling. Because accurate sampling is crucial for making harvest decisions, tracking the content of sugar along with the concentration gives more reliable information for making harvest decisions.

Factors affecting fruit maturity

It may be useful to think of grapes at harvest as being a summation of environmental conditions for the whole of the season. Aspects such as berry size, colour, flavour, their number and how diseased they are related back to the conditions under which they developed. Decisions that were made before planting have a continuing effect on the grapes, as site selection determines how the grapes will turn out, in a broad sense. For example, the day/night temperature regime can affect aroma and colour in fruit (Kliewer and Torres, 1972; Becker, 1977; Ewart, 1987), so that grapes grown in a continental climate taste, and may look, very different to the same grapes grown in a maritime climate.

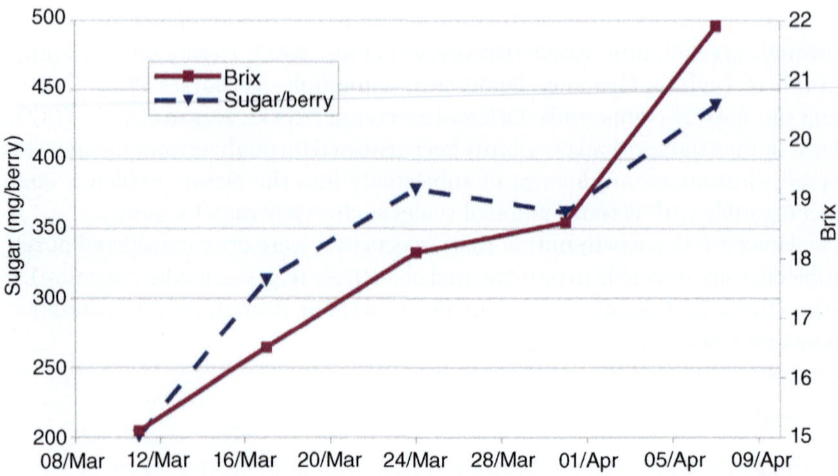

Fig. 3.15. Example of differences in trends in sugar levels when data are expressed on a per berry versus concentration basis. Content author's own.

How the vines have been managed also has a large impact, as row orientation and many canopy management decisions (e.g. shoot positioning, shoot thinning, leaf removal) can change harvest composition dramatically (Jackson and Lombard, 1993).

Some unexpected changes as a result of management decisions can also occur, e.g. the use of a sterol biosynthesis inhibitor as a spray to control powdery mildew has been found to interfere with terpene biosynthesis in 'Muscat of Alexandria' berries (Aubert *et al.*, 1998). In addition, bird netting – essential for the production of grapes in many areas – can also have an effect, slightly delaying fruit ripening (M.C.T. Trought and G.L. Creasy, 1999, unpublished results), possibly from the slight shading effect, but also from deformation of the canopy, creating shade.

There are very few things that do not change with berry development. A pivotal point of this process is the time near veraison, or between phases II and III of berry development. At this time there is a dramatic shift towards expansion, which brings with it a number of changes. The cells within the berry lose structural integrity very suddenly, resulting in a mixing of the symplast (intracellular contents) and apoplast (extracellular contents) and causing the berry to soften and become more translucent. Cell walls become thinner (Considine and Knox, 1979) and the berry skin more stretchable (Matthews *et al.*, 1987) as the berry expands and ripens.

Amino acids also change with grape maturity. These are the building blocks of proteins and so are extremely important to plant function. They are also an important grape constituent, particularly for winemaking, as the yeast cells that ferment the sugar into alcohol utilize nitrogen from amino acids in the must. Stuck fermentations are sometimes attributable to a lack of yeast-assimilable nitrogen (YAN), which is mostly that in the form of primary amino acids (especially arginine) and free ammonium ions in solution. However, yeast is not so well able to utilize nitrogen from secondary amino acids such as proline, which tends to be in higher concentrations in riper grapes (Kliewer, 1968; Bisson and Butzke, 2000). Hence, monitoring YAN in grapes can be an important predictor for possible stuck fermentations. Indeed, one can measure YAN in fruit weeks before harvest and get a good idea of what YAN will be available in the fruit before fermentation starts (Nisbet *et al.*, 2013).

The amount of primary amino acids and ammonium in grape must varies with management, cultivar and environmental conditions. For example, 'Chardonnay' and 'Pinot noir' were shown to have higher YAN than 'Cabernet franc' and 'Traminette' in New York State (Nisbet *et al.*, 2014), and vines growing in drier conditions and lower available soil nitrogen tend to have lower YAN (Wade *et al.*, 2004). This also means that cover crops, which compete for water as well as nitrogen in the soil, will tend to decrease fruit YAN. Therefore, growing grapevines organically, where there is increased competition from under vine plants, may lead to low fruit YAN issues.

It is notable that ripe *V. labrusca* grapes have much higher arginine than proline concentrations (Kluba *et al.*, 1978), potentially leading to fewer stuck fermentations. However, the amount of wine made from this species is very small compared with that from *V. vinifera* grapes.

Secondary metabolites

The components that make a grape distinctive are loosely grouped together and termed 'secondary metabolites', because they are not core to the survival of the plant (those were mentioned before as being primary metabolites). Many thousands of secondary components have been identified and many more are probably not yet recognized. Some are found in all plants while others in only a single species at a specific time in their life cycle. Their synthesis by plant cells is under genetic control. Fifty years ago, the field of chemo-taxonomy was established on the recognition that these secondary metabolites were sometimes very specific. The discovery that native American grapes and *V. vinifera* grapes had different types of anthocyanins (Ribereau-Gayon and Ribereau-Gayon, 1958) was used as a method of distinguishing between the species (and adulterated wine!). Research analysing plants for their content of secondary products was used to help differentiate them and test taxonomic and phylogenetic hypotheses (e.g. Moore and Giannasi, 1994).

Once DNA analysis was discovered, chemo-taxonomy was replaced by direct comparison of the genetic code (e.g. DNA fingerprinting) rather than the products of metabolism. This was because chemo-taxonomy was not directly determined by the DNA of the plant – there were other factors involved as well (e.g. plant age, condition and environment of the individual plant) that influenced secondary metabolite composition. Ultimately, viticulturists attempt to manipulate the secondary metabolite expression in vines as this enhances the quality (or suitability for purpose) of harvested grapes.

Within the general concept of 'secondary metabolites' are a large group of plant components called phenolic compounds. Phenolic refers to phenol, and the commonality of the group derives from a 6-carbon aromatic ring with at least one hydroxyl group (see Fig. 3.16).

Phenolic compounds are in highest concentrations at around 50 days post-bloom and then fall progressively as berries mature (Ristic and Iland, 2005). In addition, flavour compounds, or their precursors, can be synthesized in the berry at this early stage, leading to what is known as the unripe or green characteristic of developing fruit.

Some phenolic compounds are synthesized from branch pathways of the biosynthetic sequence to phenylalanine. These include the benzoic acids – parahydroxy-benzoic acid, protocatechuic acid, vanillic acid and gallic acid – all of which occur in plants. Plants alone have the full biosynthetic sequence to manufacture the aromatic amino acid phenylalanine, and animals are dependent on plants for a dietary supply of this essential amino acid. Plant reactions convert phenylalanine to thousands of aromatic (in terms of structure, not necessarily aroma) compounds, many of which occur in grapes.

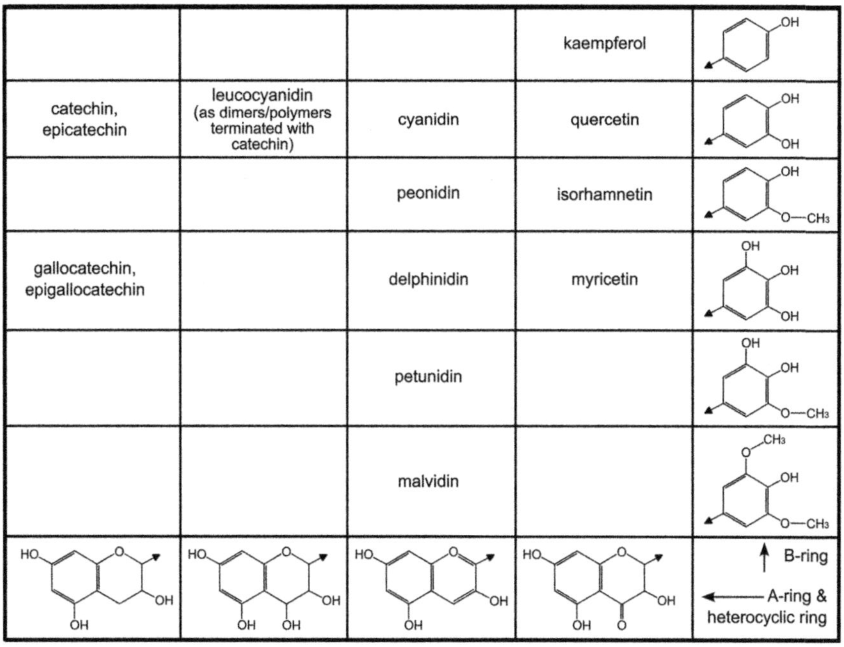

Fig. 3.16. Major phenolic compounds of mature grape berries. The A-ring substitution runs across the bottom with the trivial name of the components in the crossing boxes. Content author's own.

The products of phenylalanine metabolism include the chemical groups of non-flavonoids and flavonoids. Major flavonoid groups in grapes are tannins (e.g. flavanol oligomers and polymers contributing approximately 0.5% by weight), flavonols (0.01%) and anthocyanins (0.1%) (Brossaud *et al.*, 1999). The groups' (closely related) structures are shown on the right column of Fig. 3.16.

Anthocyanins are those molecules that contribute the bulk of red colour to grapes. The best-known flavonol is quercetin, found in high concentrations in onions, but also in significant amounts in tea, grapes and wine (Hollman and Arts, 2000). Kaempferol, myricetin and isorhamnetin are other flavonols found commonly in grapes; however, their concentration is much lower than that of quercetin (Cheynier and Rigaud, 1986). The tannins are made up of compounds from the flavanol group (e.g. flavan-3-ols), and contribute to grape astringency and bitterness.

In addition, flavonols and anthocyanins occur as glycosides (i.e. having a sugar molecule attached) for which there may be different sugars, which may also be in different positions on the hydroxyl groups and/or linked together to form polysides. Chemo-taxonomy of anthocyanins in wines was used to trap unscrupulous wineries in France that were diluting traditional varietal wines with higher-producing and cheaper-to-grow hybrid cultivars, through the different mixtures of anthocyanins in the species. This technique has also been used to detect the addition of elderberry juice (which also has different anthocyanins) to wine (Bridle and Garcia-Viguera, 1996). The phenolics are thus a diverse group, contributing to colour and aspects of aroma, bitterness and astringency in grapes as well as participating in many interactions with other molecules.

The non-flavonoids are smaller compounds that often associate with flavonoids, and are also important in browning reactions that occur in juice and wine. They can also contribute to bitterness and aroma, particularly in wines, though their overall effect is much less than that of the flavonoids. Examples of non-flavonoids would be gallic acid, protocatechuic acid, *p*-coumaric acid and ferulic acid. Gallic acid is a simple and easily obtainable phenolic compound, so mixtures of phenolic compounds are sometimes expressed on the basis of gallic acid equivalents (GAEs).

Phenolics are found in all plant tissues, hence the inclusion of leaves, rachides and stems in grapes destined for juice or wine production can alter the character of the resulting product. Within the berry, however, the non-flavonoids are primarily found in the pulp, and therefore are easily extracted into juice or wine. Of the flavonoids, the anthocyanins are the most readily apparent in red grape cultivars and are usually found only in the berry skin. There are, however, some grape cultivars that have anthocyanins in the pulp, which are known as teinturier type grapes (see Fig. 3.8). These will express red juice if the ripe berries are squeezed, quite a contrast to most red grapes. Examples of these grapes include 'Alicante Ganzin', 'Carmina', 'Kolor' and 'Pontac', which are usually used as a blending agent in wines to increase the intensity of red colour.

The flavonols are found in the skin of grapes, with almost none present in the pulp or seeds (Singleton and Esau, 1969), though green parts of the vine – such as leaves and rachis tissue – contain high amounts (Price *et al.*, 1996). Price *et al.* (1996) found that there was a strong relationship between the amount of flavonols in berries and the degree to which they were exposed to the sun. Spayd *et al.* (2002) showed that if berries were shielded from UV radiation the amount of flavonols produced dropped significantly, but not to the levels of shaded fruit, demonstrating that it is not just light that is responsible for the relationship. The connection between berry exposure and the amount of flavonols formed is such a strong one that the relative amount of quercetin (it being the flavonol in highest concentration) found in the grapes could be used as an indicator of the exposure of those grapes to the sun during the growing season.

Flavan-3-ols are found in both the seeds and skins, with there being qualitative differences between the two sources. Additionally, it is easier to extract (and perceive) those from the skins as opposed to the seeds, as the seeds are enclosed by a hardy covering, reducing the rate of transfer to the juice or wine (Singleton and Draper, 1964). Similar to flavonols, flavan-3-ols are found in the green parts of the vine as well (Price *et al.*, 1996).

Anthocyanidin pigments in grapes include these five: malvidin, cyanidin, delphinidin, petunidin and peonidin. However, they are not found naturally in these forms, but rather as the glycosylated anthocyanins: malvidin-3-glucoside (malvin), cyanidin-3-glucoside, (cyanin), delphinidin-3-glucoside (delphinin), petunidin-3-glucoside (petunin) and peonidin-3-glucoside (peonin). This is because the un-glycosylated form (aglycone) of the pigment is unstable (and less soluble in water), which is true of flavonols as well. In fact, adding on other molecules can further enhance stability, such as attaching another glucose molecule to the A-ring (creating a 3,5-diglucoside) or attaching a non-flavonoid such as caffeic, or *p*-coumaric acids (creating an acylated pigment) (refer to Fig. 3.16). A peculiarity of the wine grape 'Pinot noir' is that its anthocyanins are not acylated (Boss *et al.*, 1996; Gao *et al.*, 1997), which contributes to this cultivar's wines being shy of colour.

Malvin was first isolated from grapes in 1915 (Willstätter and Zollinger, 1915), and is the pigment found in the highest concentrations (usually 40–60% of the total, depending on cultivar) (Boss *et al.*, 1996; González-Neves, *et al.* 2001). Interestingly, red-coloured clones or sports of white grapes (i.e. 'Pinot gris', 'Pink Sultana' and 'Red Chardonnay') usually contain a different primary pigment as well as possessing non-acylated anthocyanins, like 'Pinot noir' (Boss *et al.*, 1996). This probably contributes to the lack of colour stability in products made from them.

Anthocyanin production is affected by the environment, including such aspects as exposure to sunlight, temperature and temperature fluctuations. An elegant study by Spayd *et al.* (2002) in Washington State, USA, demonstrated the individual effects that light and heat have on anthocyanin production in 'Merlot' berries. By cooling exposed berries to shaded berry temperature and heating shaded berries to exposed berry temperature, the effects of each could be readily examined. As a result, they found that exposure to temperatures around 35–40°C reduced the production of anthocyanins, but more exposure to light increased colour. Unfortunately, as the study was conducted in a relatively warm climate situation, the effect of cooler temperatures on anthocyanin production could not be determined. However, a similar study done in the same place showed that reducing the diurnal range of temperatures in that environment resulted in higher anthocyanin at harvest (Cohen *et al.*, 2008), so the average temperature is not the only factor that is important.

Other studies have shown that temperature is important to the production of anthocyanins. High temperatures of 37/32°C (day/night) inhibited anthocyanin formation as well as reducing their concentration if plants with coloured berries were transferred to that environment (Kliewer, 1977). Yamane *et al.* (2006) have also shown that high temperatures inhibit colour formation in berries, but also that the stage of development at which they are most sensitive to this is just after veraison, when the berries are starting to accumulate colour. Cultivar differences are also apparent, with 'Tokay' preferring cooler night temperatures, but 'Cabernet Sauvignon' and 'Pinot noir' able to manufacture anthocyanins at higher night temperatures (Kliewer and Torres, 1972).

We know that, by changing the degree of exposure of grapes to the sun, we can alter the amount of colour they have. Is improvement in juice or wine colour as simple as increasing the amount of colour in the grapes? Probably not: aside from extraction not being 100% during the pressing or fermentation process, while abundant anthocyanins result in a more intensely red product, the susceptibility of the juice or wine's colour to sulphite bleaching (as sulphites are commonly added to protect from oxidation and microbial spoilage) is relatively high (Somers *et al.*, 1983). Juice and wine are also susceptible to changes in colour as a result of pH shifts (Somers, 1971) and for wine, as it ages, there is a tendency for the colour to fade.

We are beginning to discover what makes a deep and long-lasting wine colour. We know that improving fruit exposure also increases levels of both flavonols and flavan-3-ols (Price *et al.*, 1995; Phelps, 1999). Researchers have been getting a handle on more of the interactions happening between these different, but closely related, phenolic compounds.

Flavonols, such as quercetin, seem to be a key factor in improving grape wine and possibly juice colour, through a process called co-pigmentation. Quercetin is only a light yellow pigment, so does not contribute much to colour by itself. However, if a small amount of quercetin is added to a solution of anthocyanin pigment (such as malvidin-3-glucoside), the result is a fairly dramatic increase in colour density of the solution (Mirabel *et al.*, 1999). Figure 3.17 shows the positive effect on absorbance at 520 nm (peak wavelength for anthocyanins) through the addition of caffeic acid, *p*-coumaric acid and the flavonol rutin (quercetin-3-rutinoside). While there is a benefit from adding these acids, the rutin provides the greatest effect in this young wine. In this case, one plus one equals three: by increasing the amount of flavonols in the grapes, more colour is visible from the anthocyanins present. This effect is somewhat short lived in wine, however, in that as time passes, the number

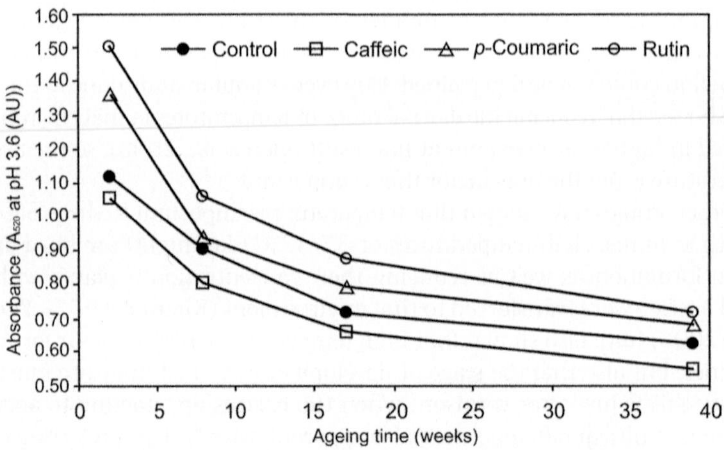

Fig. 3.17. The effect of adding various compounds to 'Cabernet Sauvignon' must prior to fermentation on the resulting wine's colour (reproduced with permission from Schwarz *et al.* © 2005 American Chemical Society J Agric Food Chem 53:8372-8381).

of anthocyanin molecules able to participate in co-pigmentation decreases (Somers, 1998). Up to 50% of the colour in a young red wine, for example, can be the result of co-pigmentation (Boulton, 2001).

Other substances have an effect on the perceived colour of anthocyanins. There are several metal ions that associate with anthocyanin molecules, such as Al^{3+}, Cu^{2+} and Fe^{2+}, which result in a slight colour shift in the pigments (Singleton and Esau, 1969). These associations may be partly responsible for the colour differences between *vinifera* and *labrusca* grapes, the latter of which have a more bluish hue (Ingalsbe *et al.*, 1963).

The flavanols are the third group of the most important flavonoid compounds. These include the flavan-3-ols catechin, epicatechin, epicatechin gallate and epigallocatechin as the most common in grapes (Souquet *et al.*, 1996). Their main importance lies in their involvement in the formation of tannins.

Tannins can be an important component of grape quality; they are phenolic compounds that are naturally found in grape skins and seeds, as well as in other plants (e.g. tree wood). A formal definition of tannins is: 'Water-soluble phenolic compounds, having molecular weights lying between 500 and 3000, which have the ability to precipitate alkaloids, gelatin and other proteins' (Swain and Bate-Smith, 1962).

Effectively, it is the ability of these compounds to precipitate proteins found in the saliva of our mouths that is important to the sensation of astringency. Increased precipitation of the saliva proteins translates into increased perceived astringency of the tannins (Gawel, 1998). In addition, tannins also may contribute to bitterness in grapes and, probably most importantly, juices and wines.

The basic building blocks (monomers) of grape tannins (also known as condensed tannins or proanthocyanidins) are the aforementioned catechin (C), epicatechin (E), epicatechin gallate (EG) and epigallocatechin. These monomers and their chains (polymers) are found in grape skins and seeds, as well as in green plant tissues such as the cluster stems (Sun *et al.*, 1999). The polymers can be two (dimer), three (trimer), four (quatromer) or more monomers in length; these short chains are sometimes referred to as oligomers. An example of a six-unit polymer might be represented by C-E-E-C-EG-C, which could also be referred to as a compound with a degree of polymerization (DP) of six. This representation is a possible conformation of a chain in that it is linear. However, connections between monomers can occur in two different ways, meaning that branched chains are also formed, creating complex mixtures of subunits and polymer configurations.

Chains are formed in a polymerization process that occurs in the ripening grape as well as during the winemaking process and wine ageing (Haslam, 1980). In the grape at harvest, skin tannins tend to have a higher average degree of DP (more subunits), while those from seeds tend to have a lower DP (Cheynier *et al.*, 1998; Downey *et al.*, 2003).

For table grapes, excessive astringency through the presence of tannins can detract from perceived quality (chewing the skins or seeds of a grape will bring out the more astringent tannins). The same occurs for juice production, which may mean the product needs to be manipulated to reduce the amount of tannins (e.g. through fining). Greater maceration of the berry skins and seeds results in more phenolic extraction, so the method by which the grapes are harvested and the juice extracted has an effect on this.

In winemaking, grapes that are left on skins following crushing release the most phenolics. Red winemaking procedures take advantage of this, as flavan-3-ol monomers and polymers are extracted from the solid parts of the grapes first by water and later more and more by alcohol, assisted by heat, as fermentation proceeds. Examination of skins and seeds would show that phenolics are more easily extracted from the former as opposed to the latter (Sun *et al.*, 1999). Therefore, the composition of the grapes has a major effect on the wine product – most perceptibly in colour because anthocyanins are extracted, but also in flavours and mouthfeel effects from other compounds.

In terms of sensory parameters, anthocyanins do not have mouthfeel characteristic themselves (Singleton and Trousdale, 1992) and quercetin may have only a slight bitter or numbing attribute (Dadic and Belleau, 1973; Price *et al.*, 1995; Vaia and McDaniel, 1996), so the extraction of these has little direct impact on juice or wine taste or mouthfeel.

However, adding complexity to this is that different types of tannins affect the product in different ways: flavan-3-ol monomers tend to contribute bitterness while the polymers tend to contribute more in terms of astringency (Gawel, 1998). In addition, all tannins are not created equal, as they can associate with anthocyanins to create pigmented polymers, and the length and complexity of tannins can have an impact on the quality of their sensory characteristics (astringency or mouthfeel). If an anthocyanin molecule attaches to a tannin chain, extension beyond the anthocyanin is not possible, so they become terminal units affecting polymer chain lengths. However, not only are pigmented polymers a more stable form of colour, but they also soften the quality of astringency compared to non-pigmented polymers, altering mouthfeel in beneficial ways (Weber *et al.*, 2013). The subunit monomers that make up tannin molecules, and the lengths and branching within the polymer, also affect its interaction with salivary proteins, affecting mouthfeel in significant ways. This is one of the reasons why research into this area is difficult – there are many and varied combinations that cannot be measured, but result in sensory changes.

Grape aromas and flavours

With the exception of table grapes, where fruit texture is a concern, the aromas and flavours that are associated with grapes are their *raison d'être*. Grapes can be found to be bland, insipid, flavourful or overpowering to the taster, reinforcing the fact that there is a tremendously large range of characteristics and intensities in existence. From the neutral taste of 'Sultana' to the fruity muscat characteristic of 'Gewürztraminer' to the foxy flavour of 'Concord' to the acidic, astringent and bitter sensation of *V. riparia*, there are flavours to meet almost any need.

FOXY GRAPE How the term foxy came into use to describe the flavour of native American grapes is not clear. It may have come about because the wild grapes that grew in North America were thought of as wild or feral by the European settlers, or that the word is a derivative of 'faux' or 'false', which was a term used to describe the grapes that came from seedlings of a desirable vine, in that they usually looked nothing like the parent. An excellent discussion on these and other possibilities for the origin of the term can be found in Pinney (1989). It seems, however, that there is no clear origin for its use.

It had been thought that the foxy character was due solely to the presence of methyl anthranilate in the grapes, but it has been shown that there is not a direct relationship between the amount of this compound and the intensity of the foxy aroma, rather, methyl anthranilate contributes more to the grapey aroma associated with them (Amerine *et al.*, 1959; Nelson *et al.*, 1977). In addition, methyl anthranilate has been found in *V. vinifera* wines (cv. 'Pinot noir'), albeit at low concentrations (Moio and Etievant, 1995), and recent research associated 30 or more chemical compounds associated with the foxy characteristic (Moriyama *et al.*, 2024).

However, being able to quantify and classify flavours and aromas is difficult, due to each person's own interpretation of how to describe them and their particular sensitivities to aromas. The analytical capability to separate and identify individual aroma compounds is expanding rapidly. As with other qualitative characteristics, we are moving from an era of using our senses to measure these aspects of grapes to a more quantitative one, such as through the use of gas chromatography (GC), GC linked with mass spectrophotometry (GC-MS), GC tandem mass spectrometry (GC-MS/MS), where another mass spectrometer is placed after the first, and gas chromatography ion mobility spectrometry (GC-IMS). The value of the human nose has not been lost to science, however, as the use of GC-olfactometry – where an instrument separates the compounds in a sample and a person measures them by rating each one's intensity and describing the quality of smell (via some descriptor, such as fruity, like a rose flower, etc.) as they emerge from the GC – is now common. This technique, in combination with parallel analysis of the sample by GC-MS, can lead to a much better understanding of how each compound matches with a smell and contributes to the overall aroma of a grape.

Aromas arise from many different chemicals, and in some cases the collective effects may be different to the individual ones. For example, researchers in New Zealand found that when ethyl octanoate, which contributes red cherry characteristics to 'Pinot noir' wine, is mixed with ethyl decanoate, which by itself contributes no aroma, the result was a black cherry aroma (Tomasino, 2011). Additionally, one aroma compound can mask the perception of another, making prediction of how a mixture of compounds will be perceived almost impossible. It should be noted that these synergistic and masking effects could be due to molecular interactions, but more likely they are due to perception by the human subject (Ishii *et al.*, 2008).

Additionally, many aroma compounds are present in grapes but are not odour-active, meaning that there is something that blocks our ability to detect them. In many cases this blocking is through glycosylation of the compound, and removal of the glucose from the molecule makes it readily detectable. The process of heating crushed grapes or grape juice enhances this process, as well as the presence of yeast during a fermentation, as they excrete enzymes that will cleave off the sugar (Delcroix *et al.*, 1994). Therefore, the spectrum of aromas and flavours in wines often greatly exceeds that found in the grapes themselves, or even the juice made from them.

In terms of primary grape aroma compounds that have been studied in detail, two classes stand out: terpenes and methoxypyrazines.

TERPENES Terpenes are a family of chemicals that (generally) impart fruity characters to grapes and are all generated from the common precursor geranyl pyrophosphate. The three significant classes of terpenes are monoterpenes, sesquiterpenes and C13 norisoprenoids.

Commonly found examples of monoterpenes are linalool, geraniol, nerol and citronellol, which are found in higher concentrations in cultivars such as 'Muscat', 'Sauvignon blanc', 'Gewürztraminer', 'Riesling', etc. In many cultivars, the monoterpenes dominate their ripe grape characteristics. The smells they are associated with are tropical fruits, orange, rose, floral, etc. They are found in both the flesh and skin of grapes (Wilson *et al.*, 1986) and in two main forms: free (which includes some polymers called polyols (Williams *et al.*, 1980)) and glycosylated (described above). The former are sometimes referred to as free volatile terpenes (FVT) and the latter potential volatile terpenes (PVT).

The monoterpenes accumulate as berries mature, though the FVTs and PVTs can do so at different rates. Increasing fruit exposure generally increases terpenes, though PVTs tend to be more affected than FVTs. Fruit that is not well exposed to the sun tends to have lower monoterpene levels, as do those grown in warmer areas (Ewart, 1987; Belancic *et al.*, 1997). The latter can result in grapes losing varietal characteristics when grown in excessively hot climates.

Interestingly, terpenes are also involved in a signalling response from plants. For example, grapevine leaves damaged by Japanese beetles release a number of aromatic compounds, including limonine and linalool (Loughrin *et al.*, 1997), and these compounds affect the behaviour of pests, such as the grape berry moth and light brown apple moth (Cha *et al.*, 2011; Suckling *et al.*, 2012).

Sesquiterpenes come from a similar synthetic origin as monoterpenes, having 15 carbon units as opposed to the monoterpenes' 10, and were not originally thought to be important contributors to grape aromas. In 2008, Wood *et al.* reported that rotundone, a sesquiterpene found in high concentrations in white and black pepper, was responsible for the aroma in some Australian 'Shiraz' (syn. 'Syrah') grapes and wines, particularly those from cooler climates or seasons.

Rotundone is found mainly in the grape skin, and accumulates post-veraison, the timing of this dependent on season (Caputi *et al.*, 2011; Logan, 2015).

Although it has been reported to be in higher concentrations in seasons with less heat accumulation (Caputi *et al.*, 2011; Logan, 2015), in other research higher amounts were found in a warmer season (Geffroy *et al.*, 2014), leading those authors to speculate that water stress may have been related to the increased amounts. Although the effects of water stress have not been fully elucidated, further research strengthens the link between heat and low rotundone concentrations, as Zhang *et al.* (2015) showed that the greater the length of time air temperature was above 25 °C, the less rotundone found in the fruit and that berries in shaded portions of the canopy, or shaded parts of clusters, had higher concentrations.

Cultivars where rotundone can have a sensory effect include 'Syrah', 'Mourvédre', 'Durif', 'Vespolina' and 'Grüner Veltliner', to name a few (Wood *et al.*, 2008; Mattivi *et al.*, 2011). There was considerable variation between 'Syrah' clones in one New Zealand study (Logan, 2015), meaning that if the peppery aroma is desired, there are a number of ways to enhance rotundone in the fruit. However, one study has noted that 40% of their sensory panellists were anosmic to rotundone (Gaby *et al.*, 2020), suggesting that using this aroma as a marketing tool may be problematic.

The third class of terpenes are the C_{13} norisoprenoids, derived from carotenoids (which are generally orange-coloured pigments, e.g. as found in carrots and tomatoes that assist in the harvest of light energy in photosynthesis and, importantly, protecting the photosynthetic reaction centres from damage caused by high light intensities, Griffiths *et al.*, 1955; Baker, 2008). An example of this type of aroma compound is beta-damascenone, which has cooked apple- or raspberry-type characteristics.

Levels of norisoprenoids in grapes increase with light exposure and environmental temperature, and are significant in such cultivars as 'Syrah', 'Cabernet Sauvignon', 'Chardonnay', 'Semillon' and 'Muscat of Alexandria'. Compounds in this group have also been found to be important contributors to 'Pinot noir' aroma, such as beta-damascenone and beta-ionone (berry and violet aromas) (Fang and Qian, 2006). As carotenoids are made only in the skins of berries, and because norisoprenoid concentrations have been found to be higher in wines made with skin contact (Cabaroglu *et al.* 1997), it is likely that the pulp of the berry contains little or none of these compounds. Norisoprenoids in the grape tend to exist in the glycosylated form, meaning that until the grapes are fermented, the compounds have little sensory impact.

Some connection has been discovered between the levels of a carotenoid compound and those of 1,1,6-trimethyl-1,2-dihydronaphthalene (TDN), which has been associated with an atypical ageing (ATA) character in wines, particularly Rieslings (Marais *et al.*, 1992). As levels of the carotenoid compound go down, levels of the TDN released from the grapes increase, and high levels of TDN in wine relate to a kerosene-like character that appears in young affected wines, but is more associated with older, aged wines. Although ATA is more strongly linked with a different chemical (2-aminoacetophenone) in wine of most regions, TDN appears to be a good indicator compound for ATA that can be measured (Henick-Kling *et al.*, 2008). ATA is associated with increased temperatures and light exposure in the vineyard, and vine water stress is also implicated, perhaps because vines under these conditions tend to shed leaves and expose their fruit more.

METHOXYPYRAZINES The previously discussed aroma compounds tend to be associated with the ripe end of the flavour spectrum, and their concentration tends to increase with time. The last group we will discuss, the methoxypyrazines, are associated with unripe grapes, and amounts tend to decrease during grape ripening.

Isobutylmethoxypyrazine and isopropylmethoxypyrazine are the most commonly found of the methoxypyrazines (MP), and are important in cultivars like 'Sauvignon blanc', 'Cabernet Sauvignon' and 'Semillon'. They are impactful components in grapes, as their high concentration in unripe grapes has been thought to be a deterrent to animals that might otherwise want to eat them before the seeds are viable (Guilford *et al.*, 1987), and their presence can go a long way to masking the riper aroma compounds that might be present. In winemaking it is usually undesirable to have much MP in the grapes as it is associated with aromas such as green capsicum or canned peas, which do not harmonize with well with the desired berry fruit aroma of ripe grapes.

This is very much a stylistic issue, as in some cases the green character has been found to be beneficial, as per the New Zealand 'Marlborough Sauvignon blanc' style. In this, the maritime climate has conspired to result in a mix of ripe aromas (e.g. from certain thiols, which are sulphur-containing compounds some of which have aromas like passion fruit) and unripe (from, for example MP and hexenols) that has defined a new international style quite apart from the traditionally accepted style that was mostly fruit driven (Parr *et al.*, 2007).

One of the reasons why these compounds are of such importance is that the human sense of smell is very sensitive to them. Isobutylmethoxypyrazine can be detected at concentrations as low as 2 ng/l, or two parts per trillion (Buttery *et al.*, 1969).

There has been much research into the factors that affect the rate at which it degrades as grapes ripen, primarily increased light and temperature (Marais, 1994; Allen *et al.*, 1997; Hashizume and Samuta, 1999; De Boubée, 2004), but there is less research on what factors influence how much is created in the grapes in the first place. Research done in Texas showed that increasing light and temperature pre-veraison also decreased MP, together and independently (Plank *et al.*, 2019), meaning that there would be less MP to degrade during ripening. It has also been suggested that methoxypyrazines are imported to the grape berries from the leaves (De Boubée, 2003), which, if true, would put another layer of complexity on its management. Research using bell peppers (*Capsicum annuum*) was unable to demonstrate the movement of MP through phloem (Zamolo *et al.*, 2023), but it remains a possibility.

Methoxypyrazines are also important in some regions because of an insect pest. The multicoloured Asian ladybird beetle (*Harmonia axyridis*) was introduced to many areas around the world as a biological control (biocontrol) agent for aphids, as which it has been fairly successful; however, it has a tendency to find refuge in grape clusters near harvest. Because they contain a potent-smelling compound (2-isopropyl-3-methoxypyrazine), even the inclusion of as few as one beetle per kilogram of fruit can be enough to taint a wine, even less depending on cultivar and wine style (Galvan *et al.*, 2007).

In all cases, colour, flavour and aroma constitute some of the most important attributes that grapes and grape products can have. Compounds contributing to aroma (and flavour, but since the two are so closely linked, we refer to aroma only) were evolved in plants to attract pollinators or seed dispersers, and so are called secondary metabolites. These are produced only when the plant requires them, so we see them in flowers (to attract pollinators) or in fruit approaching ripeness (to attract animals that may move the seed around). In grape, flowers may or may not be scented since, though they do have nectaries, it appears that they are not always active (Pratt, 1971). Therefore, pollination by insects may or may not be of importance. However, of primary importance to consumers of grapes and grape products are those aroma compounds made in the ripening fruit.

Environmental/Climatic Influences

As with many organisms, the environment has a major impact on how the grapevine grows. Since the vine is a perennial plant, it exists under a set of varying conditions and, in a vineyard setting, the plants experience variation through both space and passage of time. Growth and development of the vine (and thus composition of the fruit) is modified by the environment.

In order better to discuss how the environment impacts grapevines, it is useful to classify it. However, it is important to remember that the environment is an integrated system without specific boundaries and, just because we split it up into different categories, it does not exactly exist in that manner.

In winegrowing, climate has been split into three levels: macroclimate, mesoclimate and microclimate (Smart, 1982). Starting with the largest scale, macroclimate (also called regional climate) is defined as the general climate pattern as may be determined from a central recording station. A scale that could be associated with it is tens to hundreds of kilometres (Smart and Robinson, 1991).

The types of factors associated with macroclimate are: (i) latitude (which influences sunlight hours, among other things); (ii) altitude; (iii) temperature and altitude versus latitude (e.g. for every increase in altitude of 150 m, there is a decrease in average temperature of 1°C); (iv) topography (large scale, such as mountains/landscapes); (v) length of growing season (time between the last spring frost and the first frost of autumn); and (vi) quality of growing season (factors such as humidity, presence of relevant insects and diseases).

One step down from there is mesoclimate (also called topoclimate or site climate). Every factor that affects macroclimate also influences mesoclimate, but the scale being discussed is smaller, in the order of tens of metres to kilometres. Factors here include differences in elevation, slope, aspect or proximity to large bodies of water. These affect vineyard temperature, wind, rain, relative humidity, frost-free period (down to 150 days, with 180 days preferred; Tukey and Clore, 1972), row orientation and slope (aspect and inclination), which can reduce frost risk, increase warmth and usually has faster-draining soils and increased air movement.

Soil type may also have an effect; for example, light-coloured soils contributing reflected light in an albedo effect.

The smallest scale of climate is microclimate, also known as canopy climate. Like mesoclimate, microclimate is a subset of the larger scale climates above it, so every factor that affects macro- and mesoclimates also influences microclimate. More precisely, microclimate is the climate within and immediately surrounding a plant canopy, which means a scale of millimetres to metres.

Factors associated with microclimate include: (i) bunch and leaf exposure and their temperatures; (ii) localized photosynthetic differences (e.g. sun-flecks); (iii) irrigation dripper zones; (iv) fertilizer patches; and (v) plant water supply necessary for cell expansion/growth and stomatal function. Much of what the viticulturist does to vines on a seasonal basis relates best to the microclimate.

Many people use the term microclimate when they are referring to mesoclimate: as in, 'This vineyard has a wonderful microclimate'. If the subject has a scale of tens to thousands of metres, then it should be referred to as mesoclimate.

Trying to quantify climate

Once we have classified climate to help us understand it, we could also use some method of measuring the effect climate has on the growth of grapevines. A popular method of doing this is known as calculating the growing degree days (GDD) for a location.

This has its origins in France, where accumulated heat was used to predict harvest dates (Angot, 1883). Later, Professor Bioletti of the University of California, USA used temperature and proximity to the ocean to divide California into six different growing regions for planting table and raisin grape vineyards in the early 1900s (Bioletti, 1929). Later researchers found that this was not precise enough for the production of wine grapes, and set about improving the methodology used to create the regions. Using heat summation as a basis, Amerine and Winkler (1944) performed an extensive study using temperatures in areas already successfully growing wine grapes and quality parameters of the wines being produced. As a result of this, they defined five growing regions within California depending on the number of GDD accumulated. Each region produced grapes with distinct properties, even when comparing the same cultivar (Winkler *et al.*, 1974).

A GDD in this case is defined as the number of degrees the average temperature for a given day is above 10°C. This can be calculated on a daily basis and the results added to determine the number of GDD for a time period. In some cases, there may not be access to daily temperature data, so monthly average temperatures can be used, e.g. the monthly average temperature in °C minus 10°C, multiplied by the number of days in that month, with this calculation for each month summed over the growing season.

The threshold of 10°C is used for grapevines because there is little biological activity in plants below this temperature. An example of calculating the GDD for the month of February: if the average temperature for the month was 19.6°C, then GDD = (19.6–10)*28 = 269. Note that negative values of GDD are not included in the growing season summation. For a cool-climate season, the data look like that presented in Table 3.3. Data are usually presented in relation to the growing season, or generally April through to October in the northern hemisphere and October to April in the southern hemisphere, though in areas with a longer growing season (time between killing frosts), such as the eastern coast of New Zealand, it can be longer. In hot climates grapes are usually harvested long before the autumn weather, so it's more helpful to discuss GDD from the beginning of the growing season to typical harvest date.

Although access to monthly temperature averages is reasonably common, calculations done on a daily basis will better reflect what the vines experience and, similarly, those that integrate information on hourly, or even smaller, intervals are also available and are claimed to be the most accurate way of forecasting the suitability of a site for the production of a certain wine style (Smart, 2003). However, the availability of data with that sort of resolution is fairly limited, especially in a new region that is being evaluated for grape growing.

In comparing GDD calculations, we must know on what basis the data were calculated. For example, calculating the data on a daily basis gives a slightly different value for the same time period (see Table 3.3). In addition, because plants will develop whenever temperatures are above the 10°C threshold, data from the months surrounding the traditionally defined growing season

Table 3.3. Growing degree day (GDD) calculation for the 2016–2017 southern hemisphere growing season, Lincoln, Canterbury, New Zealand. GDD calculated on a monthly and daily basis are presented for comparison.

Month	Average temperature (°C)	Days/ month	GDD/ month	Cumulative GDD (monthly)	Cumulative GDD (daily)
September	10.4	30	12	12	30
October	12.0	31	62	74	102
November	14.1	30	123	197	225
December	15.7	31	177	374	402
January	17.2	31	223	597	626
February	17.2	28	209	806	827
March	14.9	31	152	958	979
April	12.3	30	69	1027	1051
May	7.1	31	0	1027	1057

may be appropriate. In this case, for example, the GDD for the location given in Table 3.3 is 18 GDD higher for September due to individual days that were greater than 10°C and in May there were a few days warmer than 10°C, but because it was cooler later in the month, the average is below 10°C. These late-season warm days can be important to those cultivars that are harvested at the very end of the season.

Gladstone (1992), in his book *Viticulture and Environment*, took a more detailed and physiological view of GDD and added several modifications to suit the grapevine, such as limiting the effect of very hot temperatures, as photosynthesis – and therefore vine productivity – stops as the environment becomes too extreme.

While 10°C has been settled on as a standard, it is not necessarily the most appropriate threshold temperature to use (as discussed earlier). Moncur *et al.* (1989) suggested that base temperatures for budbreak should be 4°C and for leaf appearance 7°C, based on studies with grapevine cuttings. However, for most uses, the base temperature of 10°C is adequate for the bulk of the season and, importantly, comparable with much of the data already available for many areas.

There are a number of other methods of estimating the suitability of an area for grape growing, including the mean temperature of the warmest month and the Latitude Temperature Index, the latter developed by Jackson and Cherry (1988). Every method has its advantages and disadvantages, so it is important to know some of the limitations of whichever system is used.

Having an historical database of information related to the climate at any given site can be very useful when it comes to predicting when important phenological stages of grapevines (such as budbreak, flowering, bunch closure, etc.) will occur. Keeping records of these stages, along with seasonal weather information, can help to make better management decisions, saving money and time.

The Vine as a Perennial Plant (Carbohydrate Partitioning)

Because the vine is a woody perennial, it is important to know something about how it survives from year to year, and possible interactions of this with management of the vines.

In the dormant season the vine survives by metabolizing stored energy. Starch is the primary storage form of carbohydrates in the permanent parts of the vine (Mullins *et al.*, 1992), and it is solubilized by conversion to glucose and fructose whereby it can be used for respiration and maintenance of the plant cells. Because there is no photosynthetically active tissue at this time, the vine is totally dependent on the stored reserves, not only to survive until the next growing season but also to invest in early shoot and leaf growth, which will then be able to produce photosynthates than can sustain further growth and, ultimately, contribute more stored carbohydrates.

As such, factors that affect the ability of the vine to make or store carbohydrates will potentially affect how the vine will grow in subsequent seasons. For example, if a vine is over-cropped, much of the carbohydrate the vine produces will go into ripening the fruit. Edson *et al.* (1993) demonstrated this with potted vines by adjusting crop load and looking at the effect of that on total vine dry weights. Figure 3.18 shows that the total amount of dry weight the vines produced (which is a measure of the vine's ability to create dry matter, see discussion on vine capacity in Chapter 6) was similar (and not statistically different) across crop loads. Petrie *et al.* (2000b) demonstrated this as well, but with the added treatment of leaf removal. Results of that experiment showed that total vine dry weight was proportional to the leaf area present regardless of the presence or absence of crop, but that the presence of fruit drew away dry matter from the vegetative parts of the vine leading to much reduced leaf, root and shoot growth. Yet other researchers (Bennett *et al.*, 2005) have shown this in the field, where defoliated vines had much less soluble carbohydrate in the trunks and roots at the end of the season, which had follow-on effects in terms of the fruitfulness of the vines in the year following defoliation. In warm climates, where vines can photosynthesize for weeks or months following harvest, late defoliation still has a significant impact on return performance (Holzapfel *et al.*, 2006), highlighting the perennial nature of the vine and the importance of stored carbohydrates.

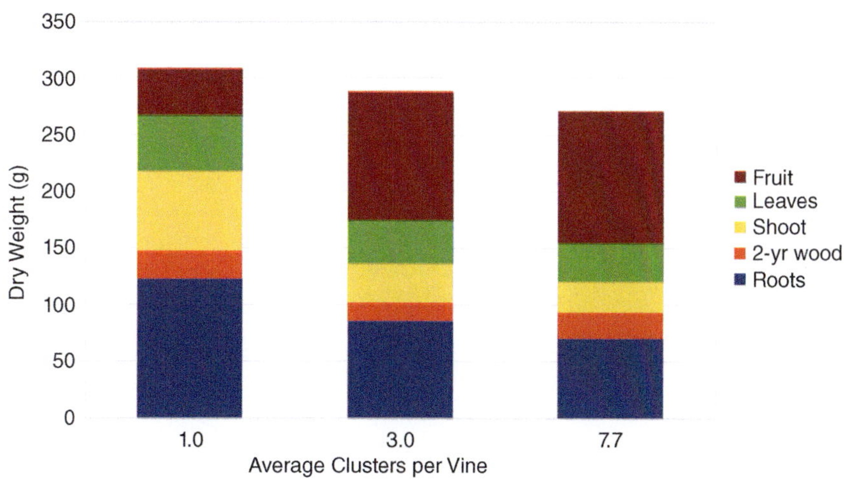

Fig. 3.18. Effect of different crop loads on total potted vine dry weight and its partitioning at the end of the experiment. Despite large differences in the amount of fruit between treatments, there is little difference in the total vine dry weights (data re-plotted and used with permission, from Edson *et al.* © 1993 American Society for Enology and Viticulture. AJEV 44:139-147).

Balancing vegetative and reproductive growth

The discussion above highlights the importance of finding the right balance between the growth of vegetative and reproductive parts of the vine, given that the vine has a fixed capacity, or ability to make carbohydrates, in any given season. If there is too much crop, then the vine will decline in terms of shoot and leaf growth, and also have less energy to survive the dormant season and produce early-season shoots and roots in the following season. If there is too little crop, then vines will have excess growth that will have to be trimmed and the vineyard will not be sustainable economically.

The balance between the vegetative infrastructure and the commercial crop will vary depending on location, climate, type of grape being produced, management, etc. Therefore, there is no way accurately to predict what the numbers will be that make up the balance, and even from year to year that balance will vary, along with the weather and presence or absence of disease (and, to a lesser extent, the age of the vines).

In fact, the viticulturist's job is never finished: it is akin to walking a tightrope when there are unpredictable winds – constant, and intelligent, adjustment is necessary.

Climatic Requirements

4

Abstract: Grapevines are a hardy and adaptable plant, allowing grape production in diverse climates, but there are limits. Vitis species vary in their tolerance to dormant season below-freezing temperatures, but as budbreak approaches, all suffer damage at temperatures not much below 0°C. Because Vitis evolved in temperate areas, special management techniques are needed to grow them commercially in tropical areas. In a practical sense grapevines have the same basic environmental needs as any other plant during the growing season, though soils have a significant impact on how vines grow due to depth, water and nutrient holding capacity and pH. The impact of this is implicated in the concept of terroir, though it depends on which definition of that term is being used. Climate change is actively changing the influence of terroir on vine growth and is a point of disruption even now.

In many areas, winter low temperatures define which grapes can be grown. The limits can be modified by lake influences, slope and many cultural practices. In other areas, the climate is too mild for easy commercial production of grapes, as their growth habit in these conditions encourages uneven cropping. Management decisions such as choice of specific cultivars, training system, ground cover, delayed pruning, burying vines in the soil, use of mulches and vine health factors such as crop load and insect and disease control (leaf health) are tools through which grapevines can be grown in less than ideal areas, potentially overcoming environmental limitations.

This chapter reviews some of the physiological requirements of grapevines with regard to their environment, with an eye to commercial production. Vines can survive almost everywhere, but managing them to the point where they produce a viable and economic crop in extreme climates can be the challenge.

Cold Hardiness

Almost every climatic region grows grapes. Grapes are grown in extremely hot regions and very cold regions, but for a long time it has been known that some

DOI: 10.1079/9781800627062.0004

cultivars perform better in warm areas and others better in cooler areas. Cool versus cold climate locations produce different characters in the same type of grape grown, but cold climates create the additional challenge of grapevine survival through the winter period. Therefore, a realistic acknowledgment of the viticultural potential of a region is required: while it may be possible to grow grapes in a site, that does not mean that it is possible to make a living out of it.

Low temperature is one environmental factor that presents significant challenges, but also potential advantages. Cold climate viticulture is defined by the winter low temperatures in the region, a parameter that is a major concern of site selection. Data exist on the historical low temperatures in most regions likely to be planted to grapes. Figure 4.1 shows the 100-year minimum temperature probability for Geneva, NY, USA. In this example, in half of the past 100 years the low temperature was –22°C or less, which is a pivotal temperature for native American grapevine survival. This information can be used to assess the degree of risk a prospective grower might be willing to take to grow a specific cultivar, though climate change means the risks are changing significantly.

The effects of cold temperatures in the dormant season are not usually apparent until the following spring, when buds do not grow, or grow weakly and then stall. Shoots may fail to develop normally in the spring if one or more of the shoot primordia have been killed by the low temperatures (see Fig. 4.2).

Figure 4.3 (top) shows a photograph of a cold-damaged vine well after budbreak. Its shoots developed normally for a short while and then faltered. Closer examination showed that the cambial layer in the cane had been killed

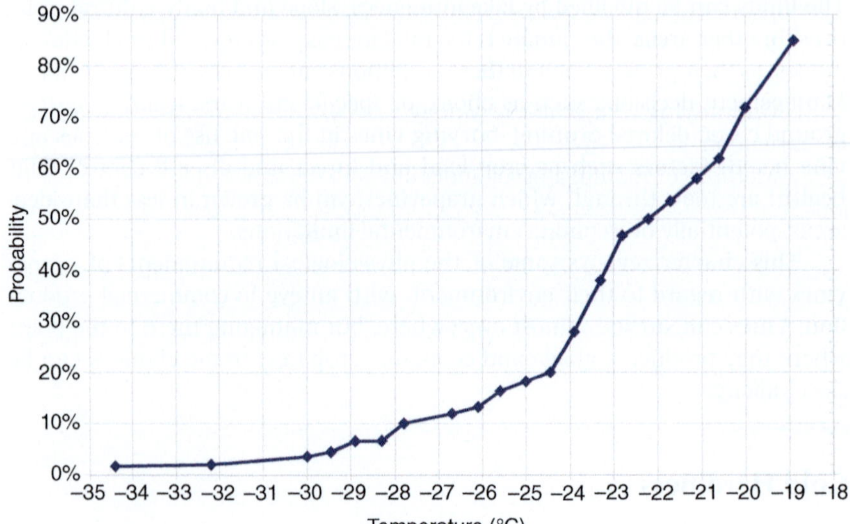

Fig. 4.1. One hundred-year probabilities for different levels of winter minimum temperatures for Geneva, New York. Content author's own.

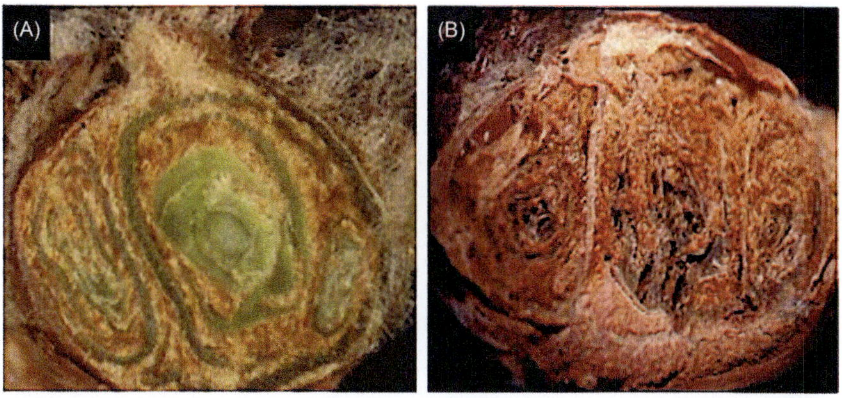

Fig. 4.2. Cross-sectional cut through two dormant grapevine buds. "A" is a healthy bud showing live green tissue for all three preformed shoots. "B" is a bud following severe cold temperatures where the preformed shoots have all been killed (reproduced with permission from Mills *et al.* © 2006 American Society for Enology and Viticulture. AJEV 57:194-200).

during the winter (Fig. 4.3, bottom). Since the xylem had not been affected, early shoot development was normal but, since there was no functional phloem in the cane, carbohydrates were not able to be moved to the developing shoots.

Grape species and cultivars differ in their adaptation to low winter temperatures (Pool *et al.*, 1990) and, as one might consider cool climate viticulture being defined by summer temperatures and frost during bloom, cold climate viticulture is directly associated with the minimum low temperature reached in the winter dormant period. Grape species have dormant survival ranges from a few degrees above freezing to less than 40°F below zero (–45°C) (Pierquet and Stushnoff, 1980).

Grapevines enter winter by gradually developing hardiness (the ability to withstand temperatures that the growing plant could not survive). Grapes acclimate to winter temperatures at different rates: some harden off more rapidly than others and some lose cold hardiness in the spring more quickly than others (Wolf and Cook, 1992). Not all grape buds on a vine are killed at the same low temperature, so some proportion of the buds may survive to produce a crop or, failing that, allow the vine to survive and grow another year.

Grape buds that have had time to develop hardiness generally survive lower temperatures, so therefore the more basal buds (for all cultivars) and buds on canes that stop growth earlier (on early-maturing cultivars) tend to survive lower temperatures. Any environmental event that shortens the leaf activity season, such as an early frost, will delay the development of maximum hardiness (Howell, 2001). A gradual change in temperature, as usually occurs during the transition from summer to autumn to winter, is most desirable in maximizing cold hardiness, as periods of warmer temperatures can deacclimate the buds and make them more susceptible to cold damage.

Fig. 4.3. Top; a healthy vine flanked by two with poor growth in early spring. Bottom; when the bark was cut away from the weak vines, it was clear that the phloem and cambial layer had been killed (brown tissue under bark), causing the poor shoot development. Content author's own.

It is possible to measure the killing temperature of grape buds during development of hardiness and in the depth of winter through a freezing isotherm procedure (Wolf and Pool, 1987), which has been used to predict the potential damage caused by a freeze event. With this knowledge, viticulturists can then adjust their pruning levels to try to compensate for the damaged buds. Canes may also be collected and their buds examined microscopically for damage (Goffinet, 2004). Percentage bud survival can then be determined and pruning to a proportionally greater number of nodes delayed until the danger of cold temperature is reduced. As grapevines have compound buds at each node position (with primary, secondary and tertiary buds enclosed within one set of bud scales) and the degree of hardiness of the buds varies, the effects of some freeze events can be overcome at pruning.

The possibility of damage from low winter temperatures can also be managed. The first defence (aside from selecting a site in an area that does not become so cold) is in choosing grape cultivars known to tolerate the expected low temperatures of the region. The second defence is to protect the vines from damage by the low temperatures through avoidance. One way of doing this uses the same technique as used for spring frost mitigation (Chapter 6), as long as an inversion layer, where warm air is on top of the cold air near the ground, exists. Fans can mix the warmer upper air with the colder lower air and prevent the temperature from reaching as low as it might have otherwise done. However, if no inversion layer is present, there is no possibility of warming up the air near the vines.

Another method of managing winter cold damage is to leave multiple trunks, frequently three to five, each a different age so that if some are killed (a selection of trunk ages spreads the risk of damage) others can provide fruit during the next growing season. As the trunks of grapevines are just as vulnerable to winter cold injury as the canes and buds, if they are damaged they are susceptible to the development of crown gall (see Chapter 10), which can result in vine death during the subsequent growing season.

A more labour-intensive method is covering or burying the vines or parts of the vines. Trunks or canes trained to a low wire, or those trained to the fruiting wire, can be taken off, laid on the ground and covered with soil using a grape hoe or plough (Fig. 4.4). The soil provides insulation from the cold air temperatures, but is also kept warm by the large mass of soil beneath. Straw mulch applied around the trunks (Fig. 4.5) or over canes laid on

Fig. 4.4. A dormant vineyard in Ningxia, China. Vines are trained such that the trunks can be laid on the soil surface and covered to protect them against cold temperatures, as well as physical damage, from for example, blowing sand. Content author's own.

Fig. 4.5. Straw-covered table grapevines in Japan. The bundle of dried plant material is supposed to limit heat loss from the plant tissue, but also prevent warming in the sunshine, which can lead to cambial tissues re-activating and becoming more susceptible to freezing temperatures. The uncovered trunks in the background are fruit trees, which are hardier than the grapevines. Content author's own.

the ground has also been used successfully for this purpose (Zabadal and Dittmer, 2003). This ensures at least a minimal number of bearing canes for the next season, although there can be a reduction in bud viability and cluster number as a result of leaving the canes under the straw (Goffinet and Martinson, 2007).

Another approach does not protect the bearing canes but buries the vine crowns, assuring that new growth can arise from the crown. However, if there are killing temperatures, there will be a delay in getting a crop off the vines as new shoots will need to be trained up on to the trellis. Because burying protects the crown from cold damage, infection by crown gall-causing bacteria is also limited, compared with methods leaving the canes exposed.

In the methods involving burying of vines, if they are grafted it is possible that scion rooting may take place (Fig. 4.6). This is where the scion develops roots from above the graft union and, as these tend to be quite vigorous, they can take over from the rootstock roots, resulting in an own-rooted vine. For this reason, it is important to remove all soil from near the graft union following the winter period.

To summarize, grape production in cold climates is accomplished by: (i) avoiding damage, usually by growing extremely hardy cultivars or by burying vines; and (ii) managing the damage by delayed pruning or leaving extra buds based on bud survival counting.

Growing Vines in Tropical Areas

At the other end of the scale, vines are also grown in areas that do not experience temperatures even close to freezing. This creates some problems of its own, not because the grapevines have trouble surviving the hot temperatures, but rather because they do not produce grapes at a single time, which means batch processing of grapes is effectively impossible.

Grapevine growth is indeterminant, meaning that as long as conditions are favourable, the shoots will continue to develop (much like a tomato vine). If a grapevine is left in this state, shoots will elongate *ad infinitum*, with the basal part of the shoot turning woody and laterals developing at node positions. Fruit would still be produced by the vine, but largely from lateral shoots that eventually would be growing at a range of stages, with reproductive development spanning flowering to over-ripe fruit. This kind of situation is not compatible with traditional production practices, where grapes are harvested at one time

Fig. 4.6. An example of advanced scion rooting caused by the vine being planted too close to the soil surface. The rootstock is still living, but large roots from the Aramon scion have emerged from above the graft union. Content author's own.

and batch processed. In addition, since many disease management practices are designed to coincide with certain phenological stages, control of pathogens such as botrytis becomes problematic.

So much of the challenge with growing grapes in perpetually warm climates has to do with trying to synchronize vine growth and defend it from the wider range of pests and diseases that are present.

Several methods are used to coordinate vine growth. Drought stress, severe pruning and/or leaf removal of vines can provide the stimulus necessary to start growth of the compound bud over that of the lateral (Lavee, 1973), but these methods are labour-intensive and potentially harmful to long-term vine health. Application of chemicals such as hydrogen cyanamide (Shulman *et al.*, 1983), calcium cyanamide (Kubota and Miyamuki, 1992) or preparations of garlic paste (which contain volatile sulphur compounds, Vargas-Arispuro *et al.*, 2008) also work to enhance budbreak (Kubota *et al.*, 2000; Botelho *et al.*, 2007). Once bud development has been synchronized, commercial management of the vines becomes much more viable.

Frost Tolerance

Frost in this sense refers to below-freezing temperatures that occur near or after budbreak, as the vines emerge from dormancy. Green tissue, being much higher in water content, is more susceptible to freezing than woody tissues. This is illustrated by controlled-environment experiments where grapevine buds at different stages of emergence were tested for the minimum temperatures they could survive. For example, 'Pinot noir' tissues were tested where the threshold temperature for significant damage progressed from −14°C for dormant buds, −3.4°C for swollen buds, −2.2°C at budbreak and increasing to −1.1°C for shoots with three leaves emerged (Gardea, 1987). As shoots grow they have less and less tolerance to cold temperatures, so date of budbreak (which varies with site and cultivar) is an important factor in susceptibility to frost events.

The effects of freezing on green tissue is devastating, causing rapid cell death followed by visible translucency and then browning (Fig. 4.7). This is followed by drying of the damaged tissue (Fig. 4.8). As with winter cold injury, not all shoots are equally affected, so damage can be patchy, in unpredictable patterns and with symptoms not appearing until well into the growing season.

Temperatures can also cause changes to shoot development without actually killing the tissue. Figure 4.9 shows changes to leaf chlorophyll that are caused by near-lethal freezing temperatures in the early part of the growing season. These are somewhat similar to symptoms of some types of herbicide damage.

Site selection is the best tool for managing frost risk, as it is costly to prevent and recover from. More information about frost management is given in Chapter 6.

Fig. 4.7. Frosted shoot and berries soon after the event. Tissue browns quickly and then dries out if the weather is fine. Content author's own.

Heat and Light

In terms of a requirement for light and heat there is little needed for a grapevine to grow in the wild. In low light situations the vine grows upward rapidly and, once it reaches bright light it fruits, with the focus being the bringing of whatever seeds are in the fruit to maturation and then rendering the fruit enticing enough for something to take it away.

In the commercial sense, however, there is a need for both in order that the vines have enough opportunity to produce carbohydrates that will go into both sustaining the vine from year to year and producing enough crop with adequate compositional parameters to keep the manager's business

Fig. 4.8. Spring frost-affected grapevine bud and shoots some time after the event. New growth from a secondary shoot can be seen growing in both images. Content author's own.

Fig. 4.9. Yellowing of leaves caused by early season sub-lethal cold temperatures while the leaf tissues were developing in the bud. In some cases, it can be severe (left) or more subtle, resulting in slight translucency of the blade (right). Content author's own.

sustainable. Integral with this is the choice of cultivar, as some require more heat (and indirectly, light) to ripen to the point where they produce the flavour and aroma characteristics that are expected by the consumer. So, while there is a minimal requirement in terms of heat and light for the growth and survival of grapevines, if the goal is to produce a commercial crop the requirements become more complex.

The use of growing degree days (see Chapter 3) or other heat accumulation tools is helpful in determining whether the selected cultivar and end use will match with a particular location. Ultimately, however, knowledge of what the end product is expected to be, how the vine interacts with the climate at the proposed location and to what extent management interventions are willing to be used are all necessary in making the most accurate assessment as to the success of the enterprise.

Water Use

As for heat, the requirement of the vine for water depends on its situation. A vine can survive on very little water, but it will not be able to sustain a crop in doing so. Addition of fruit into the equation means a greater canopy area is needed to ripen it, which requires a supply of water to allow the stomata to be open and photosynthesis to occur.

Many grape-growing regions of the world rely on natural rainfall to supply water to their vines, but many are also reliant on water brought to the vines. If there is no mechanism to supply vines with water, the amount and quality of the crop is dependent on rainfall and soil water-holding capacity. If the viticulturist has a dry ripening season, then water application becomes a (potentially useful) management tool. A balance must be struck between applying enough water to maintain vine photosynthesis and fruit development without applying too much, resulting in overly vigorous vines that require excessive management.

Typical values of water use efficiency (the tonnes of dry matter produced per megalitre of water used by the plants) range from 10 to 30 for grapevines grown for wine production, though the value depends much on how the irrigation water is applied to the vines. Flood irrigation brings values down, while deficit irrigation results in higher values (see Chapter 6).

Too much water available to the vines is also detrimental to quality grape production and efficient management. Vines will grow to luxurious levels with ample water supply, to the point where fruit can be well shaded, which reduces quality (Smart *et al.*, 1988). To correct this requires many more management interventions, such as shoot thinning, hedging, leaf removal in the fruiting zone and there are carry-on effects in the dormant season, the vines requiring much more time to prune. Hence, managing the amount of water the vines have access to is important for efficient commercial production.

Soils

Grapevines can grow and produce crop hydroponically or in sand culture, provided that water and appropriate levels of nutrients are supplied with it. If soil serves its primary purposes – anchorage and reservoir for water and an appropriate range of nutrients (see Chapter 8) – there is little direct effect of soil on grapevine growth and even fruit composition, though there is debate on this issue (Bodin and Morlat, 2006; Coipel *et al.*, 2006; Huggett, 2006; Van Leeuwen and Seguin, 2006).

However, the interaction of physical characteristics of soil and the origin of its parent material can have quite important effects on the magnitude of grapevine growth. For example, the chalk soils found in the Champagne region of France have a desirable combination of high water-holding capacity yet good drainage, porosity and permeability that makes them well suited to grapevine growth and grape production (Huggett, 2006).

The proportion of gravel in soil can have an effect on the root to shoot ratio (the dry mass of a vine's shoots divided by the dry mass of its roots), with high-percentage gravel soils reducing shoot growth relative to root growth (M.C.T. Trought, G.L. Creasy and G. Wells, 2008, unpublished results). This could be an advantage in maintaining optimal fruit exposure with less canopy management.

Soil properties like water-holding capacity are recognized as being important, but perhaps less so are how the soil makes the water available to the vine, and how the roots explore the soil volume. Anecdotally, clay soils have a reputation for producing higher quality wines, and in drier areas, this could have something to do with: (i) those soils having large reservoirs of water, but perhaps (ii) that they tend to dry out slowly from the surface to the lower depths, meaning that vines are getting water from deeper in the soil profile as the season progresses. This means that different roots are doing the gathering, and the soil solution has different nutrient status: an interesting idea that has not been explored as yet.

The factors discussed so far are fairly well connected with soil pedology, the study of soil in its natural environment, which focuses on how soils are formed, how they appear and how they can be classified. The other main study of soils is edaphology, which is all about how living things interact with the soil. Soil is a matrix for life, and it makes perfect sense that other living organisms in the soil will interact with the plants growing in it. Examples of soil-borne organisms are bacteria, fungi, nematodes, springtails, worms, beetles, and vertebrates like moles and mice. Mycorrhizae have been discussed already as a beneficial fungus in the soil, but there are hundreds or thousands of other organisms that could benefit grapevines, as well as those that are neutral or could harm them. The result of this 'soil food web' is an ecosystem that provides many benefits (decomposition and making nutrients available being significant ones) and is much more resilient to challenges (Barrios, 2007), which is why managing a vineyard now encompasses managing the soil and the living things in it (see Regenerative Viticulture section in Chapter 7).

Terroir

A discussion of the environment and grapevine interaction must include aspects of terroir, which is a term coined by the French and for which there is no direct English translation. Perhaps as a result of this, there is much confusion as to exactly what it means, so it has ended up meaning different things to different people (Moran, 2006).

It has been attached to other food products (such as cheese), but is of much interest in the wine industry. Aspects of terroir include territorial, ecophysiological, oenological, identity, promotional and legal (Moran, 2006). Analysed in such detail it is an exceedingly complex topic and therefore one of great discussion.

At the crux of it all, however, is the idea that aspects of the environment (which can take into account the physical and social) end up being associated with a unique style of a product. The style may be associated with flavour or aroma, marketing/branding or any number of other characteristics. In business, having a point of difference is key to marketing your product and retaining customers. Being able to say that your product is from a unique terroir is useful as a marketing tool as it provides a point of difference. The concept of terroir should be able to be applied to grapes in all forms, not just wine, although with the latter there are additional layers of nuance added to the product during winemaking.

One difficulty in developing the idea of terroir arises from the necessity that products coming from an area considered to have a similar terroir should have some underlying common characteristic, usually some aspect of flavour, aroma, and (in the case of wine) mouthfeel. For grapes that are as heavily modified after harvest as those used in wine, the influence of the winemaker also comes into play in determining the characteristics of the product. Thus, defining a region as having a recognizable terroir is usually more than just agreeing that it comes from a specific area or land: it also depends on those making the wine having some common thread such that the wines that are made from that area do have a recognizable trait, which can form the basis of market differentiation. It is therefore easy to see that the concept of terroir can be the subject of many and lengthy debates!

Climate Change

Alterations to the atmosphere's composition brought about by human activity is resulting in an increase in global temperature, which, in turn, affects many, many other processes on earth, both biological and non. The magnitude and rapidity with which these changes are taking place is a matter of some debate but it is inevitable that, within the life of most vineyards, the conditions under which they were designed and developed will not be the same. Therefore, understanding what climate change will bring is imperative when planning for the future.

Two of many global changes that are occurring are a rise in the average temperature and an increase in carbon dioxide (CO_2) concentration in the atmosphere. Both have an impact on the basic process of photosynthesis. The rate of photosynthesis increases with an increase in temperature (up to a point), and higher CO_2 levels mean that RuBisCO, the enzyme catalysing the conversion of CO_2 to carbohydrates, will operate more efficiently. Thus, on the basis of this, we would expect to see that plants will be able to produce more biomass than they were able to under the earlier conditions. Note, however, that with an increase in temperature there is also an increase in the rate of respiration, which counteracts photosynthesis. In addition, when one thing changes, others can change in complicated ways. Take, for example, that as global temperatures rise there will be increased respiration from soil organisms, but the level of increase is altered by historical levels of soil moisture (Hawkes *et al.*, 2017). This makes modelling its effects more complicated.

Despite all this, it is the changes to regional climate that have the most practical effect on grape production. The limits of grape production are moving northward and southward in the northern and southern hemispheres, respectively (van Leeuwen *et al.*, 2024), which was reported to have been happening in Germany decades ago (Schultz, 2000). Dates of budburst, flowering and veraison all moved forward by two (budburst and flowering) and three (veraison) weeks between 1965 and 2003 in north west Germany (Duchêne and Schneider, 2005) and harvest dates in Australia were earlier by up to 2.4 days per year over a span of 14 years (Petrie and Sadras, 2008). More recently, Italian researchers reported that in the 36 years from 1986, phenological stages advanced by 20 to 30 days, with increasing risk of damage from spring frosts (Faralli *et al.*, 2024). This last note confirms that this process is a two-edged sword, because as cooler areas become warmer and more acceptable for grape growing, warmer areas also get warmer and can get too hot for grape growing. van Leeuwen *et al.* (2024) report that around 90% of low-lying viticultural areas in the Mediterranean and southern California could be too hot for commercial production by 2100. At this point, it seems inevitable that some people's livelihoods are going to be uprooted in the next few decades.

Despite that threat in the future, right now we know that climate change is associated with an increase in the number of extreme weather events (Goodess *et al.*, 2003; O'Gorman, 2015) with severe impacts on grape production: spring frosts that occur later after budbreak can wipe out grapevines and their crops; flooding can physically damage vines; wind can break shoots and longer times between rainfall events can cause non-irrigated vineyards to fall out of production or reduce the area that can be farmed, etc.

Importantly, there are impacts on the quality of grapes grown as well. With increasing temperature there are changes to fruit ripening, sugar and acid levels and, importantly, flavour and aroma compounds as well as the relationships between them. Areas that are considered as having a hot climate will no longer be able to grow grapes, and those that are considered cool will no longer be able to produce the wine styles they were well known for (Jones *et al.*, 2005; Hannah *et al.*, 2013). In short, the landscape for wine production is changing and will continue to change drastically.

Another aspect of the environment that has changed in some areas is the amount of ultraviolet (UV) radiation that reaches the earth's surface. Normally, the ozone layer in the upper atmosphere filters out the short-wavelength UV radiation but, in areas affected by ozone layer depletion, more of this high-energy radiation affects humans and plants alike. Typical responses to more UV radiation are smaller and less green leaves, shorter internodes and greater growth of lateral shoots (Barnes *et al.*, 1990; Keller and Torres-Martinez, 2004).

Plants do have mechanisms to cope with higher incident UV radiation, such as the formation of flavonols, alkaloids, waxes and free radical scavengers, or mechanisms to repair the damage caused to components of the photosynthetic pathway, DNA, membranes, etc. (Jansen *et al.*, 1998; Kolb *et al.*, 2003), but the impact of incident UV has already been shown to have effects on fruit amino acid content and aroma compounds (Schultz, 2000; Martínez-Lüscher *et al.*, 2014; Šuklje *et al.*, 2014; Song *et al.*, 2015). In addition, the interaction between different types of stresses, for example, UV and water deficit, are not yet well described, so further research is needed in these areas. If terrestrial UV is to increase in the future, a distinct possibility with climate change (Barnes *et al.*, 2023), viticulturists should be aware of the potential effects on grapevine growth and wine style and use management techniques to avoid them.

Vineyard Establishment

5

Abstract: The most important decision made in growing grapevines is where they are going to be planted because of the many factors that cannot be changed afterward: climate, soil, slope and aspect of the block. Site planning should take into account vine needs as well as logistical ones, like irrigation, frost control, vine spacing, equipment turning spaces, access roads, buildings etc. Vines may need to be ordered well in advance due to the many steps needed during propagation, and rootstock choice is a critical decision. Pre-planting site preparation should be thorough and completed before vines are delivered. Vine planting can be done mechanically or by hand, but it is critical to keep the soil around the vines well watered during the establishment phase. The first years should concentrate on developing the root system and not cropping. Creating the vine's structure on the trellis is another primary goal during this time.

Where to place your vineyard and how to bring it into production are critical questions in the process of developing a viable and quality grape production system. It is therefore important not to be hasty at this stage, as a mistake then will stay with the vineyard for the long term. Growing grapes is a high-cost venture so, to minimize risk, as much information as possible should be gathered at every opportunity. In this chapter, the process of putting in a vineyard is followed from first thoughts to making sure that the vines get off to a healthy start, ready to produce as much high-quality crop as possible for the life of the vineyard.

Site Selection

Probably the most important decision that is made for a vineyard is where it is situated. The site influences many factors that affect how the grapes will grow, how easy they will be to manage and what kinds of grape products can be successfully produced from it. It is easy to see, then, why this decision should be thoroughly researched and well considered before choosing the spot: an inferior site results in an inferior vineyard producing inferior grapes.

DOI: 10.1079/9781800627062.0005

In choosing the best site, then, it is important first to consider the market. Without knowing what the market wants to buy, there is no justification in putting vines in the ground. While this is not a book on market research, it must be emphasized that this information is vital to the longer-term success of any vineyard. It may be possible to get this information from a marketing body associated with the grape industry, or from a winery or packing house. From them it should be possible to get some indication of what the market is likely to want to buy by the time your vineyard is in production.

Once the market has been defined, then what needs to be produced and, therefore, where it needs to be produced can be determined. For example, if a market need for 'Sauvignon blanc' of a particular style has been identified, land in a region that can produce that style in an economic quantity should be sought. In doing this, the site with the lowest risk is a vineyard that is already known to consistently produce high-quality grapes. As another example, if one were to want to produce 'Champagne', then land within the Champagne region in France must be used. However, these sites also cost the greatest amount of money, if they can be purchased at all! In some cases, there will be reasons to look at planting in untested sites, where the risk is higher but the cost of the land is less.

The concept of matching vineyard site to grape product is an important one to work toward. This will also narrow down the potential areas that need to be investigated, and provide a basis through which to evaluate potential sites. Chapter 3 reviewed the environmental factors that are important to vine growth and productivity, and in site selection the pros and cons of a particular site must be evaluated with regard to the target outcome (i.e. grapes matching a particular set of quality criteria, refer to Table 2.2).

Now information about what will be required of a site in order to produce the target product can be gathered. For example, if the market demands 'Flame Seedless' table grapes, then a vineyard site that will be capable of growing, sustaining and reliably ripening the 'Flame Seedless' cultivar is needed – there is no point in looking at a site that is too cold, does not have enough water or does not have access to the labour required to allow its production, etc.

To make the best decision about a potential site, the requirements of the cultivar must be determined and then compared with those that are available at the potential site. Common factors to be considered are heat accumulation, winter minimum temperatures, water availability and soil characteristics. There are modifiers that a site may also possess, such as a slope (related to air drainage and frost incidence), the slope's aspect (influencing sun exposure and temperature), proximity to a large body of water, elevation, etc.

It is difficult to say which of these many factors are the most important, as each may be essential in a particular aspect of grape production. A discussion on the most important factors follows.

Climate

The basic parameters the vines need, as laid out in Chapter 4, apply here. As temperatures are so critical to grapevine survival and production of the desired cultivar and composition of the grape, having access to long-term weather information at the site makes it much easier to make a confident decision about the site's suitability. Often, however, sufficient weather information is lacking, which increases the risk associated with developing that site. At a minimum one should be confident that enough heat will accumulate at the site to produce grapes of the desired composition, as well as the knowledge that damaging winter, spring or autumn temperatures do not occur with a frequency that will mean economic ruin.

Large bodies of water can also assist in changing the mesoclimate. As water stores a lot of heat, in the winter it can increase air temperatures nearby and also act in causing air to circulate, which can prevent the accumulation of freezing air and grapevine damage (Newman, 1986; Shaw, 2001). In the spring, relatively cool air coming off the water can delay budbreak, meaning less risk of damage from early-season frosts.

The amount and timing of rainfall is relevant to the need for an irrigation system, or to aid vine spacing and rootstock decisions. Rainfall occurring close to and during harvest season can be particularly detrimental, as it can cause berry splitting (Fig 5.1), degrade fruit quality (e.g. through decreased fruit sugar and flavour concentration or increased disease incidence) and hinder human and machinery traffic in the vineyard.

Soil

Soil is an important consideration for site selection because, like climate, its basic properties cannot be altered much. The soil serves several purposes for

Fig. 5.1. Example of berry splitting following rainfall. In this case, the Cinsault berries were whole before the cluster was handled: a slight disturbance was all that was necessary to cause the skin to break from the pressure. Content author's own.

the vine. At its most basic, it anchors the plant to the ground. However, it is also the medium through which the plant obtains the vast majority of its water and nutrients. Chapter 2 discussed what roots are, but here how roots interact with the soil is examined.

Grapevines can grow in many and varied soil types. There is no one soil type that is best for grapevines. Indeed, with the use of rootstocks a single cultivar, if grafted, can have roots that are completely different, and even then there can be particular rootstock/cultivar combinations that are best suited to certain sites. With this in mind, when evaluating soils it is best to confirm that the most deleterious characteristics are not present. These include flooding, impervious layers, excessively low or high nutrient levels or pH and insufficient organic matter. Soil testing (for nutrients and water-holding capacity at a minimum) is an integral part of evaluating a potential vineyard site and should be performed in a comprehensive manner (Thomas and Schapel, 2003). To some extent, each of these soil-based problems can be mitigated with pre-planting management decisions.

Interacting with these are factors such as rainfall and/or water availability on the site. Irrigation water may be lacking or expensive. If the site has ample water supply for irrigation, then very sandy or gravelly soils can work well, whereas if there is not enough water for irrigation then they would not be suitable. This is another demonstration of how complex and interrelated the components of the system can be.

So, from the overall point of view, it is very easy to identify potential locations for vineyards. What becomes more difficult is to decide which soils match which rootstocks and cultivars for which end uses. Regional soil maps, though widely available, are rarely detailed enough to provide enough information for rootstock/cultivar planting decisions. As soils can vary considerably over a short distance, having a soil expert walk the area and take soil cores as necessary is a valuable exercise, as is having a look at the soil profile in a soil pit (see Fig. 5.2). Additional information can be gained by remote sensing techniques (Smith, 2002), such as aerial photography (both at visible and infrared wave lengths), satellite imagery or electromagnetic (e.g. EM-38) (Reedy and Scanlon, 2003) or ground-penetrating radar determination of soil water content (Huisman *et al.*, 2003).

The water-holding capacity (WHC) of the soil is important in assessing a potential vineyard site. Together with the amount of annual rainfall, this provides the vines with their water supply for growth. The maximum amount of plant-usable water a soil can hold is determined by its field capacity, which is the water that is retained in the soil a few days after heavy rains (leaving air spaces as gravity pulls water down the profile), and its permanent wilting point, where plants can no longer extract water from the soil. The values for these will vary depending on the soil type, with sandy soils not holding very much water but almost all of it being easily extracted by plants. High-clay soils will hold lots of water, but a larger proportion of it will be held too tightly for the roots to extract it. High WHC is not necessary for vine growth, but there

Fig. 5.2. Soil pit used for evaluating a potential vineyard site. Note that several distinct horizons are visible in this clay-based soil. Content author's own.

must be an alternate supply of water for the vines if the WHC will not be a sufficient buffer against natural rainfall events.

Since plant rooting depth changes the volume of soil from which water can be taken, usable soil depth is also an important consideration. Increased depth means more potential for water and nutrient supply, which has effects on grape quality parameters (Coipel *et al.*, 2006). Note that, though vine roots can penetrate to significant depths and find water, there may be little in the way of nutrients available at those depths due to the lack of proper soil to carry them. Changes to soil depth or even texture across a site, as well as the presence of claypans (which slow water movement through the profile and inhibit root growth), are also causes of vineyard variability, which can complicate management of the vines due to differences in vine performance caused by the soil (Trought, 1997). The installation of tiles to drain excess water from the soil, as with perched water tables caused by impermeable layers, is important in soils that waterlog in the winter or spring as vines prefer well-drained soil (see Fig. 5.3).

Fig. 5.3. This wet spot in a springtime vineyard is a problem for machinery and human traffic, as well as keeping vine roots in anaerobic conditions. Content author's own.

Soil pH is relevant because of its influence on the availability of essential nutrients. At low pH, macronutrients such as phosphorus and potassium become less available, while at high pH, microelements like iron and zinc become scarce (Marschner, 1986). *V. labrusca* vines are known to be more tolerant to acidic soils than *V. vinifera* (Himelrick, 1991), probably because the former species evolved in an area with soils of lower pH. Soil pH can be altered through the addition of lime (increase) or gypsum (decrease), but these are temporary measures as the soil pH will tend to revert to its natural state over (long stretches of) time (Himelrick, 1991). Soil texture has an influence on this as well, with clay soils being more resistant to pH change, and lime, in particular, being very, very slow to move through the soil profile. Nutrient status of the soil can also be changed, and soil organic matter is a large contributor to soil WHC as well as nutrient carrying capacity (see Chapter 8 for further discussion of nutrients).

If a vineyard is being replanted there can be issues with carry-over effects (Westphal *et al.*, 2002). These can range from depleted soil nutrients, nematode or insect populations to pathogens or herbicides left in the soil and infected remaining plant roots. Proper soil testing and preparation is essential to avoiding these problems.

Slope and aspect of land

In many areas, higher-quality grapes can be grown on sloping land (see Fig. 5.4). In cooler climates there is a benefit in terms of water and air drainage/frost

Fig. 5.4. A vineyard along one of the Finger Lakes in New York State. The slope helps with air movement and the large body of water circulates air and adds heat into the mesoclimate in winter. Content author's own.

risk, as well as increased sun exposure if the slope faces the noon-time sun. This increase in the interception of solar radiation heats not only the air, but also the soil, which creates benefits for vine growth through the season.

However, sloping land also brings difficulties with machinery access and vineyard uniformity. There are many examples of very successful vineyards on both flat and sloping land but, in areas that are pushing the boundaries of eco nomically viable sites, slopes are generally regarded in a positive way.

Other factors

Other factors to take into account include land cost/availability/zoning, likely pests and diseases, proximity to existing industry infrastructure, presence of skilled and unskilled labour, accessibility to markets or shipping ports, reputation of the region, pollution concerns from neighbouring activities or previous cropping, water availability and quality, natural rainfall's timing through the year etc. As some vineyard activities – such as bird-scaring, frost-fighting, early morning fungicide spraying, etc. – may not be expected by neighbours, it is best to resolve these issues before planting the vineyard to avoid conflict.

Knowledge of other plants and their phenological stages in relation to grapes can also help predict the performance of vines not yet planted. If there are reliable data for time of budbreak, flowering and ripening in grapevines and other crops, including ornamentals, in one region, the relationship between phenological dates of those other crops and grapevines can be used to predict when grapes will reach certain stages of development. For example, if it is known that in one region grapevine budbreak coincides with first bloom of the 'Red Delicious' apple, then if that same apple is growing in another area that is being evaluated for grape production and the date that the apple trees bloom is also known, this leads to a good predictor of when budbreak will be for grape in that area. Advance knowledge of all the important phenological stages can be estimated using this technique, leading to a better risk assessment for the type of grapes needed to be produced. If there is the possibility that this kind of detailed information is available, it should be sought out.

In scouting new vine-growing areas, the use of remote sensing (e.g. aerial photographs) to look for vegetation differences, global positioning systems for accurately marking out sites and elevation determination, and geographical information systems to help organize the data into a useful form and plan out blocks, is increasingly common (Smith, 2002; Filippini Alba *et al.*, 2017). Be aware, though, that there are limitations to what remote sensing can achieve, so it is prudent to ground-truth at least some of the information.

Being thorough and methodical in investigating a site is vitally important, but predicting its potential to produce quality grapes, even with the increased level of knowledge and experience available today, is still quite difficult. However, it is worth this effort to try to get it right since, once the vineyard is planted, alterations to many aspects of production become almost impossible. It is helpful to remember that management inputs can overcome many deficiencies a vineyard may have, but this may result in decreased fruit quality or increased site costs through the life of the vineyard.

At the end of all this, there are two questions to answer: (i) are there one or more factors associated with the site that will prevent it from growing the grapes the market needs? (ii) If so, then what grapes could be grown, how best should they be managed and will it be economic?

The alternative is to investigate another site. However, if the site is appropriate, then it is time to move to the next stage of development, determining how the site will be planted out.

Site Planning

A vineyard site does not have to be rectangular, flat and with uniform soils. If a site is like this, then planning out the vineyard is a much simpler exercise! Most blocks require at least some compromises to be made in order to juggle all of the requirements of a vineyard. In addition to the grapevines themselves, provision may need to be made for headlands, access roads, irrigation ponds, buildings, washing pads, storage sheds, break rooms, offices, etc. All of these take up valuable space, and so their placement should be planned to take greatest advantage of the land.

To assist with planning this, it is useful to go back to the information gathered before and decide where the best locations will be for the grapevines. Land that is less suited to grapes is better used for other supporting structures such as buildings or access ways. It is common to develop some types of vineyards on slopes and, indeed, there are often a number of advantages to doing so. It is important to use the soil information to help with the placement of blocks, as erosion and potential vigour of the area are two important factors that impact on future management.

The block shape and size may be determined by a number of factors that may or may not be controllable. It can be defined by soil changes, access roads, source of irrigation water, row length, slope, etc. For many, minimizing the amount of variation within a management block is a desirable goal (Long, 1997). Thus, planning out blocks so that their management will be as uniform as possible helps in obtaining consistent results from season to season.

Row orientation for most vineyards typically run somewhere near north to south to allow for even ripening and sun exposure on both sides of the canopy. However, east–west-running rows can be beneficial in raisin production, for example, if the grapes are to be dried in between the rows, they obtain sun exposure for most of the day. In some cases, rows will run along the contour of a hill, to minimize erosion or facilitate use of machinery (see Fig. 5.5). Row orientation can also be changed to account for site-specific

Fig. 5.5. An example of a vineyard where the rows are contoured along the hillside instead of running straight. In this case, the slope is so extreme that the hillside has had to be terraced so that there is easier machine access. Content author's own.

factors, such as in areas with more rainfall, running the rows in a direction that enhances morning sun exposure in the canopy to help dry out the fruit. Likewise, maximizing canopy exposure in the mornings rather than in the afternoons can be beneficial, as the latter tend to be hotter. This can be an advantage in a climate that is a bit too hot for the cultivar being produced, as it lowers fruit temperature and modifies the development of flavour and aroma compounds.

Preparing the site

Prior to planting, you may need to clear the land. If possible, do this in the season before you intend to plant and establish a cover crop in the meantime (see Fig. 5.6). This may help add nitrogen and organic matter to the soil, reduce pest (e.g. nematode) and disease (e.g. root rot) populations and, in particular, allow the control of problem weeds before the vines are established. If the land has been used for other crops, it is worth checking to see whether there could be carry-over effects from those to grapevines. This might be due to past chemical use, or in some cases, to microorganisms or their products that build up in the soil and hinder the establishment of the vines (Westphal *et al.*, 2002).

While there are no obstructions, this is the stage to make any soil changes, such as ripping to break open claypans or remove large stones (Fig. 5.7), terracing, adjusting nutrients or pH, etc. or to improve drainage through the laying of tiles. If soil is low in organic matter, incorporating

Fig. 5.6. Mixed plants on a future vineyard site. Planting a cover crop prior to vineyard establishment can decrease pest and weed problems in subsequent years, modify soil nutrients and structure and, in some cases, provide an interim cash crop. Content author's own.

Fig. 5.7. Heavy machinery with a two-tined attachment designed to pull large stones out of the soil. A similar attachment with one tine would be suitable for breaking up compacted soil layers and improving grapevine root penetration. Over time, the tendency will be for the layers to re-form, but by then the roots will have grown down as far as they need to. Content author's own.

compost into the soil that the vines will be planted in will assist with quicker establishment through the addition of nutrients and increased water holding capacity. The interest in using biochar (charcoal made from plant materials under low oxygen conditions) has grown and experimentation with it incorporated into soil with or without compost has been trialed. Thought it can have benefits in terms of sequestering CO_2, applying to an existing vineyard is difficult and can have mixed results (Schmidt *et al.*, 2014; Genesio *et al.*, 2015). The best benefit might come from incorporation into the soil that the vines will be planted in.

A word of caution on the practice of contouring, where the soil surface is altered from its original form. While it is tempting to shift soil from high spots to low spots when preparing a vineyard site, be advised that: (i) there may be local laws regulating this, and (ii) it is extremely rare for the practice to result in a uniformly healthy and productive vineyard. Research into this area is scarce and long-term effects of the practice are not known (Hughes and Wilde, 1989; Sharp-Heward *et al.*, 2014), but research demonstrates negative effects in terms of erosion (Bazzofi *et al.* 2006), soil organic matter, depth, and most importantly, vine growth (Martínez-Casasnovas and Concepción Ramos 2009).

If a source of water is needed at the site, then a well should be dug to ensure water supply before plans are finalized. If the yield of the well is below expectations, then a redesign of the vineyard may be necessary to accommodate it (e.g. incorporation of a storage pond, change of rootstock choice or a reduction in the number of vines grown). If water is necessary for grape production in an area under consideration, then purchase of the land should be conditional on the availability of water.

Shelter from wind

If the area where the vineyard is to be established has high winds, it is worth considering establishment of windbreaks (see Fig. 5.8). Compared with vines growing without protection from the wind, those sheltered by windbreaks exhibit: (i) better budburst; (ii) larger leaves with more stomata; (iii) more clusters per shoot; (iv) longer internodes and longer shoots; (v) higher pruning weights; (vi) increased capacity to carry and ripen a crop; and (vii) increased phenolic content in the fruit (Lomkatsi *et al.*, 1983; Dry *et al.*, 1989; Bettiga *et al.*, 1996; Smart, 1999; Neel, 2000).

Sheltered vines are also less susceptible to damage caused by whipping of the shoots and breakage (usually snapped off at the base or tips broken by hitting foliage wires early in the season). Loss of shoots can result in a potential 10–15% loss of productivity (P. Evans, New Zealand, 2000, personal communication), and removal of the tips causes growth of multiple laterals, crowding the canopy.

Studies that have compared the productivity of sheltered versus non sheltered vines tend to show around a 15% increase in yields (Lomkatsi *et al.*, 1983; Bettiga *et al.*, 1996). This could be due to higher rates of photosynthesis because the chance of stomatal closure due to water loss is lower (Freeman *et al.*, 1982), and/or that leaves tend to be larger in areas with less wind. One South Australian study (Dry *et al.*, 1989) suggested that yield gains are from

Fig. 5.8. A living windbreak in Canterbury, New Zealand. In this case, the trees in use are poplars. Another effect of windbreaks is evident here, too: the vines closest to the trees are noticeably smaller than the ones further in due to the trees competing for light, water and nutrients. Content author's own.

increases in flower cluster number rather than from percentage fruit set or berry size, although gains in percentage fruit set could be envisaged in other situations.

Ideally, a windbreak is permeable to the wind so that the air moves both through and over the top of the obstacle (Cleugh, 1998). This slows the speed of the air in the vineyard and also provides protection from the fastest air, which is shunted over the shelter. The latter is provided for a distance of up to 20 times the height of the windbreak (Marshall, 1967), so windbreaks need to be high enough to justify the area of land that they take up. However, with their use comes maintenance costs, competition with grapevines for water, nutrients (Cleugh, 1998) and even sunlight due to shading, especially at latitudes closer to the poles (Abel *et al.*, 1997).

Windbreaks can be made of artificial woven nets, but more often they are grown. Judicious use of windbreaks can help in areas that are frost prone by preventing cold air from moving into a vineyard area from above (deflecting it away), though this precludes the use of deciduous plants in their make-up. However, deciduous tree shelters better allow cold air movement through them in the dormant season, which can be helpful in frost-prone areas as well.

Windbreaks do require extra effort and cost to establish and maintain (for example, to avoid effects on nearby vines, the soil in the area in between should be ripped every few years), take up land area that could be used to grow grapes and also can be a nesting site for pest bird species; however, in certain areas these drawbacks are countered by increased production and greater vine productivity.

Vine planting density

One of the early decisions that will need to be made in planting a vineyard is: at what density will the vines be planted? Across the world there is the whole gamut of spacings, from very high density, e.g. 1 m between vines and 1.1 m between rows (for example, Château Lafon-Rochet, Saint Estèphe, France, has a planting density of 9000 vines/ha, which works out to this spacing) to low density 3 × 3 m systems (1100 vines/ha) or even greater spacings for some dry land vineyards. How is the vine spacing determined?

In earlier times, vine spacing was determined by who, or what, was going to be working the soil and the vines. In days when people did all the work, it is said that vines were planted at densities that equated to tens of thousands of vines per hectare. Later, when animals began to be used for tilling the soil, vines started to be grown in more distinct rows or grids, with the spacings determined by how wide the animals were. More recently, especially in the New World areas, row spacing has been determined by the width of machinery that was available to work the soil and plants (see Fig. 5.9), which were generally large-scale agronomic crops.

Fig. 5.9. An example of a vineyard with wide row spacing, necessitated by the desire to use large machinery. Here, weeds are being controlled with an in-row mechanical hoe. Content author's own.

Another factor involved is the availability and value of the land. In Europe, land was in short supply given the number of people it needed to support, hence it was used intensively (a higher plant density means greater productivity per unit area). When people moved to the New World land was in abundant supply, so wider row spacings were seen as being convenient.

From the point of view of the vine, however, things are a bit different. There is the notion that to cope with high predicted vigour, planting density should increase because as the number of vines per unit area goes up, the resources available to each vine (e.g. water, nutrients) decreases, thus leading to less vigorous vines. This appeared to work in the Old World areas, such as practised in many Grand Cru vineyards, and so was emulated in many other, newer, grape-growing areas. However, the result in these new areas was some times very different, in that the vines remained very vigorous, and eventually declined into vinous messes producing poor-quality fruit (Figure 5.10). It is thought that this was caused by the different soils that are typical in many of the newer areas, in that they tend to be more nutrient rich than the farmed-for-centuries soils in the Old World and, in many cases, the soils are much deeper, leading to luxury levels of water and nutrients being available to the vines, despite them being packed tightly together.

So the high vine density to combat vine vigour hypothesis proves not to be a rule, and brings us back to the question of how to determine planting density.

Fig. 5.10. Vineyard with close row spacing, but high fruiting wires and therefore tall canopies. Self-shading here is limiting the productivity of the vines as well as having a negative effect on fruit composition. Content author's own.

One of the current theories is that between-row spacing should be determined by the availability of equipment used to work the vineyard, and within row vine spacing is the tool to really influence vine vigour. Closer vine rows result in more efficient use of the land area (meaning that if vines are cropped to the same level per unit of vine row, the total tonnage of fruit per unit of land for a narrow row-spacing vineyard is greater than that from a wide row spacing vineyard), so the goal should be to have the closest row spacing that is still compatible with available equipment.

Within-row vine spacing should depend on the predicted vigour of the site (and rootstock/cultivar combination) and how it is managed. Again, there are many different vine spacings used in practice, and there will be no one spacing that is ideal for any given situation.

For most vineyards using discrete rows of canopies, row spacing will approximate a 1:1 ratio with canopy height, which ensures that between-row shading is minimized. As shown earlier, Fig. 5.10 demonstrates one of the problems with close, but tall, rows. Therefore, if the vineyard has very close row spacing, the canopy height must be low to prevent between-row shading, and so the height of the fruiting wire must be low, too. Working with low fruiting wires has challenges in terms of equipment and labour, as the vines are so close to the ground working with them is very difficult. Row spacing is also influenced by trellis type used, with divided-canopy systems requiring wider rows.

Rootstocks

The choice of rootstock is another pre-planting decision that lasts for the life of the vineyard, and so should not be taken without careful consideration. Rootstocks allow vines to perform better in soils of a particular type (e.g. high pH or very deep and fertile) or in those with an undesirable infestation (e.g. of phylloxera or nematodes). Because the producing part of the vine (the scion) is dependent on the rootstock for all things coming from the soil, there can be a considerable effect of rootstock on vine performance.

These effects can be categorized as being direct or indirect (Striegler and Howell, 1991). An example of a direct effect would be root production of plant growth regulators (e.g. cytokinins) that change shoot growth or vine fruiting. Other direct effects would be the ability of the roots to take up water and nutrients and to store carbohydrates and nutrients.

Indirect effects centre on follow-on changes caused by the direct effect: for example, increased uptake of water and nutrients will cause more luxurious vegetative growth and shading, which can lead to poor fruit quality. The fact that a cultivar on a particular rootstock has lower Brix, higher acid and less ripe flavours may not be caused by the rootstock itself, but the fact that the rootstock influenced crop load and canopy density, which led to the sub-optimal fruit composition.

The parentage of a rootstock has an influence on its characteristics. Historically, examples of commonly used species for the production of root stocks included *V. rupestris, V. riparia, V. berlandieri, V. champini* and *V. vinifera*, while some were selections of a particular species. Most rootstock crosses were performed in the mid to late 19th century in response to the introduction of phylloxera to Europe. Unlike the producing cultivars used, rootstocks are still commonly dioecious (having separate male and female plants), since the regular production of grapes is not a concern for them.

Selection criteria for rootstock breeding programmes were originally phylloxera tolerance (in that all roots support some size of population of the insects) and, importantly, ease of grafting to *V. vinifera* scions. Hence, *V. vinifera* was used in some crosses, though its susceptibility to phylloxera was the reason the other grapevine species were being used. AxR1 was one such rootstock developed (a cross of *V. vinifera* × *V. rupestris*), which performed well in Californian trials due to its high productivity (Lider, 1958). However, despite the fact that its resistance to phylloxera was already in question in Europe, as included in the report by Lider (1958), the grape and nursery industries used it widely as it did perform well in the field and grafting houses (Wolpert, 1992). However, over time and through the process of natural selection, a more aggressive strain of phylloxera was able to overcome the tolerance of the rootstock and cause the vines to decline. Eventually the vines had to be pulled out and replaced, at great cost in time, money and lost production (Himelrick, 2001).

The reason the rootstock failed was that its *V. vinifera* parentage meant that it tolerated a higher population of phylloxera, which led to more opportunities for a beneficial (from the point of view of phylloxera!) natural mutation to occur.

This eventually resulted in the appearance and identification of the new strain, or biotype (Granett *et al.*, 1985). Ironically, it was noted by Wolpert (1992) that Perold (1927), described the similar downfall of AxR1 in South African vineyards due to phylloxera. In that 1927 publication he postulated that a 'new biological race of phylloxera has evolved' on the AxR1 roots, which predated the first confirmation of its occurrence by 58 years. There are now many different biotypes of phylloxera characterized (indeed, there is now PHYLLI, an online international database for grape phylloxera to house information about them), cautioning the grower again not to rely on a single rootstock, and to be informed about the potential problems that could appear during the life of the vineyard. Notably, research has found that 5C Teleki is susceptible to at least one known biotype (Clarke *et al.*, 2023).

There has been, and will continue to be, no doubt, considerable debate over the merits of grafted versus ungrafted vines, particularly in wine circles. In most cases, this is a moot point, as without the use of rootstocks, vines would not survive in soils with phylloxera. In addition, there are other good reasons to use a rootstock, already mentioned, that will allow vines to be grown in places they otherwise could not. However, there is no getting around the fact that the grafting process changes the vine. A longer-term study in New Zealand has found that own-root, but also grafted, vines are consistently poor performers in terms of vegetative development in comparison to their grafted counterparts (G.L. Creasy, unpublished results). This is contrary to most studies, which report that own-root (ungrafted) vines tend to be more vigorous than grafted ones: the graft union itself is likely responsible for a change in vine performance.

Rootstock choice

Rootstocks that impart greater vigour to the scion are often used in situations where high crop loads are needed, and those considered to be devigorating used in areas with higher predicted potential vine vigour. However, there are no dwarfing rootstocks available for grapevines in the same way that there are for apple trees, for example. So, although there is a range of rootstock performance, giving recommendations for a given site is highly problematic, as each will perform differently according to their environment, the scion that is grafted on top, management and the end use of the grapes that are to be grown.

In the case of rootstock influence on the growth of the scion, a major management influence is through the use of irrigation. In a dry region, rootstocks with a narrow root angle send roots further down into the soil profile and tend to find more water. Hence these rootstocks perform better in non-irrigated areas compared with shallower-rooting rootstocks. If irrigation is part of management, then both types of rootstock have the ability to perform well (McCarthy *et al.*, 1997). This highlights the importance of determining how a vineyard will be managed prior to planting as well as the challenges in interpreting the results of rootstock trials from around the world.

A look through the literature shows that the best performing rootstock changes according to these factors, so the final choice of rootstock should be determined with the best local knowledge available. In looking through much of the published results, some rootstock characteristics are reasonably common, however, and information for some rootstocks is presented in Table 5.1. It must be emphasized, however, that actual performance varies as described above, and before ordering vines, sufficient time should be put into research as to how rootstocks may influence vine growth at the proposed vineyard site.

It is worth noting that there is probably not a single, *right*, rootstock for any particular vineyard. It is particularly difficult to choose the best rootstock without having tested it at that site, and a rootstock may or may not perform well in producing the end product desired or under a certain management regime. For these reasons, it is best to choose several rootstocks to plant out a vineyard to spread the risk. If the vineyard needs to be redeveloped in the future, the best performing rootstocks can then be used.

Propagation

Grapevines can be propagated from seeds, cuttings, layers, grafts or by meristem culture. Sexual propagation, as for other plants, is through the formation of seeds. The genetic complexity of the genus *Vitis* assures that the vast majority of grapevines grown from seed will be new and unique plants. Therefore, all named cultivars of grape are propagated asexually to ensure that the progeny have the desired properties of the original plants.

Growers can choose to buy their vines from a nursery, propagate their own vines asexually or even produce vines for sale. The choice depends on the availability of labour and plant material, what kind of vines are required (e.g. own rooted or grafted) and the advisability of establishing an additional enterprise. Given the specialized nature of grafting grapevines and the need to maintain disease-free status of the plant material, purchase from an established producer of vines is entirely justifiable.

Layering

Difficult-to-root cultivars can be propagated by layering (Fig. 5.11). Long canes still attached to the mother plant are partially buried in the row where a new vine is wanted, with the end of the cane protruding from the soil, usually supported by a stake. The buried part of the cane roots readily and, when the new vine has developed a sufficiently large root system to be independent, the connection to the mother plant can be cut and the vine left to fill a gap in the vineyard row (or dug out and planted elsewhere). Layering is a commonly used technique for filling empty locations in an own-rooted vineyard due to its ease and the fact that the young vines get assistance during establishment from the mother vine.

Table 5.1. Summary of generalized rootstocks characteristics from a range of parentages.[a]

Rootstock	Parentage	Vigour	Drought resistance	Tolerance to lime	Tolerance to phylloxera	Tolerance to nematodes	Tolerance to salinity
Freedom	1613C × *champini*	Moderate–high	Moderate–high	na	High	High	Moderate
Harmony	1613C × *champini*	Moderate–high	Moderate	Moderate	Moderate–high	Moderate–high	Moderate
SO4	*berlandieri* × *riparia*	Moderate	Low	Moderate	Moderate–high	Moderate–high	Moderate–low
420A	*berlandieri* × *riparia*	Moderate–low	Low	Moderate	High	Moderate	Low
5BB	*berlandieri* × *riparia*	Moderate	Moderate–low	High	High	Moderate–high	Moderate
99R	*berlandieri* × *rupestris*	High	Moderate–high	Moderate–high	High	Moderate–high	Moderate–low
110R	*berlandieri* × *rupestris*	Moderate	High	Moderate	High	Moderate	Moderate
1103P	*berlandieri* × *rupestris*	Moderate–high	High	Moderate	High	Moderate	Moderate–high
Ramsey	*champini*	High	Moderate–low	Moderate	Moderate–high	High	High
Riparia gloire	*riparia*	Low	Low	Low	High	Moderate–low	Moderate
44–53	*riparia* × (*cordifolia* × *rupestris*)	Moderate	High	Moderate	High	Moderate–high	na
101–14	*riparia* × *rupestris*	Moderate–low	Moderate	Moderate	High	Moderate	High
Schwartzmann	*riparia* × *rupestris*	Moderate	Moderate–low	Moderate	High	Moderate	Moderate–high
3309	*riparia* × *rupestris*	Moderate–low	Moderate–low	Moderate	High	Moderate–low	Low
St George	*rupestris*	High	Poor	Moderate–high	Moderate–high	Moderate–low	Moderate
1616	*solonis* × *riparia*	Moderate	Moderate–low	Low	Moderate–high	High	na
AXR1	*vinifera* × *rupestris*	Moderate–high	Moderate	Moderate–high	Low	Low	na

[a]Sources of information: Anon (2007); Anon (2008); Arbabzadeh and Dutt (1987); Carbonneau (1985); Corso and Bonghi (2014); Downton (1985); Hoskins *et al.* (2003); Lider (1958); McCarthy *et al.* (1997); Pouget and Delas (1989); Smith *et al.* (2017); Southey (1992); Southey and Jooste (1991); Stafne and Carroll (2006); Whiting (2004); Whiting and Buchanan (1992); Whiting *et al.* (1987); Wooldridge *et al.* (2010).
na, not available.

Fig. 5.11. Grapevine layering, which uses a long cane still attached to a vine as the basis for a new plant. Top left: the vine has been allowed to grow a long shoot in the previous season. Bottom left: the cane is left attached to the vine, but buried with the end sticking up out of the ground and staked. Top right: Lateral growth is trimmed off. Bottom right: admiring the finished product. Content author's own.

Cuttings

A cutting is a piece of a parent plant that will develop into a new plant when provided with an appropriate environment. Cuttings can be made from dormant canes or green shoots, although in practice most are from the former (for details of the latter, see below).

If own-rooted vines are needed, easy-to-root cultivars can be planted directly in the vineyard. Dormant unrooted cuttings can be placed into holes made with a steel rod. It is important to remember that the cuttings will need access to adequate water and freedom from weed competition and animal damage in the early stages, as they have to develop an entire root system. Alternatively, the cuttings can be grown in a nursery for a season, dug out in the winter and planted in the vineyard as dormant rooted cuttings.

Grafting

The previously discussed methods produce own-rooted vines, normally only used in phylloxera-free regions or with phylloxera-resistant cultivars. Most vines planted in major grape-growing areas are grafted due to the soil conditions, presence of a soil pest or to take advantage of rootstock effects on scion growth.

The majority of commercially produced vines are propagated by bench grafting. A rootstock section of about 35 cm long is cut from dormant vines of the selected rootstock mother plant and all buds removed either manually or by a disbudding machine, which uses stiff-bristled brushes to remove the buds. Note that increasing the length of the rootstock section improves subsequent shoot and root development, likely due to greater carbohydrate storage (Larrey *et al.*, 2025). The scion section is cut from healthy vines of the selected cultivar in lengths of approximately 35 cm for ease in handling but, before the actual grafting takes place, they will be cut into individual nodes.

If cuttings are not to be used immediately they should be treated with a botryticide and covered with a moist sterile media, like perlite, and stored in high-humidity refrigerated area near, but not below, 0°C, as the fungicide used may reach toxic concentrations if the water on the vines turns to ice (Nicholas *et al.*, 1992). These cold and moist conditions prevent desiccation of the cuttings, development of disease organisms and slow respiration in the tissues.

Before grafting the rootstock and scion, cuttings are soaked in water and the scion wood cut into one-node sections with the longest portion of the inter node below the bud. These are matched in diameter with a rootstock cutting and the ends slotted with a special blade, giving a complementary 'V'-shaped or omega (Ω)-shaped cut. The two pieces when fitted together should provide optimum contact between the cambia of the rootstock and scion, which will then 'knit' together the two pieces of wood. They may or may not be wrapped with special tape or dipped in wax to keep the union from drying out or coming apart too easily (Fig. 5.12).

Fig. 5.12. Waxed cuttings, being incubated to allow callus to form at the graft union. These cuttings are packed in perlite-filled boxes in a warm and moist environment. Content author's own.

The joined cuttings are placed in a moist medium (e.g. perlite, sphagnum moss, sand) and incubated at approximately 28°C for up to 4 weeks to encourage growth of cambial cells, which can be seen externally as undifferentiated cells (callus) at the graft union, and roots at the cutting's base (Fig. 5.13). When the grafts have healed the cuttings are unboxed, the graft union checked for integrity and then are planted outside, closely spaced. Maximum rooting will occur if the vines are well irrigated and the soil covered with plastic to prevent weed competition. Typically, the grafted vines are grown for a full season in the

Fig. 5.13. Grafted vines in the nursery, after incubation at 28°C. The buds start to push and shoots form during this time, adding a bit of colour to the nursery. Fresh callus tissue can be seen growing from under the black tape in the center-bottom of the photo. Content author's own.

nursery, and in the winter dug out mechanically by undercutting, graded as to quality, roots trimmed and then bundled for delivery. Although they can be stored, the best results are from planting as soon after digging as possible.

Another grafting technique commonly used to change cultivars in an established vineyard is field-, or top-grafting. In this procedure, a small portion of a dormant cane (that includes a bud of the cultivar desired) is cut so that its cambium can be closely matched to the cambium on the trunk of the parent vine (Fig. 5.14). Methods of doing this include T-budding or use of a cleft graft. Top-grafting has the advantage of enabling a vineyard to be converted to another cultivar quickly with very little loss in production, due to the existing and established vine root system backing up the newly grafted scion. However, it also has the disadvantage of its success rate being highly dependent on the weather and skill of the team doing the grafting (Nicholas *et al.*, 1992). A poor take of grafts necessitates another round of top-grafting in the following season, and even then, can result in incomplete vascular connections that may not become evident for some time afterwards.

Fig. 5.14. Left: top-grafted vine, showing the developing new shoot and the white tape used to keep the cambia of the cane and trunk closely associated as well as decreasing water loss. Right: what an established top-grafted vine looks like. Content author's own.

Whatever method is used, it is essential that the plant materials' disease and viral status is known to be as clean as possible. While the different plants used to source the grafting material may grow well, when they are matched together the interaction of viruses and/or fungal organisms in them can result in later vine failure (May, 1994).

Other methods

Green, softwood cuttings can be rooted in mist beds in a greenhouse. An advantage of this is that many cuttings can be rooted in a small space, but it does require special facilities and the softwood cuttings are more susceptible to disease than dormant cane cuttings (Dirr and Heuser, 1987). This method can also be used to produce grafted vines: The rootstock is rooted in a mist bed and then a section of the scion cultivar grafted to the rooted understock. Afterwards, they are grown in pots under controlled conditions until acclimated and then planted directly in the vineyard.

A new cultivar or clone that is originally selected from a single plant can be propagated asexually to provide a small number of vines for further testing. The rapid production of large numbers of vines for commercial planting can then be accomplished by growing some of the vine tissue in sterile culture through micropropagation. Undifferentiated callus cells are grown from a small sample of tissue from which many shoots are induced in culture tubes. Each shoot is harvested and transferred to sterile test tubes containing growth media where roots are induced The plantlets are acclimated to greenhouse conditions and then transplanted. These 'test tube plants' can be further acclimated to nursery conditions and then planted in the vineyard.

It is notable that vine health and, increasingly, pedigree, is an important consideration in the production or purchase of vines. Vine stock that has any of a multitude of diseases can result in low vigour and performance or, at worst, complete failure of vines (see Chapter 10), so using certified stock is a worthwhile investment.

Planting

Many of the usual vineyard structures, such as end posts, intermediate posts and wires, may or may not need to be put in prior to planting. The advantages of doing so are primarily those relating to cash flow but, in practical terms, it depends on how you plan on planting and training the young vines to the wire. Most mechanized methods of planting vines are incompatible with trellising structures, for example, so will need to be erected afterwards.

However, it is vitally important that all relevant site preparations have been completed prior to the vines going into the ground. If the site is not ready, it is best to delay planting until it is. Vines have small root systems and will not survive unless the soil has been prepared, weeds controlled and water supplied directly to them. An example of ensuring the latter would be matching irrigation emitter position with vine position. Many modern irrigation laterals have built-in emitters spaced at equal intervals along the tube. If these do not match closely enough with the vine spacing, some vines can be left with the nearest emitter being too far away to supply water to the small root system, leading to poor vine establishment or vine death. It is worth noting that irrigation tubing shrinks when cold water is passed through it, leading to a substantial change in overall length (and thus position of emitters relative to the vines) for long runs; this should be accounted for when laying out irrigation laterals. Tying the ends of the irrigation lateral to the far post can prevent the shifting of where the water is deposited, but it is best to do this when the lines are in use, so as not to put too much tension on the tubing when it contracts.

Vines should be handled properly to ensure good establishment and a minimum of young vine deaths. Prior to planting, the soil should be thoroughly watered so that the soil is moist but not wet. Common ways to plant vines are with a mechanical furrow planter (Fig 5.15), an auger or waterjet (Fig. 5.16), or by hand with a shovel. If unrooted cuttings are used, or dormant grafts that the roots have been trimmed off of, a hole can be made in the soil with a rod, the cutting inserted and then heeled in (Fig 5.17). Note that trying to establish a vine without roots is especially difficult in dry climates, as water needs to be

Fig. 5.15. Mechanical vine planting. Rooted cuttings are placed into rigid holders mounted on a wheel, which rotates and deposits the vine in a furrow that is closed after placement. This set up also injects water alongside the vine to get soil particles settled around the roots. Content author's own.

Fig. 5.16. Top: An augered hole for planting a vine. Note the friable sides of the hole, which will allow vine roots to grow unhindered. Bottom: Example of a recently planted vine with use of a waterjet. The abundance of water helps settle the soil closely around the roots. Content author's own.

delivered exactly at the plant so it is accessible to the emerging root hairs, and the soil cannot be allowed to dry out or death of the small roots will occur.

The use of tree planters is not recommended, as correct positioning of the roots is difficult: roots should not be bent upwards in the hole or furrow, as this discourages root growth and can result in root diseases, such as blackfoot (see Chapter 10) becoming established. If roots are properly placed with the use of a

Fig. 5.17. Hand planting of rootless vines. Vines are produced normally and the roots trimmed prior to delivery. The "T" bar is used to make a hole into which the vine is placed and heeled in afterward. The ropes are color-coded for planting the vines at various spacings. Content author's own.

mechanized system, then it is the most economical method of getting the vines in the ground and has the option of water being injected with the vine at planting. Waterjets are another method, and have the advantage of settling the soil closely to the roots and also providing water right where it is needed at planting; however, this method may not work in all types of soils (Hoag *et al.*, 2001).

In soils with a high clay content, the use of augers is not recommended due to possible glazing of the sides of the hole. This effectively creates a barrier to root growth beyond the hole, leading to stunted vine growth.

Once planted, the vines should be watered immediately and often, depending on the condition of the soil beforehand. Although these guidelines are general good practice, it is preferable to follow the recommendations of the vine supplier as closely as possible to avoid any questions if there turns out to be a problem with vine survival as the season progresses.

Record keeping is also important, as any last-minute changes to the original plans may be easily forgotten. Ensure that information as to cultivar, clone and location are all recorded in a safe place and perhaps even on the end posts, to avoid confusion by workers (and companies buying the fruit) later on.

Vine Establishment

There has been considerable debate in viticultural circles about the best way to get the vines established and on to the trellising system. One of the basic decisions to be made is whether or not to train up the vines in the first year. The sprawling method (see Fig. 5.18) allows the vines to grow wherever they like on the ground, and little is done to them in the first season, aside from sprays to control disease. Advantages are no cost of vine training, and there may be increased photosynthetic productivity of the vines compared with those trained up and with a single shoot, as the leaf area is greater with sprawl. This is beneficial as the vine has more carbohydrates available, which means more shoot and importantly, root growth. Also, the trellising system does not have to be installed before planting, which delays expenditure and creates less work in pre-planting preparations and increased flexibility in moving about the vineyard (e.g. for cross-cultivation). The downsides are that it is nearly impossible to control weeds when they are intermingled with the grapes, disease control can be problematic and the quality of canes produced may not be as great (e.g. possibly not as straight or as long as vines trained upward). To counter the weed problem some have used polythene, which controls the weeds and has the additional benefit of conserving moisture for use by the vine (Fig. 5.18). However, disposal of the plastic in later years can be an issue.

Fig. 5.18. A recently established vineyard showing the vines without any supports, leading to a sprawling growth habit. This vineyard is also using plastic mulch, which conserves moisture and reduces weed competition. Disposal of the plastic, especially once it starts to degrade, is a problem, however. Content author's own.

The alternative to allowing the vines to sprawl is to train them up. This can be achieved through the use of individual stakes for each vine, or installation of end assemblies, the fruiting wire and a fraction of the total number of intermediate posts, providing something to tie or secure other vine support to. Training the vines up is thought to be beneficial because the vine grows more quickly upward than downward (Kliewer *et al.*, 1989; Schubert *et al.*, 1999), it is easier to control disease and it allows the vine to be trained to the fruiting wire in the first season (if growth is sufficient). This is desirable because future growth will be better exposed to light and have the framework for substantial shoot development, leading to more photosynthesis. The downsides are increased labour costs for removing unwanted shoots and training, that the vine may not have the same carbohydrate production as vines that are allowed to sprawl (due to lesser leaf area) and, since they are up to the wire in the first year, there could be the temptation to leave a crop on the vines in the second year, which may have consequences on the establishment of the vine due to its effects on carbohydrate partitioning (see Chapter 3).

In addition, there are ways to help the vines grow up to the wire and reduce some management requirements. Vines can be individually staked or string can be tied from the vine (some part of the vine that will not have any growth distal to it, to prevent girdling of the cane/shoot) to the fruiting wire. In either case, it will be necessary to fasten the vines to the stake or string to ensure the selected shoots grow upwards, which is a materials and labour requirement.

Vine shelters

Vine shelters range from short plastic or cardboard sleeves (in the latter instance, empty milk cartons have been used), to narrow tubes of some kind, usually made of plastic, that are slipped over or wrapped around the vines (Fig. 5.19).

Simple plastic or waxed cardboard sleeves (e.g. milk cartons), perhaps 30 cm tall, are commonly used and are effective at protecting the vine from herbicide and perhaps some rodent damage. They are inexpensive, but do not provide all the advantages of using taller structures.

Taller plastic sleeves, ideally rigid, encourage shoots to grow up and straight to the wire, provide protection against herbicides and herbivory of the young plants, and increase the rate of growth of vines so that they reach the trellis wire more quickly. The environment inside the structure is warmer and more humid, and also protected against the effects of wind (Wample *et al.*, 2000). Warmer temperatures increase the growth rate in the spring, when average air temperatures tend to be cooler. The greater humidity and reduced air movement places less water stress on the vines, allowing them to grow unhindered.

Disadvantages of shelters in general include: (i) their cost (although rigid ones can be used on more than one set of vines); (ii) the extra labour required to open and close them again when working on the trunks; (iii) the possibility

Fig. 5.19. Montage of different types of vine shelters. Flexible shelters (bottom left) will need at least two stakes to prevent the wind from moving it and damaging the young vine shoots. Content author's own.

of temperatures exceeding damage thresholds on hot and sunny days; (iv) the requirement of the presence of a fruiting wire or stake to steady the sleeve; and (v) perhaps more spindly growth inside the shelter (Due, 1990; Wample *et al.*, 2000; Olmstead and Tarara, 2001).

Regardless of the method of training used, it is important to train the vines to have a straight trunk, in-line with the row. This improves the load-bearing capacity of the vine (thus reducing the load on the trellising system) and allows for mechanization that relies on specific placement of the trunks.

Young vine care

The goal for establishing young vines is to maximize vegetative growth, which includes the growth of the root system. To enhance this, the irrigation and weed management programmes should ensure water availability and minimize weed competition. Many newly planted vineyards will have bare ground to avoid competition from weeds, and soil moisture monitoring should be practised to keep water readily available to the vines in the volume of soil where their roots are positioned (i.e. shallow in the soil profile). Diseases should be monitored and controlled as needed.

Young vine trunks need protection from herbicide sprays because, not only will green tissue take them up, but the thin bark on the canes will also be able to absorb the chemicals. Hence the use of vine guards of some sort is desirable. For mechanical weeding, robust stakes are needed to trigger the weeder's sensor.

Any vines that do not survive should be identified quickly and replaced after a likely cause for their death has been determined. For example, if the irrigation emitter was faulty or too far away from the vine, it should be replaced before the new vine is planted. To attain maximum vineyard uniformity, it is best to replant failed vines as soon as possible in the same year, as long as the vines will be able to grow enough to survive the dormant season. Extra vines can be planted in between others at the ends of rows and used for replanting. This ensures the replants are as strong as the neighbouring vines when repositioned.

If multiple shoots grow from the young vine, the strongest-looking one should be retained to form the trunk. This is done to focus the energy of the plant into a single shoot, ensuring that it grows as much as possible in both length and girth. Two (or more) can be left if there is risk of shoot damage, for example, from passing equipment.

When the vines reach the fruiting wire there are the options of training the shoot along the fruiting wire in one direction or cutting off the shoot tip above the wire, allowing lateral shoots to be trained along the wire in both directions. This sets the vine up for development of its head later on, during pruning.

During the first dormant season, vines should be pruned back to a point that is relative to the amount of growth in the previous season. If the vines have grown a lot, several to many nodes may be left on the vine but, if it has only just grown past the fruiting wire, it may be pruned to just above the wire with the cane tied straight against the wire (Fig. 5.20, left). It is important when tying vines to remember not to bind them in such a way that later growth will be girdled (Fig. 5.21).

If vines have not grown much at all, there is the option to prune them back to the most distal healthy bud and tie them with string up to the fruiting wire or against a stake (Fig. 5.20, right) or, if in a tall vine shelter, left on its own. This ensures that the next season's growth will start from as close to the wire as possible. It is not recommended to 'two-bud' (i.e. cut a healthy cane back to a spur) vines after their first season unless the growth was minimal. The thinking around two-budding is that the buds at the lower positions will have stronger growth the following year, but as long as the most distal bud retained is healthy, it will grow just as strongly and the shoot will reach the fruiting wire sooner.

Fig. 5.20. Left: tying a cane distal to the last node to prevent girdling in the subsequent growing season. Note that the end node has been cut through diagonally so that the bud has been removed, but there is still part of the node (and tendril in this case) there to prevent the tie from slipping off. The topmost bud that will grow in the following season is just below the fruiting wire. Right: upward growth that young vines make during the season should be preserved by leaving the topmost viable bud so that the shoot will reach the fruiting wire as soon as possible. Here, the vine is fixed to a stake that is tied to the fruiting wire. Content author's own.

Vines that have been left to sprawl in the first season should have their strongest healthy shoot retained and tied up to the wire as already described.

If vines have grown well and already have been wrapped on the fruiting wire, it may be a good time to disbud the trunks. If this procedure is done well, it will only need to be done once – for the life of the vineyard. Figure 5.22 shows how this is done – all of the bud must be removed from the cane/trunk to prevent leaving any buried buds in the wood. Please note, however, that once a bud has been removed in this fashion, it will never form another shoot from there: the viticulturist must be certain that it will never be needed in the future. A general rule of thumb is to leave three to four buds below the fruiting wire so that if the vine head needs to be re-established, there will be a source of shoots to do it. Note that this practice may reduce sap flow through the trunk as the vascular system is disrupted (see Chapter 6, "Pruning Style"), so the trade off between reducing labour requirements and vine performance should be taken into consideration.

Fig. 5.21. Left: Cloth tie left from securing the young vine's cane to a support at vine training that was not removed. Note the swelling above the tie, which was caused by carbohydrates not being able to be transported to the roots. Centre: Girdle that resulted from leaving the tie in place. Right: A trunk that has been partially girdled because the string used to help train it to the fruiting wire was not removed. Content author's own.

Fig. 5.22. Disbudding a young vine trunk. Secateurs are used to cut the entire bud out of the nodes along the lower part of the trunk. This prevents future, and unwanted, shoot development, but may have negative impacts on sap flow through the trunk. Content author's own.

It is not possible to completely de-bud around the graft union, as there are already many older buds there, so there should always be a source of new trunks. In climates with very cold winter temperatures, for example, these buds are valuable for training new trunks, should there be freezing damage.

Second season

In most cases, the management goal for the second season is the same as the first: to maximize vegetative growth, including growth of the root system. To ensure this, any fruit that the vine may set should be removed as soon as possible. The easiest time to do this is early in shoot development, when the flower clusters are easily seen and flicked off quickly by hand. Removing shoots growing from the base or the trunk of the vine (also known as de-suckering) should also be done to concentrate growth in the remaining shoots.

In certain environments, it may be possible for the vines to carry a small crop in their second year of growth. Usually, this would be in warm climates where the vines had made a lot of growth in the previous season and there was ample development of both shoots and roots. In areas where this is not possible, carrying a crop will be a drain on the vine, resulting in poor vine performance in future seasons. Crop load, therefore, can become a tool for managing vine uniformity in a new vineyard. The most vigorous vines can have a small crop left on them to reduce their rate of vegetative growth, which allows the weaker vines to better compete and exploit the soil and canopy area.

In addition, it is important to replant any vines that have not survived as soon as possible, again, to ensure that there is as little developmental difference between them and the more established vines. Follow-up on this is vital to these vines' development: they will be at a disadvantage compared to the older vines unless they are given additional water and protection from weed competition. If this is not the case, they may never perform at the same level as their neighbours.

What looks like the beginnings of a proper canopy may appear in the second season, which ensures that the plants are investing lots of carbohydrate energy into shoot and root growth.

Irrigation and disease and weed control are still important in this year. Powdery mildew can easily appear on young vines as they grow rapidly, and maintaining adequate fungicide coverage can be difficult. Any disease on the vines will reduce the amount of energy they can invest in other parts of the vine structure and will have an impact on future vine performance, so therefore should be controlled.

By the time the second dormant season comes around, vine pruning will take a lot more time than in the first. Strong canes should be trained along the wire, and the goal for the third season is to consolidate the development of roots and capacity to store carbohydrates, especially in a cool climate growing area. The number of nodes left on the vines should depend on, and be proportional to, the amount of growth the vines have put on in the past season: larger vines should have more nodes left on them than smaller ones.

In instances of less vigorous growth, canes can be cut back to just above the wire and tied for a straight trunk and weaker vines similarly treated as for the first year, so that a proper vine head can be developed. Spot fertilization or investigating why these vines have done poorly is worthwhile to ensure they catch up with their neighbours.

Third season and on

During the third season after planting management should prioritize vine root establishment as much as possible by encouraging leaf growth and only having crop on vines that need to be reigned back so that less vigorous vines can catch up. The presence of fruit on vines should be seen as a viticultural tool rather than a product to be sold. If the vine warrants a commercially viable crop, that's great, but don't be tempted if growth isn't strong enough. Restraint at this point will pay dividends later.

In the dormant season is the opportunity to really establish the structure of vine head, so care is needed at pruning to choose appropriate spur locations and leave a number of nodes that matches each vine's vigor (refer to Chapter 6).

By the fourth season the vines may look mature, but even though the shoots may have filled the canopy, the root system is still exploring the volume of soil underneath. Caution is necessary not to over-crop the vines and physiologically-speaking, it is better to err on the side of too little than too much.

If all has gone well, by the fifth season the vines should be more or less mature, but close observation is still required to ensure that vine growth is robust each year.

Seasonal Management

6

Abstract: Many operations are necessary through each season of vineyard management. Pruning sets the vine up for the coming season and should be done with care. The balance between vegetation and crop should be considered: various ways of evaluating vine balance and vine capacity are presented to help make decisions around pruning severity. The most common pruning types are cane and spur, each with its own advantages and disadvantages, and there are many types of trellising systems adapted to grape production for different markets. Vineyard floor management uses plants for improving equipment access, biodiversity and soil nutrition. Cover crops have an influence on frost control, for which active and passive methods are discussed. Vine canopies should be porous for best fruit maturation and disease control, and if irrigation is installed its management is key to maintaining desirable canopy density. Harvest timing and method have a large impact on fruit quality.

Once a vineyard has been established there is ongoing management to keep it in top producing form. In many ways this process starts at pruning with the process of matching the vine to the trellising system, as this sets the vine up for producing fruit in the following year. There are then many other management interventions that can be used during the growing season, including those dealing with the canopy, cover crop, water application and other methods that fine-tune the grape product. Finally, determining when and how to harvest the crop caps off a season of preparation.

Pruning and Training

Pruning and training the vine are two of the most important aspects of quality grape production, whether it be for table, wine or other uses. How the vine is pruned and trained interacts with where the vine is growing, the proposed end product and management style. The two practices are firmly intertwined.

In essence, through pruning and training, the vine is shaped and manipulated to fit within the volume of trellis that has been made available for it.

DOI: 10.1079/9781800627062.0006

There are two main types of pruning that can occur: summer and winter. The purpose of summer pruning (removal of green shoot parts during the growing season) is mainly to open up the canopy, prevent self- or between-row shading or just to tidy up the appearance of the vines.

The substantive work in pruning happens in the dormant season, when a decision has to be made as to how much, and which parts, of the previous season's growth must be removed (see Fig. 6.1).

The aims of pruning are to:

- Establish/maintain the vine in the desired form.
- Select nodes that will produce fruitful shoots.
- Regulate shoot number/crop load.
- Regulate vegetative growth.
- Produce fruit of the target composition.

Vine balance at pruning

A key concept in thinking about pruning is balance. In the case of most grape growers, the balance is in preparing the vines to 1) produce an economic crop, but also 2) keep the plants from spending too much of their energy ripening that crop at the expense of their ability to grow in the following season (a consideration in the cropping of any perennial plant). Refer to Chapter 3 for a discussion about the carbohydrate balance of the vine and how it is related to management.

As the pruner leaves more nodes on a vine, they leave not only more growing points (shoots), but also potential fruit (all the clusters on those

Fig. 6.1. A partially pruned row of grapevines in Oregon, USA. Note how much material is removed in the process. Content author's own.

shoots). Thus, pruning has a significant influence on the ability of a vine to grow and to produce a crop.

Let's look at this in more detail. If we approach a vine in the dormant season, we see canes that were the shoots from the previous season. On each of those canes are a number of nodes. At each of those nodes is a compound bud that contains three preformed shoots. Each of these shoots is capable of growing and each may, or may not, have flower cluster primordia on them. In most cases, only the largest of those preformed shoots will grow in the spring. So, if a vine was not pruned, it would seem like a very large bush, with small shoots coming out at all angles from every node, each potentially carrying fruit (see Fig. 6.2). In that growing season, the vine will be overcropped and likely not be able to ripen the fruit to a commercial level.

If we prune a vine back to just a few nodes, in the spring we will have a very limited number of shoots coming out, as well as having a limited amount of crop – ideally there would be the same number of shoots as nodes left, with each shoot having some crop on it. In this case, we have pruned off a lot of the potential vegetative and reproductive growth of the vine. In the spring, it will have available a much smaller leaf area with which to photosynthesize and make carbohydrates and, perhaps more importantly, it has a very limited fruit

Fig. 6.2. Unpruned vine in Upstate New York showing the many nodes left and the small shoots growing at an early time of the season. This vine suffered from winter freeze damage, resulting in patchy death of the cambial layer. A new trunk would need to be trained from the base to get a healthy, productive vine. Note that large number of berries in relation to the leaf area available. Content author's own.

load. The result of this will be rampant growth of those shoots, and in practice, the pushing of many buried buds (left over from previously removed canes or shoots) and other preformed shoots in the compound bud (see Fig. 6.3). This is not a desirable situation due to the crowding of shoots and leaves leading to shading, difficulty in controlling disease, and poor fruit maturation.

So not pruning a vine is not what we want, and pruning it too hard is also not a good thing. So the question then is how many nodes is appropriate to retain in a given situation? Some of the most widely used guidelines relating to pruning decisions are published in the book *Sunlight into Wine* (Smart and Robinson, 1991), where the balance between shoots and crop can be estimated

Fig. 6.3. The head of a vine showing the growth of many (all but one of the visible shoots) unexpected (non-count) shoots. These lead to high canopy density, which delays fruit development and encourages disease. Content author's own.

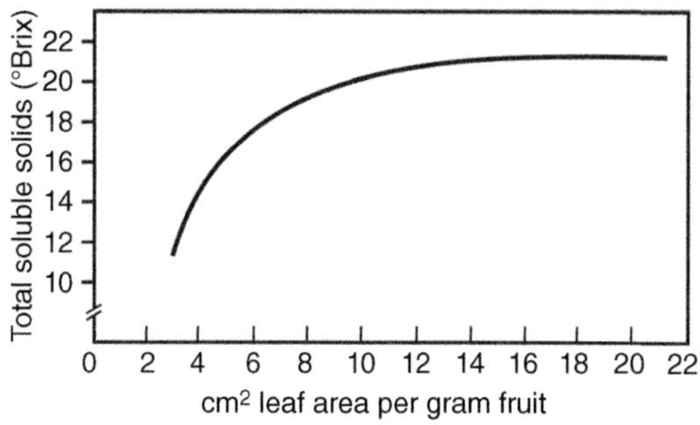

Fig. 6.4. Effect of a range of leaf areas per gram of fruit on fruit Brix after 4 years of treatment application to 'Thompson Seedless' vines. No appreciable increase of Brix is obtained at the very high values (Used with permission, from Kliewer and Dokoozlian © 2005 American Society for Enology and Viticulture. AJEV 56:170-181).

through examining the fruit weight to pruning weight ratio. This concept can be used to help make practical pruning decisions.

For the fruit weight to pruning weight ratio, a value of 5–10:1 is said to represent a vine that is in balance – that is, the amount of fruit it is producing is appropriate to the amount of canopy the vine has. This concept is not new, having been first proposed by Ravaz in 1911 (as cited in Howell, 2001), and indeed, the ratio is often referred to as the Ravaz Index. The principle that there is a balance between vegetative and fruit development was explained by experiments that looked at a range of crop loads and leaf areas on vines and came to the conclusion that approximately 5–14 cm^2 of leaf area was needed to ripen each 1 g of fruit (see Fig. 6.4), depending on the trellising system and environment (Kliewer and Antcliff, 1970; Kliewer and Dokoozlian, 2005). Crop weight is straightforward to measure, and as there is a relationship between shoot growth and leaf area, it is easy to make the jump from that to using pruning weight, which is the summation of shoot growth over the season, as a proxy for leaf area. It should be noted that the 5–10 ratio for the Ravaz Index and 5–14 cm^2 of leaf area per gram of fruit values were developed in moderately warm to hot climates - in cooler climates the former could be around 3 and the latter towards the upper end and beyond 15.

Another tool for considering balance is through the calculation of approximate dry matter production of a vine: (fruit weight × 0.25) + (pruning weight × 0.55), where the numbers that are used as multipliers are roughly equal to the proportionate dry weight of the fruit and dormant canes, respectively. This value can be thought of as an overall measure of the capacity of a vine (Winkler *et al.*, 1974), which is related to its ability to grow vegetatively or produce grapes.

Knowledge of vine balance and capacity can be used to make actual pruning decisions, but before we get there, some terminology: in pruning we will refer to count nodes and non-count nodes. Count nodes are those that are left on purpose during pruning *and* which we expect to produce one fruitful shoot. In practice, the first count node on a cane is not the first visible node, as the compressed nodes at the base of the cane (which can be defined as with less than 5 mm of internode below them) do not normally push during budbreak. The shoot that arises from a count node is referred to as a count shoot and, in normal circumstances, this will be the primary shoot, which bears the most potential fruit.

A non-count node is any node that is not a count node. For example, the compressed nodes at the base of a cane are non-count nodes, as are those that are embedded in the trunk, head or cordon of a vine (buried buds). Similarly, shoots that arise from non-count nodes are termed non-count shoots or canes (see Fig. 2.5). In addition, secondary or tertiary shoots that may arise from a count node are also considered non-count shoots, as they are not anticipated to burst in the spring: Figure 6.3 shows only one count shoot despite the large number of shoots present.

Types of pruning

The principal methods of pruning are cane and spur (also known as cordon) pruning. With cane pruning, the fruiting shoots for the next season come from a length of the previous season's shoot, which is secured to the fruiting wire of the trellising system (see Fig. 6.5).

Cane pruning typically uses up to four canes of 8 to 15 nodes in length on the vine, depending on trellis, vine spacing and cropping aims. In most cases, one or two replacement spurs (which is basically a short, two-node cane) are

Fig. 6.5. Examples of cane-pruned vines. In this case, there are two fruiting wires and three canes per vine. Content author's own.

also left below the head, which is to facilitate cane choice the following year. The vines in Figure 6.5 have their heads at the centre of a bilaterally trained vine, but the head can be to one side on unilaterally trained vines (typically used for close vine plantings, where there is only one cane tied down). The goal is to choose canes that are: (i) in a good location to retain the shape of the vine; (ii) well-exposed in the previous season, to give maximum fruitfulness; (iii) have good periderm formation; and (iv) not excessively thick or thin. The latter two parameters are associated with the cane's ability to withstand cold while dormant and increased fruitfulness in the following growing season, respectively.

Advantages of cane pruning are that vine fruitfulness is maximized and the number of potential non-count nodes on the vine is minimized, because there is less permanent wood on the vine and therefore less surface area from which non-count shoots can arise. As some cultivars are inherently less fruitful at the basal node positions on the shoot (Swartwout, 1925; Winkler, 1926; Buttrose, 1969), cane pruning generally results in higher potential crop loads compared to spur pruning.

Disadvantages of cane pruning include: (i) blind budding (nodes along the cane where no shoot develops), especially with long canes; (ii) limitations to between-vine spacing (in that the distance between vines is restricted to around twice the length of a typical shoot during the growing season); and (iii) uneven shoot growth along the cane. The latter phenomenon has been called the End Point Principle by David Jackson (1997), where shoots at the end of a cane tend to grow more vigorously than those in the mid-portion of the cane (see Fig. 6.6). Extreme examples of this arise when there is no shoot development in the middle parts of the canes, but there is at the ends and bases, which also demonstrates the Trunk Proximity Principle, where shoots tend to develop more vigorously near the base of the canes (Jackson, 1997). Cane pruning can also be more difficult when training workers to make the best decisions, resulting in loss of some of the advantages of cane pruning in the first place, and it generally take more time per vine to complete due to the need of wrapping and tying the cane to the wire.

Spur pruning does away with continued use of canes by establishing a permanent arm, or cordon, from a cane (see Fig. 6.7). Note that, in establishing a cordon, shoots arising from the downward-facing bud on the original cane are removed, leaving every alternate and upper node positions present. Downward-pointing shoots do not grow so vigorously (Kliewer *et al.*, 1989) and, if shoots from both nodes were used, it would enlarge the fruiting zone, making some management practices more difficult. To counteract the loss of shoots from these positions, canes arising from the remaining nodes are cut back to (typically) two-node spurs so that the number of shoots per metre of fruiting wire is the same as with cane pruning. Note that, when training a cane that is destined to become a cordon, it should not be wrapped as tightly as if the vine is cane pruned every year. Too many revolutions around the wire can result in girdling of the cordon as it increases in girth (see Fig. 6.8).

Advantages of spur pruning arc that it is easier to teach unskilled people how to do it and it is possible to partly mechanize the process using cutting bars (see Fig. 6.9), which means that the crew that goes in afterwards has much less brush-pulling to do. Spur pruning also results in the vine having more permanent vine material, which has been implicated in fruit quality (Howell, 2001). Disadvantages are: (i) vine fruitfulness is limited (because the basal nodes on canes are typically not as fruitful); (ii) spurs can die out over time due to injury, disease or other problems, leading to barren sections of canopy; (iii) the fruiting zone can creep upwards as seasons pass if there is a tendency for pruners to choose the distal nodes on each of the spurs (this can result in very uneven or high fruiting zones); (iv) the cordon needs to be renewed periodically due to damage, disease, etc.; and (v) there is a tendency towards a higher number of non-count shoots due to the larger amount of older wood.

Other factors to be aware of in spur pruning relate to the canes that were used to set up the cordons. Spur spacing depends on the internode length of the original cane, so initial cane choice is important. If internodes are too long,

Fig. 6.6. Example of canes showing signs of the End Point Principle and Trunk Proximity Principle. For the former, growth of shoots tends to be more vigorous at the ends of canes compared with that at the mid-sections and for the latter, shoots near the head of the vine grow more vigorously than those in the mid-cane region. Content author's own.

Fig. 6.7. Example of a spur-trained vine, showing two-node spurs. Note that each node position has a spur, which will lead to twice as many shoots per metre of row compared to a cane-pruned vine. Ordinarily, every other spur, usually those on the bottom of the cordon, is removed. Content author's own.

then spurs will be spaced too far apart and cropping will be limited. If they are too close together, there will be shoot shading, dense canopies and a host of other problems that are related (see the Goals and tools for canopy management section, on page 166). For most situations, keeping replacement spurs near the head of the vine is vital, because it is from those that renewal canes/cordons can be sourced. To avoid the upwardly creeping fruiting zone, pruners must be aware that shoots from the lower of the two nodes need to have preference over the upper. The low potential fruitfulness of spur-pruned vines can be overcome somewhat by altering the length of spurs retained, though this does have detrimental effects on the resulting canopy unless remedial measures (such as shoot thinning and leaf removal) are taken during the growing season.

No matter what the pruning system, there remain some tips to smooth the pruning process. If movable foliage wires are used in the trellising system, be sure to move these down before budburst as then they will catch the shoots as they are raised, reducing hand tucking after. If any trellis posts or wires have broken, your best chance to fix them is at pruning, before any tying down of canes has taken place.

Pruning style

In recent times there has been a lot of pressure on minimizing costs of production, which, of course, leads to changes in how vines are managed. Pruning is a very time-intensive operation and so is a prime target for potential savings, but since pruning sets the foundation of the vine up in terms of structure and

Fig. 6.8. Cordons where the wire has become embedded into the plant. Eventually, the performance of the cordon will decline due to restriction of nutrient flow and, in extreme cases, the entire cordon may die. Content author's own.

fruiting, it's not a practice to be changed without reflection. It's possible that more economic methods of pruning may be responsible for the increase in trunk diseases such as esca and eutypa (see Chapter 10). The thought is that pruning without regard to how the vine responds to repeated and numerous cuts on canes and older parts of the vine results in greater susceptibility to some fungal pathogens, which is logical because in the case of trunk diseases the infection site is open wounds. As a result of this some have been returning to a pruning style designed to protect sap flow through the wood, also known as gentle pruning, or respectful pruning.

Gentle pruning revolves around 1) minimizing pruning wounds, particularly large ones that will cause death of tissue back into the vine, and 2) pruning to preserve an undisturbed pathway for sap to flow from the roots to the most distal part of the vine (Lafon, 1921; Lecomte *et al.*, 2018). For the first

Fig. 6.9. Example of a barrel pruner, which is used to remove unwanted cane material from spur-pruned VSP-type trellis systems quickly and easily. The rotating cylinders have a sharp blade on the bottom to cleanly cut the canes off and the overlapping lobes crush brush into small sections, which can fall from the trellis. Wires pass through unharmed and the cylinders move in and out to avoid posts. Content author's own.

point, cuts should not be made very close to older wood - this is to prevent the formation of a cone of necrotic tissue that extends into the wood that disrupts sap flow (Bruez *et al.*, 2022) and provides a pathway for infection: these stubs can be removed in the following year, after the vine has sealed off the injury. If cuts are made above the older wood, a distance of 3cm for canes is suggested, because the depth of the necrotic cone is not directly related to the diameter of the cane (Bruez *et al.*, 2022). For cuts to older wood there doesn't seem to be a definitive guideline, but leave at least 3cm or the equivalent of the 1.5 times the diameter of the wood being cut (Deloire *et al.*, 2022; Galar-Martínez *et al.*, 2024), whichever is greater. Aside from this, use a pruning strategy that minimizes larger cuts and maximizes cuts to one-year old canes (Lecomte *et al.*, 2018).

To facilitate sap flow through the vine, cuts are made with regard to the vascular system in the functional xylem and phloem, making sure to retain a continuous path on one side of a unilaterally trained vine or on each side of a bidirectionally trained vine. Functionally, this means making pruning cuts on one side of the trunk or cordon so any disturbance to sap flow is limited to that side (Lafon, 1921; Deloire *et al.*, 2022).

Vines treated with gentle pruning techniques demonstrated better sap flow and greater conductive tissue area than those pruned conventionally and also showed greater biomass production (Claverie *et al.*, 2023; Galar-Martínez *et al.*, 2024). They also are associated with lesser trunk disease incidence (Lecomte *et al.*, 2018), but further research into how this pruning method affects disease development and vine performance is needed. Note also that with grafted vines, there is a rather chaotic arrangement of vascular tissues at the graft union, which cannot be avoided.

Pruning decisions

The basis for pruning grapevines is that each vine has a capacity to support growth (of shoots and roots) and ripen a crop. Each dormant season, an appropriate balance between vegetative production and yield must be found. Pruning is your first step in managing crop load and vine size, and so is an important process that requires thought and reasoning. If too many nodes are retained on the vines, time-consuming canopy and crop management practices such as shoot thinning, cluster thinning and leaf removal may need to be done during the growing season, or the fruit will not ripen sufficiently. A vine in balance with site considerations, its crop load and management decisions will have moderate growth that fills the trellis and achieves target fruit composition.

It is worth remembering that vines have evolved without something coming in and removing 90% of their season's growth each year. Vines produce far more nodes, each with a compound bud capable of producing three shoots, than they can support with the kind of growth needed to fill typical trellis systems. However, if a vine is left with a large number of nodes, it will reduce the number of growing points, effectively pruning itself. Any growth that has not hardened off and formed an adequate periderm will not survive freezing temperatures and will be lost during the winter. In the spring, a smaller number of buds at those remaining nodes will produce shoots, i.e. percentage budbreak is reduced (see Table 6.1). This (along with other factors relating to the cluster) is why, if twice as many nodes are left at pruning, there will not necessarily be a doubling of crop (Christensen *et al.*, 1994). An increase of 100% in node number results in a relatively small 60% increase in shoot number and only a 22% increase in yield. Part of this is due to the decreased percentage budbreak, but also to factors related to cluster weight and berry number.

Table 6.1. Yield components for 'Sultana' raisin production given two different vine pruning severities (data adapted and reproduced with permission, Christensen *et al.* © 1994 American Society for Enology and Viticulture. AJEV 45:141-149

Yield component	60 nodes/vine	120 nodes/vine	Change (%)
Shoots/vine	67	108	+60
Budbreak (%)	113	90	–20
Yield (kg/vine)	5.31	6.49	+22

Note that budbreak on the 60-node vines is greater than 100%, because some nodes produce more than one shoot (two or three of the preformed shoots in the compound bud push instead of one) and because there will be a greater number of non-count shoots arising from the permanent parts of the vine (from buried buds). This is related back to vine capacity, where a vine has the capacity to push a certain number of buds each year and, if they are pruned to below that level, a greater number of shoots than nodes retained will be produced. If more buds than that level of capacity can support are left, fewer buds will break, percentage budbreak will decrease and growth of those shoots that do push will be weaker.

Matching pruning to vine capacity

Each vine in a vineyard may have a slightly different capacity due to the micro-climate it experiences. Experienced pruners recognize this when they first look at a vine and judge its growth during the previous season. If there has been a lot of vegetative growth, more nodes will be left than there were the previous season, which should result in smaller shoots with less vigorous growth and greater crop load per vine. If the vine has put on very limited shoot growth, fewer nodes will be retained to encourage more growth from those canes and reduce the crop load. Crop load in both cases will assist in balancing the vines: a large vine's increased crop load moderates vegetative vigour, and a small vine's reduced crop load puts more of the vine's resources towards vegetative growth (roots, shoots and trunk). However, for very weak vines, fruit should be removed altogether.

Experience has traditionally been the measure to balancing vine growth and pruning level. However, in the 1940s, scientists in New York quantified this by producing a formula that matches the number of nodes retained to estimates of vine capacity. This method is known as balanced pruning (Shaulis *et al.*, 1953), and can be a useful tool in learning the relationship between vine size and pruning severity.

In balanced pruning, the number of nodes to be left on the vine is estimated by weighing most of the vine's canes produced in that year (more nodes than needed are left on the vine so that some can be trimmed off to reach the desired number after the measurements) and using that value in an equation. In this case, the capacity of the vine is estimated by just pruning weight. For 'Concord' in New York State, USA, the formula is $30 + 10x$ (Jordan *et al.*, 1981), where x is the weight of prunings in pounds (1 pound is approximately 0.45 kg). The first value is the number of nodes to be left for the first pound of prunings, and the second value is the number of nodes to be left for each pound after that. For example, if the vine has been pre-pruned and there is 1.5 lb of canes, $30 + (10 \times 0.5)$ or 35 nodes would be left on the vine. If the vine is larger and there were 3 lb of prunings, $30 + (10 \times 2)$ or 50 nodes would be left on the vine. Thus vines that put on a lot of growth that season will have a large number of nodes left, vigour of the individual shoots will be less and the greater amount of crop on the vine will also slow growth, resulting in a smaller vine with reduced capacity the next time it comes to pruning. For a small vine, balanced pruning allows its capacity to grow from season to season. This formula was designed for use with mature vines (note that the minimum node number is 30): for very weak or diseased vines, the formula should not be used.

Note that both the location and cultivar were specified with the formula; vines growing in different areas and for different purposes may have a different formula. For instance, in New York where balanced pruning was developed for use with 'Concord' grapes, the formula was 30 + 10x (x in pounds). Under Australian conditions (and with metric units), the general recommendation of 30 to 40 nodes/kg (for comparison, this is equivalent to 13.6 to 18.1 nodes per pound) of pruning weight has been suggested (Smart and Robinson, 1991). Since trellising (divided versus single) and vine and row spacing vary considerably from place to place, it is more informative to look at weight of prunings per 1 m of vine canopy rather than on a per vine basis, which allows for more direct comparison of values from place to place.

Other indicators of vine capacity are also used. One is to note how many 'effective' shoots grew during the growing season. This can be done at pruning time once you have learned to estimate the weight of individual canes. Smart and Robinson (1991) recommend that the ideal cane should have a weight of between 20 g and 40 g during the dormant season. By weighing a variety of canes and getting an idea of what one in that range looks like, at pruning the number of canes that fall within this range can be counted. If a cane weighs less than 20 g it is not counted; if it is between 20 g and 60 g it is counted as one. If it is greater than 60 g it is counted as two. The shoot counts are totalled up for the vine and the result is an estimate for the number of nodes to leave for the following season. If the number of nodes left the previous season was much higher than the number of 'effective' canes, then fewer nodes should be left on the vine as the vine did not have the capacity to push all of the nodes retained. If the number of nodes left the previous season was fewer than the number of 'effective' canes calculated, then many non-count or large shoots will have come out and the vine must have more capacity for growth available.

Regardless of where the vines are growing or what system you use, the concept of any formula and tradition remains sound: larger vines should have more nodes, and smaller vines fewer nodes, left on them.

Performed over a series of years, larger vines will reduce their capacity and the smaller vines increase their capacity. Equilibrium is reached when you are leaving about the same number of nodes every year. Shoot, root and trunk growth should be in balance with fruit production, which means a more uniform vineyard, facilitating vine management and resulting in an ideal balance between crop load and vegetative growth. Ultimately, this leads to higher quality of grapes and wine.

The basic procedure for cane pruning would look like this:

- Look at vine structure for any deficiencies that need to be corrected (e.g. poorly placed spurs, canes originating too close to the fruiting wire, poor cane growth near the head of the vine, etc.). The goal is to remedy these problems through pruning decisions.
- Locate those canes that are likely candidates for fruiting and those for spurs, keeping in mind the four points mentioned in the 'Types of pruning' section.
- Make the main cuts, *but*:
- Leave extra canes to choose from in case the targeted ones break when tying them down.
- Carefully remove the wood that was cut off without damaging the remaining buds (particularly important if you are pruning late in the season and the buds have begun to move).
- Make final cane and spur choices, determine how many nodes to leave (by formula or using your experience), cut to length and tie to wire, if necessary.
- When cutting canes to length, cut through nodes, removing the bud and leaving part of the node swelling, which makes securing the cane to the wire easier (refer to Fig. 5.20), and avoids the possibility of girdling the cane or cordon later.
- Remove any extra canes.
- Remove older wood that is no longer needed.
- Paint large pruning cuts with an approved fungicide or biocontrol agent to protect from *Eutypa* and other wood-invading pathogens.

It is difficult to start pruning if you are inexperienced, so if you are able to observe and discuss the practice with someone that has done it for a while, you will get a much better appreciation for the mechanics of it and the decision-making processes. Pre- and post-pruning photos of a problematic vine are presented in Fig. 6.10.

Other considerations when pruning

1. If you are in an area that is prone to frosts, prune as late as possible: bud-break is delayed by late pruning (Antcliff *et al.*, 1957; Williams *et al.*, 1985; Friend and Trought, 2007). Yields have also been shown to increase with later pruning dates (Friend and Trought, 2007), which also may be an advantage. Delayed pruning and removal of pruned wood must be done carefully if the buds are swelling – they can be brushed off easily.

2. Choose canes that were in well-exposed positions: flower cluster initiation takes place in the buds at the base of each leaf near bloom, meaning that the following year's crop is beginning to be determined almost 18 months before it

Fig. 6.10. An old vine before and after reconstructive pruning. Top: the vine has a lot of older wood around the head and the head is spread out along the two fruiting wires. After pruning (bottom) much of the old part of the vine has been cut away, leaving two canes, from which the new, much more compact, vine head will be established. Note that the large cuts have been covered to protect them against disease (see Chapter 10). Content author's own.

is actually harvested. When pruning and laying down canes, pick those that have been well exposed during the early part of the growing season.

3. Plan ahead for canopy density: overlapping canes will cause excessively dense canopies, which will lead to poor flower cluster initiation, spray penetration, air movement and fruit ripening. If the vine requires more nodes to be left than there is space available on the wire then, if possible, add another fruiting wire to lay down more canes, arch the canes (so that more nodes per metre of row are retained) or perhaps consider that the vines are too vigorous for the trellising system, in which case it should be changed (see Training and Trellising, below).

4. When is big too big? If single-canopy vines are producing near or over 1 kg of pruning per 1 m of row, it may be worthwhile considering changing to a divided canopy. Smart and Robinson (1991) have suggested that ideal values for pruning weights in single-canopy systems fall between 0.3 kg/m and 0.6 kg/m of row. If, for example, vertical shoot-positioned (VSP) trained vines are producing much over that, conversion to a divided canopy of some sort should result in more moderate growth over the larger volume of space available. This will return better-quality fruit (due to more fruit exposure to the sun, better airflow and better spray penetration), increased yields per acre and fewer management interventions, which should offset the cost of the conversion (Smart and Robinson, 1991). If vigour is still a problem (see Fig. 6.11) then

Fig. 6.11. Very vigorous vines in an Oregon, USA Scott-Henry-trained vineyard. The upward-growing shoots are shading the adjacent rows as well as themselves; the downward-trained shoots have been cut off by the mower. These vines are in desperate need of some good viticultural management! Content author's own.

other management techniques must be used to reduce it, for example, through the use of more aggressive cover crops, reduced irrigation, etc.

In general, if there is excessive shading in the canopy resulting in disease (because of poor spray penetration or increased humidity due to lack of air movement), poor fruit quality (from lack of exposure) and poor fruitfulness of the buds (due to a lack of light for flower cluster initiation), some management intervention is needed to correct it.

One additional note before we leave this section is that prunings should be disposed of properly to avoid disease carry-over into the next season, preferably mulched into small (less than 2 cm long) pieces and left on the vineyard floor to decompose and, over time, return the nutrients and organic matter to the soil.

Training and Trellising

Trellising is the science and art of arranging the vine in space. The goals are to maximize light interception, minimize the number of additional inputs (e.g. labour, material costs, maintenance, etc.), provide physical support and be compatible with the management situation, which includes machinery (Fig. 6.12). In choosing a trellising system, factors such as product end use, harvest method, mechanization, cost and labour requirements have to be considered, along with the predicted vigour/capacity of the vines at that site.

Fig. 6.12. An example of where there is incompatibility between mechanization and trellis design. An entire section of row has fallen over due to mechanical harvester damage. Trellis designs for mechanized systems should be robust enough to withstand the additional stress they cause. Content author's own.

Self- or stake-supported

The simplest training systems are staked vines or bush vines. The vine is trained to a short trunk tied to a stake on which a head with several short canes or spurs is left. Once the trunk can support the weight of the canopy and fruit on its own, the stake can be removed (Fig. 6.13). Alternately, the stake can remain and canes can be trained upward and tied to the stake (Fig. 6.14). There are other forms of self-supporting systems, such as the basket vines, in which canes are woven together to form a low bowl shape, which may encourage the capture of dew, important in areas that cannot be irrigated (Fig. 6.15). Advantages of these types of systems are low capital cost, compatibility with low vine densities suited to arid areas and the possibility of movement of machinery crossways through the vineyard. Disadvantages include low cropping capacity per hectare due to inefficient use of land area.

Single wire

Single-wire trellising systems are used for a variety of purposes and are low cost relative to most other systems, due to only one wire being needed. Here, a single fruiting wire is used and shoots arising from the cane or spurs are free to grow in whichever direction they end up (see Fig. 6.16). In vine species

Fig. 6.13. Bush vines in a Californian vineyard. Note the wide vine spacings and tyre tracks indicating cross-cultivation. Content author's own.

Fig. 6.14. Individually staked vines. Note the clean cultivated soil in this young vineyard. Content author's own.

Fig. 6.15. A trellis-less vine on the island of Santorini, Greece. Vines are woven into a basket shape to encourage collection of early morning dew, and spaced far apart so the amount of soil water available to the vine is maximized (photograph courtesy Loretta Brenner, reproduced with permission).

Fig. 6.16. These grapevines in Oregon are trained to a single-wire trellis. There is virtually no canopy management, yet fruit exposure and spray penetration are still good. Content author's own.

with upright growth habits (most of the *V. vinifera* species), this creates a high hedge open at the bottom, whereas in those species with a trailing growth habit (e.g. *V. labrusca*) the hedge form is lower and less open at the bottom. Though this system is not compatible with close row spacings, the canopy is usually quite sparse, leading to good fruit exposure, spray penetration and the potential for relatively high crop loads.

The practice of minimal (only trimming the low hanging canes of otherwise untouched vines) or mechanical pruning (trimming tops, bottoms and sides of the vines as needed) can be combined nicely with a single-wire system. Vines are not pruned by hand, but rather trimmed by machine (see Fig. 6.17). The size of the cross-sectional box to which the vines are trimmed determines the approximate number of nodes left on the vine. Crop adjustment can be accomplished through trimming the vines after fruit set, when crop loads can be better estimated (usually, the bottom part of the vines is trimmed off in this operation), or later, when the berries are larger, with a mechanical harvester (Pool, 1987; Fisher *et al.*, 1997).

Hedge-type

Vertical shoot-positioned trellis (VSP) (Fig. 6.18) is a widely used trellising system for the production of wine grapes and results in the box-shaped hedge that is seen in so many vineyard calendars. A fruiting wire is relatively

Fig. 6.17. A comparison of minimally pruned vines with hedge-type vines in the early part of the season. The minimally pruned vines have many more, but shorter, shoots than the hedge vines, which can lead to greater productivity due to greater photosynthetic area. Content author's own.

close to the ground (typically 90 cm at a row spacing of 2.5–3.0 m), dormant canes wrapped to it (the horizontal cane version of this is called a Guyot system) and shoots are trained upwards from it during the growing season. This results in a fruiting zone above the cane, which localizes the crop relative to most of the canopy. Fruiting wires can number one or two, with one to four fruiting canes being typical. Vines can also be spur pruned, with a cordon trained both ways (bilaterally) or one (unilaterally) depending on vine spacing.

Additional wires are needed to maintain the shoot growth in its vertical position, adding to the system's cost. Labour is also required to position the shoots within the wire and keep them there, as vine growth habit and wind tend to make them fall out. Cultivars or clones with an upright growth habit are much better suited to VSP, as it is much easier to keep them within the wires. Typically, trimming of the sides and tops of the canopy is also necessary, as often VSP is used in areas where vigour is greater than the system can contain. To allow for more nodes to be laid down between vines, the canes can be arched rather than tied to the horizontal fruiting wire (see Fig. 6.19), which also assists in levelling out shoot growth along the cane. Names of these variations are Pendelbogen, Umbrella Kniffen or simply Arched Cane.

Fig. 6.18. Spur-pruned wine grapes growing on a vertical shoot-positioned (VSP) trellis. The steel posts make for a narrow canopy, leading to good fruit exposure, air movement and spray penetration. Content author's own.

Training to a Sylvoz system is a way of increasing the number of nodes left per metre of row without having to divide the canopy with a more expensive trellising option. Vines are cane or spur pruned to a mid-height fruiting wire, but additional canes are left trained downward and tied to a lower wire (see Fig. 6.20). This spreads out the shoots and fruiting area – the former being beneficial for canopy density and light interception, but the latter making it more difficult to target fruit sprays and perform effective leaf removal operations.

Fig. 6.19. Vines trained to an Umbrella Kniffen-style system in the Margaret River area of Western Australia. The canes are arched over wires placed at the desired height over the vine heads. Note the use of natural timber posts and straw mulch under the vines. Content author's own.

Divided canopies

Divided canopies were developed to address the concerns of maximizing crop and accommodating high vigour. The first significant breakthrough in this area was made by Nelson Shaulis and his research crew of the Geneva Experiment Station in New York, USA (Shaulis *et al.*, 1967). The Geneva Double Curtain (GDC, Fig. 6.21) addressed the issues of increasing productivity (through better light interception per unit land area) and mechanization of 'Concord' grapes, which were primarily grown for juice production. The system is effectively two single-wire systems set side by side. Due to the trailing growth habit of *V. labrusca* vines, two distinct curtains of shoots develop (though usually with the help of some shoot positioning). GDC has been adapted for use with other grapevine species, though its best application remains with trailing growth-habit cultivars.

Other horizontally divided canopy systems have been developed, such as the Lyre system, which is similar to the GDC in that it simulates two close-spaced rows of vines while only being a single row of vines. The Lyre is so called due to the instrument-like arrangement of the two VSP-like trellising systems (see Fig. 6.22) when viewed end-on, a configuration that was chosen to optimize light interception and therefore productivity

Fig. 6.20. Sylvoz trained vines, where short canes are tied to a lower wire. Content author's own.

(Carbonneau and Casteran, 1987). However, the capital costs of establishing this trellising system are high in comparison with that of most others and mechanization is also challenging, so the perceived benefits must be worth the additional expense.

A popular divided canopy system is the Scott-Henry (see Fig. 6.23) and its derivatives (e.g. Smart-Dyson) (Smart and Robinson, 1991). Rather than splitting canopies horizontally, this system splits them vertically, with one set of shoots trained upward and one downward. An advantage of this system is that it is easy and relatively inexpensive to retrofit to a VSP system, which is useful if the eventual vigour of the vines has been underestimated at planting.

Some generalizations on divided canopies would be that porosity is increased through their use, which in part accounts for their beneficial effects

Fig. 6.21. The view down the centre of a Geneva Double Curtain (GDC) trellis, showing cane-pruned vines and the distinct distance between the two fruiting wires. Content author's own.

Fig. 6.22. Looking down the inside of a row of vines trained to a Lyre trellis. It is an advantage if a person is able to walk down between the canopies and work on the shoots. In any case, there should not be any shoots bridging the interior of the two sides. Content author's own.

Fig. 6.23. Scott-Henry-trained vines in Mudgee, New South Wales, Australia. Of note is the distinct band of light running through the middle of the canopy, which demonstrates good practice with this system, as air should be able to move between the upper- and lower-trained shoots. Content author's own.

on improving crop load and fruit composition and decreasing disease incidence. They also allow for more flexible matching of vines to potentially vigorous sites and stretch the yield: quality envelope. In many cases the cost of establishing or converting to a divided canopy system are justified given the benefits received over the life of the vineyard.

Other trellising systems

Overhead systems (e.g. pergola or parronal) are the most efficient in terms of land use in that almost 100% of the incident light is captured and used by the vines. This leads to high yields per unit area of land, but there is a cost in terms of trellis construction and difficulties in managing the system. A series of wires is strung between post supports and the vines trained along the top (see Fig. 6.24), usually with a low number of vines per hectare compared with the previously mentioned trellising systems.

Overhead trellising is used in table, wine and raisin grape production, to varying extents. Adoption of this system in raisin production is growing as a result of techniques allowing raisins to be dried on the vine rather than in between rows (Christensen, 2000).

Fig. 6.24. Top: an overhead vineyard trellis system growing 'Sultana' grapes for fresh consumption in Chile. Note the fresh shoots on the ground, which are the result of shoot thinning to open up the canopy. Bottom: A hillside 'Koshu' vineyard in Japan. Content author's own.

Munson 'T'-type trellises are used extensively for fresh grape production as well as for wine grapes and raisins. As its name suggests, a single post supports a horizontal brace on which three wires are strung up and down the rows. Shoots are trained on the top of the trellis wires and the fruit hangs below (see Fig. 6.25). Variations of this have the top brace at an angle, where shoots are trained up and along it, allowing the fruit to hang underneath, but with better access for management and harvesting. In the case of the swing-arm trellis

Fig. 6.25. A 'T'-type trellis, which encourages the fruit to hang freely under the canopy. This allows good access for sprays as well as for handling. Content author's own.

(Clingeleffer and May, 1981) the 'T' portion of the trellis also pivots, to improve compatibility with mechanization.

There are many variations on these themes for providing a structure on which the grapevine can grow, each adapted to a particular situation or need. For a review of these see Freeman *et al.* (1992) and Smart and Robinson (1991).

For those vineyards that are intended to be managed by hand, research such as that by Kato *et al.* (2006) should be of increasing interest, as it has been demonstrated that certain types of trellising systems are more user-friendly than others, in terms of risk of musculo-skeletal disorders such as those associated with repetitive strain injuries.

End assemblies

Trellises support much of the weight of grapevines and their fruit, as well as resisting movement caused by wind, rain and vineyard machinery. The structures at the ends of the rows must resist the along-the-row tension caused by the weight on the trellis, as the intermediate posts carry the strain of the vines, fruit and even the accumulated weight of rainwater on them (Mollah, 1997).

The end assemblies must transfer the tension in the wire (which is, in large part, determined by the spacing between intermediate posts (Freeman *et al.*, 1992)) to the ground and, as such, need to be more robust than the intermediate posts. The main load-bearing unit of an end assembly is called the strainer, and its ability to transfer the load to the soil depends on the trellis height, strainer diameter, leverage affected by its depth in the ground (typically 1–1.2 m), and the soil itself (Smart and Robinson, 1991; Freeman *et al.*, 1992; Mollah, 1997). Three main types of end assemblies will be discussed here that should be useful for most vineyard situations.

The tieback end assembly (see Fig. 6.26) is a system where the load is transferred though the end post to an anchor in the ground. The anchor can be a deadman (solid object buried in the ground), an auger-like device that

Fig. 6.26. Tieback end assemblies. Two loops of wire are used to link the strainer to the end post and each has a wire tensioner to adjust the strain on the posts. Content author's own.

is screwed into the ground or a post. The anchor must have enough surface area, in combination with the soil strength, to withstand the forces to be applied to it.

The wire used to tie the anchor to the end post should be equivalent in strength to the sum of all wires used in the row, where foliage wires count as half because they are not under as much tension (e.g. if there is one fruiting wire and four foliage wires, then the wire to the anchor should be equivalent to three wires) (Smart and Robinson, 1991). The tieback is cost efficient and allows for vines to grow unhindered right up to the last post of the row, but does require some additional headland for equipment turning compared with other systems.

A diagonal stay system (see Fig. 6.27) transfers the tension into the row using an angled brace, which can be made of wood or metal. The foot at the end of the diagonal component acts as a pivot for the strainer post, so that the tension attempts to twist the strainer up and over the diagonal stay. This is an effective system that is inherently stronger than a tieback one (Freeman

Fig. 6.27. Diagonal stay assemblies in a vertical shoot-positioned vineyard. Content author's own.

et al., 1992; Mollah, 1997), and also requires less space outside the last vine in the row for headland. However, due to additional materials used, it does cost more, and there is some interference for vine training caused by the diagonal stay.

The box end (or horizontal stay, H-brace) assembly (see Fig. 6.28) is regarded as the strongest, as the design keeps the force on the strainer vertical rather than horizontal, distributing it over a larger area (Smart and Robinson, 1991). Therefore, this post needs to be larger than the others. Though strong, the design is expensive to build and vines growing within the box are much harder to manage.

The failure of end assemblies is usually caused by either: (i) insufficient strainer diameter or strength, leading to breakage (see Fig. 6.29); (ii) not putting the strainers into the ground far enough, which leads to them being pulled out; or (iii) soil not being strong enough to bear the load, which leads to strainers tilting or moving through the soil (see Fig. 6.30). A thorough analysis of predicted trellis loading and soil strength, before determining the type of end assembly to use, will minimize capital outlay for establishment and prevent costly repairs and lost crop in the future.

Fig. 6.28. Many rows of box end assemblies in a New Zealand vineyard. Note that no vines are planted inside the box, due to the difficulties in managing them there. Content author's own.

Fig. 6.29. Top: an end post that is not strong enough to withstand the tension of the wires ties to it. Bottom: a free-standing end post that has failed due to inadequate strength. The adjacent row has already been retrofitted with a diagonal stay assembly in recognition of the original oversight. Content author's own.

Intermediates

Intermediate posts carry the vertical load from the vines and their crop, as well as lateral loads caused by wind and machinery. Because they do not transfer any tension (longitudinal load) that is along the wires, they do not need to be as strong as the end-assembly structures. Their spacing, therefore, is determined

Fig. 6.30. Left: an end assembly post moving in the soil. This could be due to the weakness of recently disturbed soil in this newly developed vineyard. Right: a free-standing end post that is starting to lean into the row. A tieback of some sort is required. Content author's own.

by the maximum these loads can reach in the life of the trellis. Typically, spacing is in the order of 6–8 m, but this can depend on the type of intermediate post used (see below), straightness of the vine trunks (and thus the amount of the load they can bear), typical wind loads in the area and whether or not a machine harvester is going to be used.

Post materials

For many years the standard vineyard post was made of a treated wood (typically pine), so that it does not rot as quickly. The expected useful life span for a pressure-treated creosote or chromated copper arsenate (CCA) post should be in the range of 30 or more years, which is about the same life as a commercial vineyard (vines can produce quite happily for much longer than this, but often vineyards are redeveloped for market reasons). There has been a lot of discussion and research around the impact of treated posts in the environment, and indeed, research has found that there is slow leaching of the treatment chemicals into the soil around the post (Robinson *et al.*, 2006). In addition, disposal of broken or otherwise abandoned treated posts is problematic (Helsen and Van den Bulck, 2005), so their use should be carefully considered. Because of these issues, or because you wish to establish your vineyard with organic certification, the use of other materials for posts is warranted. These include metal, plastic (or plastic-coated wood), wood, but treated with a more environmentally-friendly chemical (e.g. creosote), untreated, but hardy, timbers, and concrete.

There are a number of metal post types available, ranging from simple cylinders to those that have a 'U', 'V' or even a 'Z' shape to their end-on view. They should be galvanized to prevent corrosion and have the advantage of making for a narrow canopy, which means better light penetration and air movement. Because they are also quite easy to drive into the soil (unless it is particularly stony), replacing broken posts is also less of an issue. Whereas with a wooden post you can attach things to it at any point using a nail or staple, one of the tricks with metal posts is providing methods for fixing wires and clips along its length. This can be accomplished with spring-like coils of wire for cylindrical posts, which allows them to be placed at any location, or fixed notches or holes punched into the metal that wires or clips can be fit into.

The main disadvantage of most metal posts is their relative lack of strength compared to typical wooden posts, particularly to lateral loads. Even in relatively strong soils, because of the lack of surface area in contact with the soil, posts tend to move with, for example, strong winds, enlarging the hole they sit in. Eventually this leads to failure of the trellis. For intermediates, this can be adjusted for by increasing the depth to which the posts are driven (typically this is 60 cm) and/or decreasing the distance between posts: either results in greater cost per hectare. Another alternative is to use them in conjunction with posts that have greater lateral strength, taking some of the stress off the metal intermediates.

Plastic posts are also on the market, either from virgin or recycled materials. At present, they appear to lack the strength to bear significant loads, and their durability (especially with relation to UV radiation) is a concern. A variation on this theme is untreated pine coated with plastic, giving the strength of treated wood without the environmental risks around chemical leaching. As long as the plastic barrier is unbreached, the wood inside will not decay. The plastic is designed to be self-sealing with the use of nails and staples, maintaining the moisture barrier. Again, how well these types of post last in the field is a bit of an unknown, but they deserve consideration and testing. Wood treated with creosote behave similarly to CCA posts, but the preservative doesn't have the same longevity, so they need to be replaced sooner.

Natural timbers that are resistant to decay are also in use, and if locally sourced, minimizing the need for long distance transport. There is increasing interest in using these types of posts, and if chosen well, they can last about as long as CCA treated wood (Bush, 2011). However, there needs to be more testing in this area to determine if locally grown timber meets the strength and durability requirements, as environmental conditions affect wood density and resistance to decay (Malan and Verryn, 1996).

Concrete posts are also used in some areas, especially those where timber is scarce. Due to transport issues, it is preferable for these to be made locally, but their strength and consistency within and between batches should be ensured. In most cases, these cannot be driven into the soil, so an auger and backfilling are necessary. These posts have the disadvantage of it being difficult to attach wires to, have a relatively high carbon footprint and are generally not a recommended option.

Wires

The wire of a trellis needs to support the weight of the vine and fruit, as well lateral loads, transferring them to the intermediate posts while also delivering the longitudinal strain to the end assemblies. The fruiting wires bear most of the load, and therefore are under greater tension than, for example, the foliage wires. Most modern trellises take advantage of high-carbon galvanized steel wire, which is able to withstand much greater forces than 'soft' steel wire without permanently stretching or breaking. Some amount of wire sagging between posts is acceptable, but if you think it is excessive, check the tension on the wires to make sure you do not go beyond the wire's yield limit (the point at which plastic stretching will begin). Remember that the load on the wire will be at its maximum when it is cold (steel will contract as the temperature decreases), there is a full canopy and crop, along with any moisture on the vines, and side loads from wind and/or mechanical harvesting.

As with metal posts, without some form of protection from corrosion, steel wires would soon lose strength and break. Galvanization can result in wires that will last for 30 years, as long as the coating is not damaged to the point where it can no longer protect. This can be an issue with metal posts, which may have sharp edges in contact with wires that scrape off the zinc coating. A plastic insert can be used to prevent this kind of damage.

Wires can be made of other materials, with plastic being a useful complement to steel. Plastic (usually polyester) wire has the advantage of being a lot lighter than steel and easy to manipulate and repair breaks, especially compared to high-tensile wire. It is highly elastic as well, meaning it is convenient to use for movable foliage wires where the rigidity of a high-tensile wire would be difficult to work against (especially for short rows). One should be aware of the potential for chafing on metal posts that have sharp edges, but with proper care, they should last up to 20 years.

Vineyard Floor Management

There are many options available for managing the soil surface in vineyards. For many years the soil was regularly cultivated to discourage weed growth – a bare earth policy (see Fig. 5.9). More recently there has been a shift from this to establishing certain types of plant growth between and under the vine rows. This is probably due to increasing numbers of options for controlling non-grape plant species, changing attitudes about vineyard appearance or managing soil health or increasing awareness of the sustainability of management practices.

In terms of impact on the vines, a bare earth policy would seem to be the best option. Other plants growing around the vines compete for water and nutrients, and can also hinder air movement through the vineyard. Bare earth is a simple policy for determining success rates and results in the lowest spring frost risk (Dethier and Shaulis, 1964) in areas so affected. However, continued tillage of the soil is bad for both soil structure (increasing the chance of erosion) and microorganism populations, and herbicide use is expensive and can have adverse impacts on the environment (Doran, 1980; Pankhurst *et al.*, 1995; Roper and Gupta, 1995).

Therefore, some form of planting in the vineyard system can have beneficial effects, such as reducing erosion, dust and undesirable weed development and increasing water infiltration rates, soil organic matter, quality of soil structure, populations of beneficial organisms and vineyard accessibility (for, e.g. workers, tractors and harvesters), especially in wet conditions (Hartwig and Ammon, 2002). Cover crops can also be useful as a tool to manage soil water availability and alter grapevine shoot growth. Disadvantages include the need to prepare soil, buy seed, and plant and establish the cover crop (if the volunteer plants are not suitable). In dry areas, its use of soil moisture can be a drawback, as well as introducing competition for nutrients. A vegetative cover on the soil also lessens heat accumulation during the day and release during the night, which can increase the risk of spring frost damage (Dethier and Shaulis, 1964). Vine root development is hindered in the cover crop's rooting zone and some plants can harbour pests such as rodents, diseases, insects and nematodes.

Careful selection of an appropriate cover crop is then warranted; there are many options available. Annual plants are usually shallower rooting (Schenk and Jackson, 2002) and therefore less intrusive on the vines and, if left to seed, do not have to be re-sown. Perennials have better staying power but may be more expensive to establish.

Popular options for cover crops are grasses (e.g. brome, rye, blue and fescue) or even a grain crop, which could be harvested or tilled back into the soil. Tall cover crops can be used in a young vineyard to provide shelter from wind, as well. The ability of a cover crop to die back in the summer is usually an advantage as, when conditions are at their driest, it is not competing with the vines for water and nutrients. Cover crops must also be able to survive in between the vine rows, where there is usually no supplemental water. In areas of low nitrogen, legumes may be used as a cover crop to help manage soil nitrogen availability. More recently, flowering plants have been used in vineyards as they attract beneficial insects, resulting in fewer pests (Fig. 6.31; Landis *et al.*, 2000; Nicholls *et al.*, 2000). Generally speaking, increasing plant biodiversity through choice of cover crops leads to increased insect and animal populations, and on to improved natural pest control (a type of ecosystem service) - a benefit to the grower and the environment (Winter *et al.*, 2018).

In areas with excess available water in the soil, deep-rooted, but low profile, plants such as chicory can be planted to help draw water, and therefore vigour, away from the vines (Caspari *et al.*, 1997).

Fig. 6.31. Examples of cover crops. Left: a mixture of phacelia (purple flowers) and buckwheat (white flowers). Middle: a variety of flowering plants. Right: buckwheat seedlings. Content author's own.

In some cases, a volunteer crop is allowed to become established as a cover crop, which has the benefit of being low cost and made up of those plants that are hardy enough to survive in between the rows. However, this volunteer crop may have undesirable characteristics, such as adding (or not) nitrogen to the soil, not dying out during the summer months or not achieving sufficient ground coverage to allow foot and vehicle traffic during the wet season.

Additional costs are associated with cover crops, such as the need to mow (particularly in the spring in frost-prone areas, as a short grass cover crop has minimal impact on frost incidence) and to keep the cover crop from encroaching on the vines or growing at the wrong time (usually resulting in the need to use herbicides or mechanical weeding).

During the vine establishment phase, cover crops are still desirable, but benefits in vineyard accessibility, biodiversity etc. must be weighed against competition for water and nutrients. They should also not be used in areas of severe frost risk or where soil water is at a premium. However, once the vines are established with deep root systems, cover crops can be planted with minimal impact (see Fig. 6.32).

Treatment of the under-vine area has changed in recent years, as well. Historically this has meant the use of mulches, which can be organic or inorganic, and can result in improved weed control, water conservation, soil structure, water infiltration rates and root branching. However, they can also increase vine vegetative vigour, shelter rodents and increase costs (Penfold, 2004). The use of mulches is often worthwhile during the establishment phase of vineyard development, as it can aid vine growth (see Table 6.2, Fig. 6.33) or in areas of low water availability.

Fig. 6.32. Recently seeded grass in a 2-year-old vineyard. Content author's own.

Table 6.2. Effect of using different types of mulches on the second-year pruning weights of 'Pinot gris' vines growing in Canterbury New Zealand. Cocomulch is a combination of brown plastic and coconut fibre (from G.L. Creasy and G. Wells, 2003, unpublished results).

Pruning weight (kg)	Control	Black plastic	Cocomulch	Compost
Mean	0.071	0.116	0.307	0.281
Standard error	0.020	0.019	0.048	0.039

Mulches are taking on other roles as well, in altering other aspects of the vines' environment. Spreading of mussel shells under the vine rows (see Fig. 6.34) was found to improve 'Pinot noir' wine quality, with minimal impact on other performance factors (Creasy *et al.*, 2006). These preliminary results are very promising, and have the bonus of providing an end use for a waste product.

Managing cover crops in the under-vine area has become more widespread with increasing interest in organic management and a desire to use fewer herbicides. Whichever plant you choose to grow under the vines, there will be increased competition for water and nutrients, so management practices will need to change to account for this. Therefore, changing from a herbicide strip or cultivation, vine growth and productivity will decrease unless more irrigation is provided and there is continued monitoring, and maintenance, of plant nutrient status.

Fig. 6.33. Three-year old vines with a difference. Those on the left were grown for the previous 2 years with black plastic mulch while those on the right were the control, with obvious effects in this non-irrigated vineyard. Content author's own.

Fig. 6.34. Mussel shell mulch as part of a research trial at Neudorf Vineyards in the Nelson region, New Zealand. The mulch was found to change the radiation reflected back into the fruiting zone, improving wine sensory characteristics. Content author's own.

Many cover crops will grow tall enough to interfere with the grapevine, so under-vine plants need to be managed, usually by mowing or periodic cultivation, but it can also be done with livestock. Machinery used for this is covered in Chapter 9, but it should be noted that the timing of cultivation is important, as it is ineffective if done when the soil is wet (the plants re-root easily) or the plants are too well established (i.e. robust enough to trigger the machine's vine sensor).

Frost Management

For some growing areas, the occurrence of early-season freezing temperatures is a real risk and, in some cases, late-season frosts as well. The latter tend not to be as common as the former but, if they do occur prior to harvest, fruit can be affected directly, ruining the crop, and if not, the premature death of the vine leaves means that the only carbohydrate source for ripening the fruit and providing the vine with energy to overwinter and begin growth in the following season is that stored in the woody parts of the vine. Early-season frosts of importance are those that occur near or after budbreak, when developing and green tissues are much more susceptible to freezing damage.

There are different terms when it comes to talking about freezing events, so it is worthwhile mentioning these before talking about frosts in more detail. A ground frost is said to have occurred if the temperature at the ground's surface falls below 0°C. However, this is not usually damaging for vines as they are typically held off the surface of the ground, so a more useful measure is the air or screen frost, when the temperature falls below 0°C at a height of 1.2 m (note that this is measured in a shelter, so it is the true air temperature being recorded).

Types of frost events

To ascertain ways of managing frost, a review of the types of frost events is required. The type with the fewest options to manage is the movement of below-freezing air into an area, or advection frost (Cornford, 1938). In areas with surrounding mountains or hills covered with snow, the freezing air moves down from the force of gravity (cold air is heavier than warm air) in a katabatic flow, enveloping the lower areas. Typically, these occur in mountainous areas when there are clouds, a bit of wind and low humidity.

A second type of frost is a radiation frost, which occurs on clear and windless nights as heat from the surface of objects is lost to the atmosphere (see Fig. 6.35). If there is little or no mixing of air, the heat lost from the ground and plant surfaces collects such that a layer of warmer air (the inversion layer) develops above (Evans, 2000). Since, in this case, there is warmer air above the freezing air, a strategy for reducing the freezing temperatures near ground level is to mix the two layers.

Fig. 6.35. Shoots that have been damaged in a radiation frost. Content author's own.

Radiation frosts are more of an issue if the relative humidity is low because with less moisture in the air it takes less energy for the temperature to drop. Related to this, if the dew point (the temperature at which 100% relative humidity is reached in the air) is above or near freezing, as water transforms from the gaseous to the liquid state a significant amount of heat is released, keeping the air temperature from dropping so quickly.

Passive control strategies

The best way to deal with frosts is to avoid them through site selection, which has been dealt with in Chapter 5. Cultivar choice can also have an effect. Late budbursting cultivars such as 'Riesling', 'Muscadelle' and 'Dolcetto' are a better choice in a frost-prone area than early-budding cultivars like 'Chardonnay' and 'Pinot noir'.

Given that the coldest air is closest to the ground, a simple solution is to increase the height of the buds from the ground. Data from New Zealand suggests that a cane/cordon height of 2 m (for example, as may be found in a high GDC system) could provide an extra 4°C of warmth than the fruiting buds on a vine trained to typical VSP, at 90 cm from the ground (Trought *et al.*, 1999). If the vineyard is planted on a slope, an air dam can also be used, for example, a line of non-deciduous trees that deflects the movement of cold air from higher on the slope into the vineyard.

Active control strategies

The type of freeze event determines which control strategies are most useful. Radiation frosts have more options when it comes to preventing damage. Many take advantage of the inversion layer that develops over the vineyard by mixing the layers, for example by the use of helicopters. Larger helicopters, because they move more air than smaller ones, can protect up to 40 hectares as long as the rotation time (how long it takes to make a complete loop through the vineyard area) is kept to five minutes or less. These have the advantage of not requiring a huge capital investment (if they are rented from a third party), but the disadvantage of being expensive to organize and run when they are needed, and if there are hazards such as transmission wires running nearby, may not be safe to use. Therefore, these are an option best used where frost is not a regular concern.

Air can also be mixed with engine-powered fans mounted on a support pole (see Fig. 6.36). These are fixed high enough to draw down the warmer air from a typical inversion layer and mix it with the colder air. A properly sited fan can protect in the range of four hectares and has the advantage of being available at short notice, which is not always possible with helicopters. Disadvantages include the up-front capital costs, and as for helicopters, if the inversion layer is weak or non-existent, they are no help or can make the damage worse (Rieger, 1989). Another potentially important issue is noise, though changes to impeller design have made improvements in this area.

Fig. 6.36. A frost fan in Central Otago, New Zealand. Content author's own.

Other innovations for mixing air include: (i) portable, ground-based fans similar to the fixed ones but on an articulated arm, or ones that blow cold air at ground level up into the inversion layer, resulting in mixing (see Fig. 6.37) (Augsburger, 2000; Guarga *et al.*, 2003); (ii) tractor-mounted fans (heated or unheated); and (iii) forcing air through the drip irrigation system, which causes it to heat as well as cause air to move through the vineyard (Annabell, 2001).

Other methods are available that add heat to the vineyard environment, which are useful for both radiation and advective frosts. Burning tires, straw bales or candles are options, or the use of fuel burners, also known as smudge

Fig. 6.37. Top: A portable, articulated frost fan that can be towed into place when needed. Bottom: A forced cold air system called Selective Inverted Sumps, developed in Uruguay. The fans pictured are portable, but others can be built in vineyard low spots to access the coldest air. Content author's own.

pots (see Fig. 6.38). However, for a number of reasons these are no longer favoured. The method by which they work is sound, however. Aside from providing radiant heat, they also create a mixing of the inversion layer as the hot air rises, meets the upper layer, cools and then settles back to the earth, effectively circulating the air. Point sources of heat, if too strong, can cause

Fig. 6.38. A row of kerosene burners in a Californian vineyard. Though common in years past, they are becoming less used due to cost, the inconvenience of needing to re-fuel them and environmental concerns. The tall pole to the left has a temperature sensor on it for measuring the strength of the inversion layer. Content author's own.

such an upward rush of hot air that it travels right through the inversion layer (Trought *et al.*, 1999), causing complete loss of the warmer air, with potentially catastrophic results back on the ground.

It may not seem intuitive, but the addition of water to the vineyard system can also add heat in a useful way. The most effective systems apply water over the vines, adding heat (the heat of fusion) as the water freezes. As long as there is liquid water freezing around the plant tissues, their temperature will not fall below –0.6°C (Evans, 2000), which even green grapevine tissues can tolerate. The result of this can be significant amounts of ice formed on the vines and vineyard structures (see Fig. 6.39). Application of water should start before freezing temperatures are reached due to initial cooling, and continue until the ice is melting because if not, the heat of fusion (this time working in the other direction) will draw heat from the remaining ice and tissue dropping the temperature precipitously (Evans, 2000).

Water can be applied over the whole vineyard area, or be targeted towards the vines. Micro-sprinklers (see Fig. 6.40) limit the amount of water used compared with that of large impact sprinklers; alternatively, in-row

Fig. 6.39. The aftermath of frost protection using sprinklers. Inside the ice the vine tissues are still unfrozen due to the release of heat when water freezes. Content author's own.

Fig. 6.40. Micro-sprinklers mounted atop each vineyard post deliver water evenly over the vineyard during frost protection. Left: rotary, right: in-row, flipper-type sprinkler. Content author's own.

sprinklers can be used, which saves further on water consumption (Trought *et al.*, 1999).

Use of the methods mentioned so far is fairly simple in concept, but complex in execution. Those interested in learning more about these frost protection systems are referred to Rieger (1989), Snyder *et al.* (1992), Trought *et al.* (1999), Evans (2000) and Poling (2007).

Other methods

There are a number of other methods of note for decreasing damage from freezing temperatures. Already mentioned in this book are techniques such as keeping cover crop height to a minimum and practising delayed pruning. Another makes best use of the heat that the soil absorbs during the day (enhanced by choosing a site that has a sun-facing slope) by maximizing the amount that is released at night. To do this the soil must be free of weeds or other plants and also be moist, as this darkens the soil and also increases its heat-carrying capacity. The soil surface was traditionally packed, but at least one study has found that cultivated soil releases slightly more heat at night (see Fig. 6.41; Creasy, 2004).

Fig. 6.41. Examples of cultivated (upper) and packed (lower) soils used to assess their effect on fruiting cane temperatures during frost events (from Creasy, 2004). The influence of frost cloth tents over the canes was also evaluated, but was found to make little difference. Note how dark the moist soil is, contributing to the amount of energy that can be absorbed during the daytime. Content author's own.

Electrically-heated cables have also been either built in to vineyard systems or retrofitted. Training the vine around an electrical wire that is heated during frost events has been shown to be an effective way of preventing damage in advective and radiation frost events (Lamb, 2009). A more recent twist to this is using infrared emitters that are incorporated into a cable that can be strung out along the fruiting wires when needed. The use of infrared means more efficient heating, and it can work well with all frost types (Carey, 2024). Due to the infrastructure and cost involved if protecting a large vineyard area, however, this method will likely find only niche uses, for example, with a particularly crucial block of high-value vines.

Canopy Management

Canopy management, in simple terms, is the manipulation of shoots, leaves and clusters to manage conditions for producing grapes for a specific purpose.

Goals and tools for canopy management

The goal for canopy management is to optimize light interception by both leaves and fruit. To what level this is done depends on the purpose for which the grapes are being grown. In terms of achieving the optimum exposure, site selection is still the first step. Selecting an appropriate site has a life-of-the-vineyard impact, so choosing one that is compatible with the production goals is important. For example, planting a vineyard on a particularly vigorous site could result in much additional time and labour spent correcting the excess vigour, which shades the fruit, through shoot trimming, shoot thinning, leaf removal, etc. While often practiced in commercial vineyards, a viticulturist's goal should be to minimize or eliminate the need for these operations.

The next important opportunity to affect your canopy management is through scion and rootstock choice: both should be matched to the site (or vice versa) and its potential for vine growth, e.g. choosing a shallow-rooting rootstock for a deep fertile soil to minimize vine vigour.

Vine planting density is also an important pre-planting decision that affects canopy management. Between vine spacing that is too wide can result in vines that cannot sustain enough growth to fill the trellis. Spacing that is too close can result in vigorous growth and crowded canopies. If there is potential for very vigorous growth that cannot be managed through, for example, rootstock choice, use of a divided canopy should be considered right from the start.

Row orientation is also important. North–south-oriented rows have east- and west-facing sides and therefore the sun will spend half the day shining on one side and half the day shining on the other. East–west-oriented rows have a north and a south side to the canopy, but only one side receives direct sunlight. Fruit from the continually shaded side of canopies ripens more slowly and may be more susceptible to disease than the other side, because of the lack of sunlight (Naylor *et al.*, 2003). Therefore, fruit from E–W-oriented rows tends to be less uniform.

Cover crop, irrigation and fertilization are also management tools that influence the vegetative growth of the vine. Vines will grow most in a vineyard with bare soils and luxurious levels of water and fertilizer, so to make sure canopies fit within the trellis, vine access to water and nutrients must be monitored and managed.

Leaf petiole analyses can be used to assess the nutritional health of the vines, and soil nutrient tests can supplement this information. This information should be used to apply to the vines only those nutrients that are needed. Refer to Chapter 8 for further details on vine nutrition.

Limiting the amount of water accessible to the vines is an effective way of reducing vegetative growth. This can be either directly, by managing irrigation inputs, or indirectly, by managing between-row cover crops. Monitoring and maintaining soil moisture levels to appropriate values for desired vine growth is essential in minimizing the number of canopy management interventions that are needed.

Shoot thinning and lateral removal

In terms of active, in-season canopy management, shoot thinning is an effective tool for reducing canopy density. If a vine has been pruned to two-node spurs, non-count shoots may still arise even if the vine is pruned to approximately the right number of nodes to achieve balance. These non-count shoots generally are not fruitful, increase canopy density and should be thinned out if not needed as replacement spurs.

Shoot thinning is especially important in the head area of the vine, as there are many buds embedded in the wood that can develop into non-count shoots, crowding the renewal area.

The best time to perform shoot thinning is early in the season, when the shoots are about 15 cm long – they are easy to see and come off easily at this stage. Later in the season the base of the shoot becomes more firmly attached to the cane or spur and the shoot must be cut off, because if it is broken off the cane may be damaged. More non-count shoots can grow as the season progresses, but management of cover crops, irrigation or other techniques could be used to decrease vine vigour and prevent this. The effect of shoot thinning alone as a canopy management device can be significant (see Fig. 6.42).

With vigorous vines there can be large laterals that develop from the leaf axils. These also crowd the canopy, so if time can be taken to remove these and a few offending shoots, the results can be dramatic (Fig. 6.43).

Fig. 6.42. If practised early in the season, shoot thinning is an efficient way to decrease vine canopy density. Left: before, right: after. Content author's own.

Fig. 6.43. 'Pinot Meunier' without (top) and with (bottom) lateral removal. Content author's own.

Shoot thinning, in the ideal scenario, is a temporary fix to problems with the canopy. If shoot thinning is necessary every year the vines may be out of balance and making a more permanent change to bring the vines into balance would be prudent (e.g. a trellising change, removing or planting more vines, changing the cover crop, etc.).

Shoot positioning

Another vital aspect of canopy management for many trellising systems is shoot positioning. Often, what is desired is shoots positioned perpendicularly to the fruiting wire, either straight up (e.g. for VSP) or straight down (e.g. for half the canopy of a Scott-Henry trellis or for GDC). Shoots growing out into the row, down and below the fruiting wire (for upright trellises like VSP) and also, very importantly, shoots growing along the row, should be repositioned as early as possible since these contribute to shading and also tend to develop strong lateral growth. The presence of non-vertical shoots often result in clumping (see Fig. 6.44) and areas of the canopy that are very dense are disease hot-spots. The out-of-line shoots have implications for the effectiveness of shoot topping, as described below, and if they cannot be trained to where they should go (for example, see Fig. 6.45), it may be worthwhile removing them.

Foliage wires are an important part of many trellising systems. If these wires are lowered at pruning time, in the spring, when the shoots reach 25 cm or so in height, the wires can be lifted into place to guide and support the developing shoots. Wind moves shoot tips violently at times, so avoid raising the wire too close to them as they can be broken off, causing growth of several laterals in place of each one. Additional wires can be lifted up to support further growth, and wires can also be shifted up the trellis as needed. This lessens the amount of hand-tucking that is required. Note that, in areas with high winds, foliage wires also help prevent shoot breakage, but more, rather than fewer, will be needed on the trellis.

Fig. 6.44. Shoots in a vertical shoot-positioned (VSP) trellis that are not vertical can lead to clumping, resulting in high canopy density, fruit shading, poor spray penetration and thus increased risk of disease. Cultivars or clones with more upright growth habit are much more compatible with VSP systems. Content author's own.

Fig. 6.45. These shoots emerging from the underside of the cane will not grow vertically into the trellis, but horizontally. This leads to many more leaves and laterals in the fruiting zone – so much so that it is worthwhile removing the shoots early in the season. Content author's own.

Shoot topping and hedging

Shoot topping (summer pruning or hedging) is another temporary-fix management tool. As long as the environment is conducive, a vine will continue to grow: under ideal conditions a shoot can grow many metres in a single season. This would be enough to extend beyond the top of a typical trellis and back down to trail on the ground. Clearly, such long shoots would limit access through the vineyard as well as go against the goal of optimizing exposure of the maximum percentage of leaves and fruit.

With a well-balanced vine lateral shoots will not develop, shoot growth will slow as the vines approach fruit set and stop by the time the fruit is pea-sized or so. Shoot length at this point would not be far outside a typical trellis and shading would not occur. However, in most vineyards vines are able to grow much more than this and, to prevent shading between and within the rows, any parts of the shoots growing outside the trellis volume should be cut off.

It is important not to cut the shoots too short, however, as typically there must be 10 to 14 leaves on each shoot to support the development of a typical crop load and the rest of the vine (Howell, 2001). Most trellis designs should accommodate a sufficient amount of shoot length, even if trimming is done.

There is the distinct possibility of lateral growth after initial topping. This development should be monitored to determine when to top again: regrowth should not extend outside the trellis volume to the point where it begins to shade other parts of the vine. In some vineyards shoot topping may need to be performed many times in a season – an expensive proposition.

Proliferation of laterals can lead to a very dense upper part of the canopy where fungicides cannot penetrate. If this is the case, other methods of dealing with excess vine vigour should be used.

Leaf removal

There are a number of reasons for leaf removal on vines. One follows along from the primary goal of canopy management: to optimize the exposure of fruit to the sun. In general, this hastens fruit maturity, increasing soluble solids and decreasing titratable acidity (TA), and results in increased varietal aroma, flavour and colour (Smart and Robinson, 1991).

The amount of leaves to be removed varies with the desired end use of the grapes. For wine grapes, 60% or more exposure of the clusters to the sun is regarded as the most beneficial (Smart and Robinson, 1991). In some vineyards, 100% leaf removal in the fruiting zone is practised (see Fig. 6.46), though this can result in sun/exposure damage to some of the fruit (Kliewer and Lider, 1968). Optimum leaf removal opens up the canopy around the fruit sufficiently to obtain the benefits described, but not so much that direct sun can harm the fruit. For some production regions or end uses, less direct

Fig. 6.46. Leaf and lateral removal in a VSP vineyard. Management goals for these vines include removal of all the leaves in the fruiting zone. Content author's own.

exposure to the sun may be warranted as the grape's appearance or flavours may be changed too much.

One major outcome of leaf removal is a reduction in disease incidence, particularly botrytis (Gubler *et al.*, 1987). For this, it is often necessary only to remove leaves from the east side of a N–S-oriented row. This exposes the fruit to the morning sun, which will help to dry out the area sooner, reducing the number of infection periods. On an E–W-oriented row, it is usually best to remove leaves on the side of the canopy facing away from the sun to improve light exposure. If the canopy is wide or thick, some leaves from both sides may need to be removed to obtain a sufficient improvement in airflow and exposure.

The best time to remove leaves depends on the situation, but generally should not be performed prior to fruit set, or crop load and flower cluster initiation may be affected (Poni *et al.*, 2005). Aside from this, earlier leaf removal ensures that the berries develop in an open environment, which is beneficial as if it is left too late (e.g. until after veraison), berries may be damaged through this sudden exposure to the sun (see Fig. 6.47).

Note that in some cases, pre-bloom leaf removal is warranted. This has been found to be a good method for reducing fruit set that results in lower crop

Fig. 6.47. Berries and a shoot that have been damaged from late-season leaf removal. Direct exposure to the sun and concomitant heat have caused death of some external tissues. Content author's own.

loads, but also more open clusters, decreasing the risk of disease later in the ripening season (Poni *et al.*, 2006).

Root pruning

Most active management of vines concerns manipulation of the fruit, shoots and leaves. However, aside from the use of cover crops and control over irrigation, the roots can also be managed through pruning. Much as the pruning of canes limits vine capacity, because there is a balance between root, shoot and fruit growth, root pruning can also decrease vine capacity. In practice this is accomplished by running a strong tine through the soil down the vineyard row, breaking vine roots that are growing out towards adjacent rows (see Fig. 6.48). Ripping can be carried out intensively (two sides of each row), or less so by ripping every other row.

Fig. 6.48. An example of root pruning, or breaking roots growing out between the rows. This practice cuts back growth below ground in much the same way that canes are pruned above ground, similarly, limiting vine capacity. Content author's own.

Things to avoid

1. Wide canopies: a thick canopy presents difficulties in two ways. One, it is difficult to manage a thick canopy. Exercises such as leaf removal are difficult to perform as interior leaves need to be targeted and it may require more time to remove enough leaves if the canopy is wide. Two, it presents problems in disease control as a thick canopy decreases light reaching inside, reduces air movement, increases relative humidity within the canopy and decreases spray penetration. All of these can contribute to the increased risk of disease (Thomas *et al.*, 1988). Additionally, more fruit will be within the canopy where it will not be well exposed to light, so it will not mature as quickly or accumulate as much colour.
2. Dense canopies: for reasons of exposure and disease incidence already mentioned, a dense canopy works against the goal of a high-quality crop (see Fig. 6.49). Dense and shaded canopies also reduce bud fruitfulness the following year, resulting in smaller potential crops and a tendency for the vine to move to a vegetative cycle (Smart and Robinson, 1991). Use shoot thinning and leaf removal as temporary measures, and investigate ways of reducing vegetative growth in a more permanent way.
3. High shoot densities: a contributor to dense canopies, this commonly occurs due to over-pruning (and thus the development of many non-count shoots) or leaving too many canes on fruiting wires. This may be an indication that dividing the canopy is a better management option. Alternatively, you may be able to change your irrigation scheduling or use of cover crops to reduce the amount of water available to the vine.

Fig. 6.49. Left: the interior of a dense and wide canopy. Note that the shoot carrying the lower cluster has lost all its leaves due to the lack of light. Right: Even in relatively narrow canopies leaves can be so tightly packed so as to cause interior leaves to yellow, senesce and die. Content author's own.

4. Large gaps in the canopy: areas of the trellis not filled with canopy are effectively wasted. Gaps can be caused through missing vines or inadequate growth. Replant any missing vines as soon as practicable and follow up with additional water and nutrients during the season to allow them to better compete with their neighbors. If vines cannot maintain adequate growth to fill the canopy (both along the row and up to the top of the trellis), then ensure that the vines are healthy, receiving enough water and nutrients, etc. If nothing is amiss, then the vine spacing may have been too wide, and adjustments to other management practices – such as removing/changing cover crop, irrigation and fertilization plans, etc. – should be made to increase vine capacity.

Quantifying change in the canopy

There are many reasons why the ability to acquire hard data on canopy characteristics, such as density and fruit exposure, would be useful, for example in testing the effectiveness of shoot thinning on reducing canopy density, monitoring changes of irrigation schedules on shoot growth and fruit exposure, or performing quality control on contracted vineyard labourers, etc. In each case having some figures to back up personal observations is helpful.

The point quadrat (PQ) method of evaluating vertical canopies is one such tool (Smart, 1988). In this a thin rod, simulating a ray of light, is pushed through the canopy at fruiting zone height. Each intersection with a leaf or cluster is recorded, and the process repeated at intervals of 20 cm down the row. If the rod makes no contact with anything, then the record for that point is considered a canopy gap. This simple, if somewhat tedious, operation can yield useful data about canopy characteristics, such as: (i) leaf layer number (the total number of interceptions with leaves over the total number of insertions); (ii) interior leaves (the total number of leaves surrounded by either a leaf or a cluster over the total number of leaves); (iii) interior clusters (ditto, but for clusters); and (iv) percentage canopy gaps (the total number of gaps over the total number of insertions).

The PQ data align nicely with other means of measuring canopy density, including the use of light meters (Smart, 1988). This system's advantages are: (i) only basic equipment is needed; (ii) it is objective; (iii) it is not difficult to train people to collect the data; and (iv) it gives useful information. The major disadvantage is that it requires quite a bit of time to collect the data. However, in those instances where differences in canopies need to be quantified, the PQ method is worth serious consideration.

A variation on PQ is the enhanced point quadrat analysis (Meyers and Vanden Heuvel, 2008), which takes the data from a normal PQ assessment and manipulates it more realistically to represent actual exposure experienced within the canopy. Outputs from this analysis include, for example, the occlusion layer number (number of shade-producing contacts per insertion) and the cluster exposure layer (number of shading layers between clusters and the canopy's exterior), which recognize that both leaves and clusters can create shade. Overall, the method is aimed at better characterizing the canopy structure and shading within.

Other methods of evaluating canopy density include: (i) the optical recognition of canopy gaps – usually using a digital camera in the visible or infrared range to detect areas of the canopy through which light can pass (known as the gap fraction, Dobrowski *et al.*, 2002; Diago *et al.*, 2016); (ii) ultrasound (Palleja and Landers 2017); and (iii) LIDAR (light detection and ranging, Lowe *et al.* 2021). Advances in this area will be rapid, possibly even using a smartphone given advances in processing power and sensors. A number of canopy assessing apps are already available (e.g. De Bei *et al.*, 2016; Orlando *et al.*, 2016) that will allow the routine collection of data on canopy status in a vineyard, even as the season progresses.

The results of good canopy management are;

1. An open canopy.
2. A uniform canopy, which is easy to manage and that can produce higher quality fruit.
3. Less disease: powdery mildew development is inhibited by sunlight (Willocquet *et al.*, 1996), and botrytis and other diseases need moisture and high humidity for optimal development (Thomas *et al.*, 1988), thus opening up the canopy decreases the length of time when conditions are favourable for disease organisms as well as improving spray penetration.
4. Higher yield: better shoot exposure means less loss to disease, greater fruitfulness and a greater ability to ripen a crop.
5. Higher-quality fruit: not only can there be more fruit, but it can all be of better quality: with good canopy management, there will be an increase in percentage soluble solids, colour, fruit aromas and flavours and decreases in TA and herbaceous aromas and flavours (Smart *et al.*, 1988).
6. A higher-quality end product.

Irrigation

The management of water application to vineyards is a valuable grape production tool in many regions around the world. In some areas, natural rainfall is sufficient for commercial production of grapes, and in other areas there is sufficient rainfall even for the establishment of young vines. In those places where there is a seasonal net water deficit (rainfall and soil-available water is not enough to support vine growth through part of or a whole season), application of supplementary water is necessary. If water supply is inadequate, both the vine and the crop can suffer (see Fig. 6.50).

In some instances, water may be applied by hand, such as for vine establishment. Other methods of delivering water to vineyards are through flood irrigation and some type of tubular reticulated water system. Though common

Fig. 6.50. Pre-veraison berries showing signs of severe water stress, which will be affecting cell division and enlargement. Content author's own.

for some crops, in general overhead sprinkler irrigation is not recommended for grape production due to the adverse effect it has on disease incidence (though this type of system can be used for frost management or to delay budbreak (Lipe *et al.*, 1992)). Under-vine micro-sprinklers, however, can be used in vineyard systems as the canopy and fruit do not become wet.

Flood irrigation (see Fig. 6.51) has the advantage of not requiring great capital expense, but its limitations are that is it practical only in level areas, requires flattening of the vineyard floor and subsequent maintenance, can result in uneven water application and is inefficient compared with targeted irrigation methods (Peacock *et al.*, 1977b). A modification of flood irrigation, twin-furrow irrigation (see Fig. 6.52), uses less water than flood irrigation but has similar drawbacks (Christen *et al.*, 1995; Neeson *et al.*, 1995).

Targeted irrigation, usually using a series of pipes and tubes to deliver water to the vines, is now common in many viticultural regions. Establishment costs can be high, but there are savings in the lesser amount of water used and advantages with the precision to which water can be applied to vines.

Typically, a small-diameter tube is laid down under each row and drippers (also called emitters) – either built in or inserted at appropriate intervals – deliver water at a known rate (most modern emitters are pressure compensating, so they can be used on sloping land). Tubes can be tied to a wire or simply rest on the ground (see Fig. 6.53), though they should be fastened so

Fig. 6.51. Flood irrigation in a New South Wales, Australia 'Semillon' vineyard. The land must be levelled quite precisely for flood irrigation to be efficient, but even then, more water will be delivered to the vines near the water entry point than at the far ends of the row. Content author's own.

Fig. 6.52. A modification of flood irrigation: twin-furrow. This system uses less water, but has similar drawbacks. Content author's own.

Fig. 6.53. Irrigation lateral tube with built-in emitters lain on the ground. The short distance between emitters ensures an evenly irrigated line down each row, which is especially advantageous with newly planted vines, but note that the same amount of water from fewer emitters will travel farther down the soil profile, which may be beneficial for older vines. Content author's own.

they cannot move out between rows. Older-style drippers often required the former method, as soil particles could be sucked back up into the tube as it drained, but the modern irrigation emitters are quite robust and simply need to be kept away from machinery paths, such as under vine cultivators.

In areas with high temperatures and low water availability, irrigation tubes can be buried underground, delivering the water directly to the root zone and potentially increasing the efficiency of application (Camp, 1998). Efficiency increases using this method are not universally found, however, and there are also issues around being able to identify and repair leaks and blocked emitters.

Methods of monitoring soil moisture

In grape-growing areas that need supplemental water, investment in some means of measuring soil water will benefit management of the vineyard and optimize water use. Observing the trends of how the soil profile dries out allows advance planning as to when to schedule irrigation, which can save on water and pumping costs and maintain vine growth at a preferred rate. Adding too much water can result it passing through the root zone, unabsorbed, or multiplies the work necessary to keep the canopies in good shape: more shoot trimming and leaf removal at the very least.

At its most simple, the amount of moisture in the soil can be followed through calculations of data available in many weather reports. If the soil starts the season at field capacity (where no more water can be held by the soil), evaporation pan data (a standardized method of measuring water loss from liquid form into the atmosphere) can be converted to evapotranspiration that occurs from plants. The formula for this is

$$\text{evapotranspiration (ET)} = \text{evaporation pan figure } (E_{pan}) \times \text{crop factor } (k_c)$$

The crop factor adjustment accounts for the difference between water loss from an open pan of water and water loss from plant stomata, with values typically ranging from 0.2 to 0.5 (Nicholas, 2004). The crop factor will vary between plant types, and also with the stage of development of the plants, as small grapevine shoots early in the season transpire less water than a full canopy later on. However, if a cover crop is present, the overall crop factor may not change much over a growing season (Yunusa *et al.*, 1997).

The water lost to evapotranspiration can be calculated on a daily or monthly basis (at whatever frequency that E_{pan} data are available) and subtracted from the field capacity value, which has added to it any rainfall and/or irrigation applied (under the limit of field capacity). Figure 6.54 shows an example

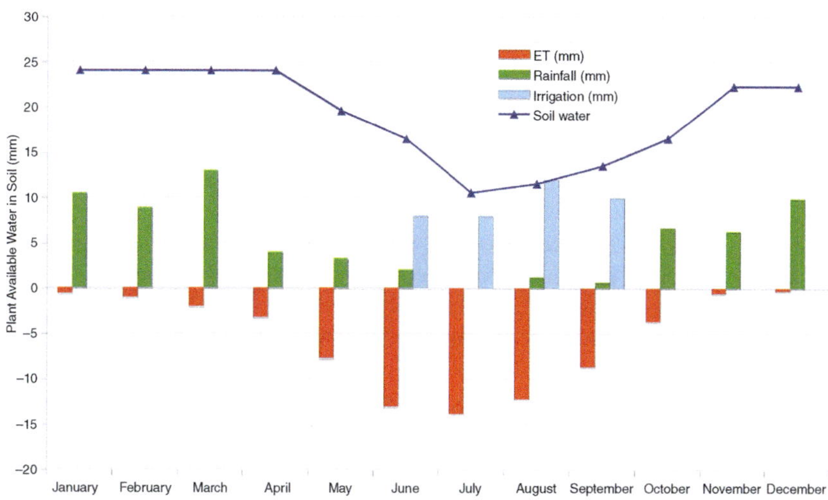

Fig. 6.54. Example of the amount of plant-available water in a soil through a northern-hemisphere growing season and how evapotranspiration, rainfall and irrigation affect it. At field capacity the soil has 24 mm of water available in the profile, but this dips to as low as 10 mm in July. Content author's own.

of a season's soil moisture as estimated through these types of calculations. In this example, irrigation needed to be applied to keep the plant-available water from dropping below zero. Even though there is quite a bit of rainfall in the winter and spring, none of this is usable in the soil profile because it is already at field capacity. Note that there is still a deficit in soil moisture at the end of the season that may not be replenished by winter rainfall: if this was the case, additional irrigation would be required in the following season.

As an indirect method of keeping track of soil moisture this has its advantages, as no equipment is needed; however, your particular site may not have the same weather conditions as the site from which the data come, which can lead to significant errors in determining the amount of water left for the vines. Unless this method is combined with periodic verification with some form of direct soil water measurement, it can only be used as an indication of irrigation needs. More precise management techniques will require other methods.

The amount of water in the soil can be measured, or the water status of the plant can be estimated. Quantifying the former can be done through measuring soil moisture tension, which is the amount of force required to extract water from the soil particles (this is the same force that a plant root has to contend with). A tensiometer (a partly filled water tube with a porous cap on the bottom and a vacuum gauge on the top) measures this directly when it is buried to the appropriate depth in the soil. As the soil dries out, water moves from the porous cap into the soil, creating a vacuum in the tube, which registers on the gauge. These are fairly simple devices that work well at higher soil moisture contents (Tollner and Moss, 1988). However, because grapevines are fairly efficient at removing water from the soil, tensiometers are not ideal for use in vineyards.

Another commonly used instrument relies on the electrical conductivity of a substance changing as the water content within it changes. Gypsum is often used as the conducting material, and a block of it is buried in the soil. As the soil dries so does the block, and the change in its electrical conductance can be recorded. These sensors are more accurate in drier soils than tensiometers (Zazueta and Xin, 1994), and have common application in vineyard situations. As an electrical output is being measured, the data can be logged and transmitted to remote stations, automating some of the monitoring process. However, the blocks have a finite life, which is shortened when used in acidic soils, and it can be difficult to convert the conductance output into actual required amounts of water to apply (Johnston, 2000). Additionally, they are not as accurate in sandy soils as compared to silt or clay soils.

Other methods measure the dielectric constant of a soil (which varies with moisture content), through capacitance or time-domain reflectometry (TDR). Capacitance is measured with a probe lowered into a tube that has been carefully installed into the soil, and TDR uses two stainless steel rods inserted precisely into the soil. A variation on TDR is time-domain transmissometry, where the same sort of sensor is used, but the transmission time of the electrical waveform sent through the probes is analysed (Fig. 6.55). Both of these methods measure only the small volume of soil immediately around the sensor and

Fig. 6.55. A time-domain transmissometry sensor (Aquaflex) being laid next to the vine row. The 3m long tape can be buried parallel to the soil surface, or at an angle, depending on where you want to be sensing soil moisture. The blue box contains a temperature sensor as well. The control box of one of these same sensors is also shown in Fig. 4.4. Content author's own.

require precise placement of the tube/rods, but they can determine volumetric soil water content (making it easier to calculate the amount of water needed) over a wide range of values (Whalley *et al.*, 2004; Plauborg *et al.*, 2005).

Perhaps the most accurate method of determining soil moisture is with a neutron probe, which measures the number of neutron particles that are reflected back to a source from the soil, with a wetter soil reflecting back more neutrons. As such, this method also gives an indication of volumetric soil water content and has the added advantage of being able to measure a much larger volume of soil (15 cm radius) than capacitance or TDR methods (Zazueta and Xin, 1994). Because of this, the tubes into which the probe is inserted do not have to be so precisely installed, leading to more consistent measurements from tube to tube. However, the cost of these machines is high, taking readings is a slower process (up to 32 s each), the unit must be operated manually (so no remote, real-time collection of data) and, as a radiation source is used, the operator often must be licensed. As a result of the convenience factor, as well as being able to gather real-time data, there is a tendency for growers to use other methods (Nicholas, 2004).

Methods of monitoring plant water status

Other than measuring what is in the soil, the water status of the vine can also be measured which, since it directly measures the plant rather than inferred plant status from the soil status, may be more useful. For many years a pressure chamber has been used to do this in research (Scholander *et al.*, 1965; Waring and Cleary, 1967; Ritchie and Hinckley, 1975), and now the technology is well within the grasp of many growers (Choné *et al.*, 2001; Williams and Araujo, 2002). With this method, a leaf or stem is cut off the vine and the distal end placed into a chamber that can be pressurized (see Fig. 6.56). The stem or petiole of the leaf is left protruding from the chamber and the pressure inside the chamber increased (usually from a compressed gas cylinder). Since the xylem of plants is under tension the fluid inside retreats distally when it is cut from the plant. As pressure is applied to the stem or leaf blade, the fluid in the xylem is forced back up the xylem to the cut surface. When this happens, the pressure inside the chamber is equivalent to the tension that was inside the xylem before the stem or leaf was removed. The greater the tension in the plant, the less available water and the greater the water stress.

Because there is a large variation in the water potential of plant tissues over the course of a day (Hardie and Considine, 1976), the time at which the measurement is taken must be standardized. Traditionally, pre-dawn measurements were used as this was the time at which the plants had the most available water (i.e. were under the least amount of water stress). Some research suggests that midday measurements are more accurate (Williams and Trout, 2005),

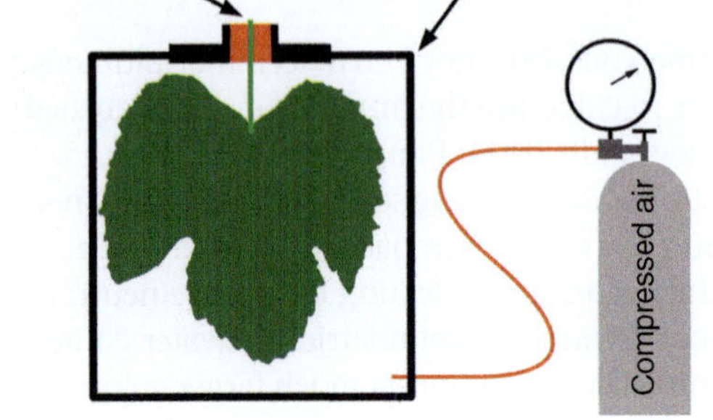

Fig. 6.56. Diagram of a pressure chamber as used to evaluate plant tissue water potential. Content author's own.

but others propose that any measurement times are indicative (Williams and Araujo, 2002) or that there are varietal differences in the response to water stress, which limit its broad application (Schultz, 2003). Choné *et al.* (2001) compared taking measurements pre-dawn or midday on stem and leaf tissues and concluded that midday readings from stem samples were most indicative of actual vine transpiration.

Other measurements are possible. As the level of tension in the xylem vessels increases (i.e. the water stress in the vine is increasing) water will be drawn out of the other plant tissues, causing a decrease in stem diameter (Klepper *et al.*, 1971). This change can be measured through the use of accurate transducers and related to vine water status (Goldhamer and Fereres, 2001; Intrigliolo and Castel, 2007).

Water movement through the plant can also be measured through water loss via leaves (stomatal conductance; Escalona *et al.*, 2002) or from water movement up the stem (using, for example, heat-pulse methodology; see Fig. 6.57) (Green and Clothier, 1988). Both technologies are available for use in commercial situations.

Fig. 6.57. An early sap flow meter attached to a vine trunk. The device applies a pulse of heat to the trunk and measures the amount of time it takes for the heat to travel up the trunk to the detector. The time is related to how fast the sap is flowing and, therefore, to the rate of water used by the canopy. Content author's own.

Scheduling water application

Once the level of vine water stress can be measured, a system needs to be devised through which the optimum time for application of water can be derived. The simplest strategies involve addition of water when a threshold has been reached. For example, if neutron probe results show that the soil moisture is reaching the point where vines will be unable to draw water from it (or some other threshold set by the vineyard manager), then water is applied. With this system the vines are always in a situation where there is no water shortage – often an advantage where maximum quantity is desired. However, in those situations where yield is not the main focus, or where vegetative growth needs to be managed, a strategy that leaves the vines under some water stress is often employed.

Two examples of this are the use of regulated deficit irrigation (RDI) and partial rootzone drying (PRD). RDI was first used on fruit trees (Chalmers *et al.*, 1981) and involves giving the plants less water than they use in ET. This reduces shoot growth because vegetative growth is more sensitive to water stress than reproductive growth (Coombe and McCarthy, 2000). A basic RDI programme for grapevines would be to maintain water availability through fruit set, limit until veraison, and then provide adequate water during ripening (Nicholas, 2004; Fereres and Soriano, 2007). This results in more efficient water use and limits berry size and vine yield (with the idea of improving wine quality) in comparison with a full-water control.

PRD is a type of RDI, but using spatial separation of water application rather than just lowering the amount (Dry, 1997). Water is applied to one side of a vine and the other side is left to dry out, whereupon the watered side is reversed and the process is repeated through the season. This results in a reduction of shoot growth, attributed to the production and transport of abscisic acid (ABA) from the roots to the shoots (Stoll *et al.*, 2000). ABA has been found to decrease the stomatal conductance of the leaves, leading to reduced photosynthesis and shoot growth. There has been extensive research into the potential of PRD to maintain grape quality while not affecting yields or vine health; however, on a use of water basis, there appears to be little advantage in using PRD over RDI (Gu *et al.*, 2004), especially considering the additional infrastructure that is needed to make PRD work.

Effective measurement of vine water status is essential for strategies that employ near to detrimental soil water deficits, as slight under-application of water can result in severe vine stress and loss of productivity, or even death (see Fig. 6.58). As most vineyard soils are not uniform, in practice irrigation often means some vines are over-watered and others under-watered, leading to uneven growth and cropping. Developing a finer network of soil and/or plant water sensors and controllers for delivering water on a vine-by-vine basis (a segment of precision viticulture) will be future big advancements for vineyard management, and that future is not that far off. Sensors that could

Fig. 6.58. Top: chronic under-watering results in short shoots, poor cropping and premature loss of leaves. Bottom: temporary water stress at bloom resulted in the loss of leaves, shoot tips and almost all the crop. Content author's own.

form the basis of a mesh-network reporting real-time data on plant water status are already in testing phase (Lakso *et al.* 2022).

Other Management Practices

Girdling

Girdling is an old practice, as Coombe (1959) credits a report by Lambry in 1817 as the first one describing the technique. Girdling is the removal of a ring of bark 3–6 mm wide from the trunk, arm or cane basal to the cluster (see Fig. 6.59).

Fig. 6.59. A recent cane girdle (left) and two examples of trunk girdles: recent (middle) and one that was cut in the previous year, but that has since healed (right). Content author's own.

The result is that the carbohydrate materials produced in the leaves accumulate in those parts distal to the wound, including the clusters of flowers or fruit. Girdled bands usually heal over within a few weeks (Jensen *et al.*, 1981) depending on the width of the cut, but that is enough to increase fruit set if performed at bloom; or, if performed at the stage of rapid fruit growth (immediately after fruit set), to increase berry size by 10–90% (Coombe, 1959; Jensen *et al.*, 1975). However, the magnitude of the response depends on cultivar. Additionally, girdling can also hasten maturity by improving colour and sugar accumulation if carried out later in the season (Peacock *et al.*, 1977a).

Fruit thinning

Sometimes vines carry too much fruit, which could be on a physiological basis (see Fig. 6.60), or commercial basis. Thinning of all types can be used for crop reduction, but frequently other goals are desired, such as removing overlapping, or diseased clusters or fruit exhibiting delayed ripening. The main types of hand thinning are of flower clusters, clusters and berries. Hand thinning is selective, and the worker can choose to retain the best size and shape of clusters for the market, which saves time later on during harvest.

Flower cluster thinning

This can be carried out early in shoot development because the clusters are easily visible before many leaves have expanded. The early removal of flower

Fig. 6.60. Examples of overcropped young vines. Excessive fruit for the size of the vine leads to poor fruit quality, small amounts of root and shoot growth, lowered vine carbohydrate status and reduced growth in following years. Content author's own.

clusters can increase slightly both harvest berry size and Brix (Dokoozlian and Hirschfelt, 1995). Some cultivars produce three to four clusters per shoot, and usually the basal cluster is the largest. Removal of the distal clusters is beneficial for basal cluster development and for crop control of the whole vine. Although this stage is the optimum time for hand thinning because the clusters are not obscured by leaves, it is not the optimum time to remove clusters in cultivars that tend to form overly compact clusters, as this will only accentuate the problem.

Cluster thinning

This involves removal of clusters after fruit set, so berry size of the remaining clusters is not affected. The major purpose is to remove excessive crop load and balance the leaf area with the crop load. It is used in table, raisin and wine grape vineyards. Small, misshapen and excessively large clusters can also be removed, which is beneficial in table grape production. In wine grape production, dropping any crop that is lagging in maturity can also help final harvested fruit composition, and removing clusters that may be laying on top of others can help with reducing disease. Non-selective thinning can be done with mechanical harvesters, viable from around 4 weeks post-fruit set to when berries are not much more than pea-sized, using beater bars to remove clusters in certain parts of the fruiting zone (Petrie and Clingeleffer 2006). If done at the beginning of this window, there are also benefits in reducing berry number per cluster and removing bunch trash (e.g. calyptra, abscised flowers etc.), which was associated with reduced botrytis during harvest (Mundy *et al.* 2021).

Berry thinning

Removal of parts of the cluster before or after berry set is primarily used in the production of table grapes. Pre-bloom clusters can be stroked with a rubber-bristled brush to remove a proportion of florets, and after fruit set, but before the cluster gets too compact with berry growth, berries or branches of the cluster can be cut by hand. Laborious, but useful in grapes of high value. Quality improvement comes from controlling cluster size, looseness and shape. Because the rachis contains less fruit, clusters are usually less compact and the reduction in flower number (with flower thinning) results in larger berries that mature earlier (Ahmad and Zargar, 2005). In some cultivars individual branches within the cluster are removed to improve cluster appearance and reduce compactness. Some cultivars benefit from cluster shortening to produce round clusters that pack easily. Though it is very labour intensive, this can be advantageous in wine grape production as well, since the cluster is more open and there less prone to disease development.

Application of plant growth regulators

Gibberellic acid (GA) sprays have several uses in table grape production. If applied during mid-bloom, flowers are thinned and result in less compact clusters (see Fig. 6.61) and larger berries, but a second GA application at fruit set (at the time of girdling) can also increase berry size (Weaver and McCune, 1959; Kasimatis *et al.*, 1971; Harrell and Williams, 1987). However, not all cultivars respond in the same way to girdling or GA applications, so each must be studied for optimum treatment, timing and suitable response.

Vine/crop protection

In some environments, protecting the crop may be necessary, for example, against hail, heat/sun, or rain, or more often from blemishes that the consumer would notice. Temporary covers are used in some table grape vineyards where the heat from direct sun exposure might damage the fruit or be too intense for the vines to continue ripening the crop (Fig. 6.62). In other instances, a more personalized approach is used, with individual bags or caps used to keep the clusters dry and preserve the natural waxy bloom on the berries (Fig. 6.63).

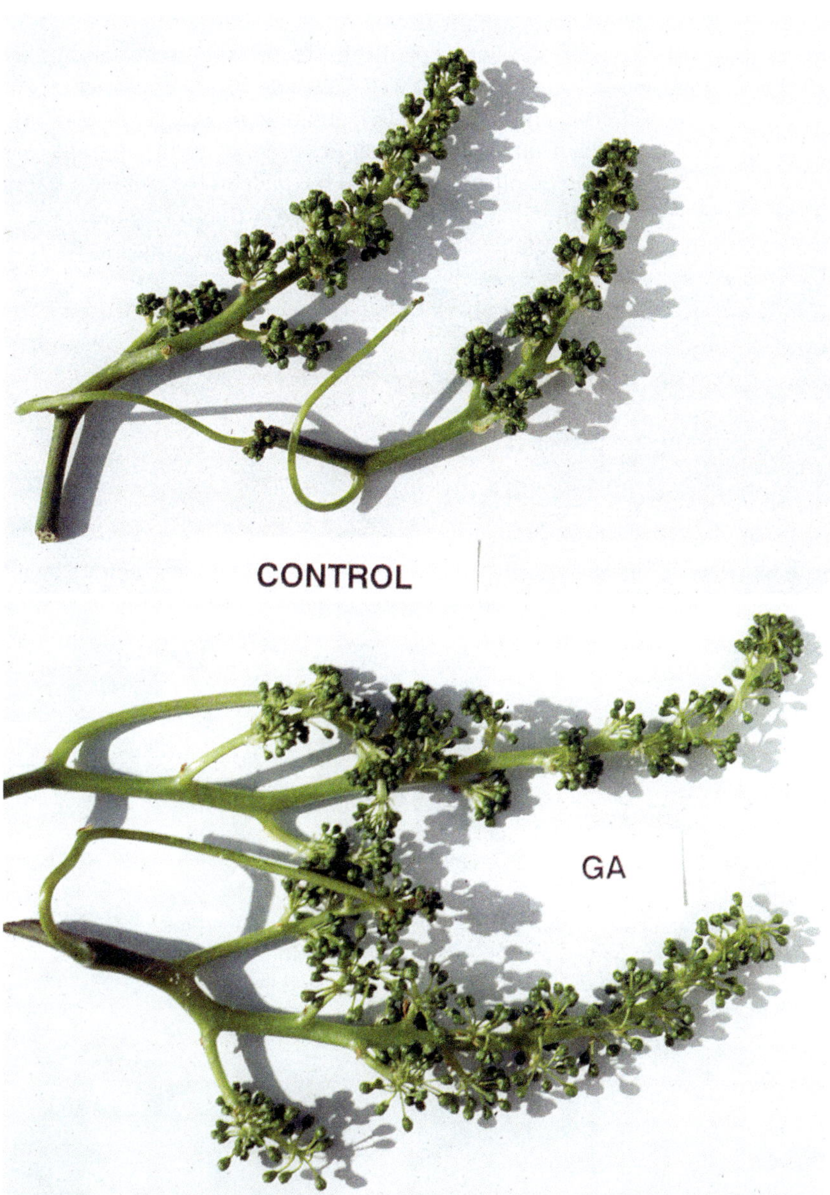

Fig. 6.61. The effect of a pre-bloom gibberellic acid (GA) spray on 'Pinot noir' clusters. Compared with the control, the treated clusters are longer and each flower has a longer pedicel, leading to a less compact cluster. Content author's own.

Harvest

The quantity of fruit available at the end of the season is made up of a number of factors that have been mentioned up to this point. To bring all the pieces of the puzzle together, a discussion about yield components us helpful. The crop

Fig. 6.62. Row covers in an Australian table grape vineyard. Content author's own.

Fig. 6.63. Cluster protection in Japanese vineyards: paper bags or paper 'hats'. Content author's own.

harvested is made up of the area of vines, the number of producing vines in that area, the cluster number per vine and the cluster weight. A diagram leading up to this is presented in Figure 6.64. The clusters per vine is decided at pruning (where the number of nodes is set and therefore potential number of shoots)

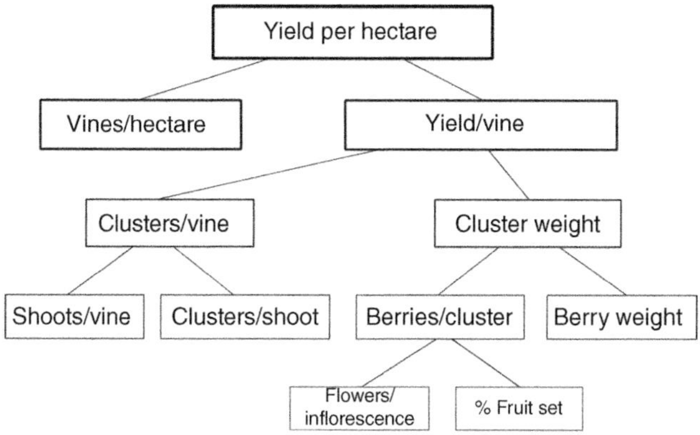

Fig. 6.64. Diagram showing yield components and how they are interrelated. Content author's own.

and how many clusters there are per shoot is influenced by a myriad of factors like flower cluster initiation in the season prior, vine capacity, frost damage etc. Cluster weight is made up of the number of berries on it and their individual weights. The former is determined by flowers per inflorescence and percent fruit set, and the latter by seed number, seasonal weather etc. There are many influences on these components, some more permanent (e.g. climate), some highly variable (e.g. seasonal weather, pests and disease) and others under the control of the manager. The sum of all of these factors and management decisions culminates in how much fruit is harvested. The question then is, when to harvest?

As a viticultural tool, harvest timing has perhaps the most impact on grape composition. Therefore, knowing what the grapes are going to be used for is vital in determining when and how to harvest. For example, early harvest of wine grapes means that flavour, sugar and acid profiles are substantially different, and suited to styles such as low-alcohol or sparkling wines or verjus (unfermented juice of unripe grapes, used primarily in cooking). Late harvesting of grapes can result in raisins or those suited to producing dessert wines, which have high residual sugar. The products can be very distinct, but in some cases come from the same cultivar.

Deciding when to harvest can be a combination of art and science. Weather can be a major factor, as can equipment and labor availability. Analysis of grapes can help with factors such as sugar concentrations, acidity or even color, but in many cases, especially in winemaking, aromas and flavours are paramount. Gathering frequent, and representative cluster samples leading up to harvest provides a tool for tracking quantitative data (refer to Fig. 3.15), but also gives an opportunity for sensory evaluation of the juice, to follow how aromas and flavours are developing so that fruit can be picked as closely as possible to optimum composition for the product.

Harvesting method is also of importance to product quality. For fresh consumption, delicate handling is required for premium fruit, as the bloom (waxy covering) of the grape should be preserved. Fruit are hand harvested, trimmed of unsightly fruit and placed in protective containers for shipping to market.

For some wine styles, hand harvesting is a must (see Fig. 6.65); for example, traditional sparkling wine production requires intact fruit before processing starts, or harsh phenolics may be extracted from the rachis and skins. For many table wines, however, machine harvesting (see Fig. 6.66) is acceptable as the breaking of the berry skins and their mingling with the juices can help extract desirable compounds.

Mechanical harvesting technology has progressed well, blurring the lines between hand- and machine-harvested fruit quality. No doubt this will continue, eventually leading to fully automated production systems (refer to Chapter 9).

Fig. 6.65. Hand harvesting of grapes is a labour-intensive process. Visible are hand-luggable picking bins, which hold 10-12 kg of fruit. Content author's own.

Fig. 6.66. Mechanical harvesting of wine grapes. These harvesters can be self-propelled (as pictured here) or towed by a tractor. There is a bin on board for harvested grapes to be held until they can be unloaded, or the fruit is conveyed to a trailer travelling in an adjacent row. Content author's own.

Vineyard Management Systems

7

Abstract: Managing a vineyard is largely based your experiences and skills, but also by regulations, governmental or otherwise. There are a number of internationally recognized management systems, including the European Appellation d'Origine Protégée (AOP), ISO, organic-approved and biodynamic. Others are locally organized (sustainability programmes) and some, at the moment, are more like movements (e.g. regenerative viticulture). All can help structure a business and provide assistance in the marketplace, but their foci may differ. ISO oversees an objective set of rules that revolve around business best-practice and safety, AOP is more about tradition and branding, while organics, biodynamics, sustainability programmes and regenerative viticulture concentrate on the environment. Organics, biodynamics and regenerative viticulture are closely related in their practices, while sustainability programmes tend to be aligned with more regional needs. Managing these and the increasing amount of data available to us can be leveraged through use of vineyard management services.

Aside from the management regime that each viticulturist develops for their situation, there are external third party schemes that may be beneficial to incorporate into your systems. Examples of these could be certified sustainable, organic or biodynamic production, ISO conformity, or other methods of running your business that are less formal, such as regenerative farming. This chapter will discuss these types of systems: what they are, what they promise, and what they require, and will end with the possibilities that software or cloud-based systems offer to help with managing data and your vineyards.

Appellation Systems

One such system is the Appellation d'Origine Protégée (AOP) in Europe. In it, producers subscribe to the programme, follow regulations as to how a product is grown/made, are subject to audits and as a result, they can label their products with a protected name/brand. Based on historical practices, it is now applied to new regions and products and by all accounts, is successful in the marketplace (Barjolle *et al.*, 1997).

DOI: 10.1079/9781800627062.0007

Appellation systems like the AOP can be found world-wide and in many cases are recognized internationally. Examples like Protected Geographical Indication (PGI, for a product that comes from a specific place and has some characteristic originating from that place) or Protected Designation of Origin (like the PGI but includes the environment and/or human influences on the product) are used for a number of products, including wine. Some are more localised, such as the American Viticultural Area designation (USA), Vintners Quality Alliance (Canada), or regional ones, like the Tsinandali appellation from Georgia. Most of these require the product to be made from a designated region, contain products made from a certain cultivar or combination of cultivars, and perhaps some other regulation about growing the crop and processing. However, in a general sense, the amount of control of production is quite limited compared, in particular, to the AOP system, so they are not really useable for managing an enterprise, but are more for recognition and marketing.

ISO

More recent systems include the International Organization for Standardization, or ISO, established in 1947. It provides a global standard for various procedures so that there is certainty by all its users that the result of the process meets defined guidelines. It publishes these standards, but does not enforce them - that is up to third parties. Some standards are generic, but adopted by members of the grape industry, like ISO 9001 (https://www.iso.org/standard/62085.html) which provides information to establish, maintain and improve an enterprise's quality management system, or ISO 14001 (https://www.iso.org/standard/60857.html), which does the same for impacts on the environment.

Other frameworks can be very specific, such as the International Classification for Standards (ICS), which is part of ISO. These are developed with cooperation of the industry involved and are designed to make an agreed-upon technical specification for a particular operation, something that might be useful in developing a standard operating procedure. For example, ICS has a section for agriculture and within that, one for "Agricultural machines, implements and equipment." A subset of that is for "Viticultural and wine-making equipment", and an actual standard is ISO 5704:1980 "Equipment for vine cultivation and wine making - Grape-harvesting machinery - Test methods" (https://www.iso.org/standard/11815.html). This standard comprizes:

- Technical test methods
- Machine description form
- Form for reporting test settings and condition
- Plot description form
- Worksite form
- Transport form

- Work quality form
- Evaluation of losses
- Sample determination form
- Vinification form -mechanical harvest
- Vinification form -manual harvest
- General report form

It's evident that these standards go into a great deal of detail and will require significant resources to implement and maintain.

Since there are costs to adhering to these standards (Yiridoe and Marett, 2004), what is the benefit to the business? With the example of ISO 14001 certification there are positives in terms of relationships with stakeholders and the possibility of avoiding government-imposed regulation, but also cost savings from improved efficiencies and reduced waste production (Camilleri, 2022). A survey in Italy found that 15.6% of 173 wineries in the study were ISO 14001 accredited, with larger firms benefitting financially, most likely due to the certification benefitting trade with international partners (Fiore *et al.*, 2016).

Organics

Other certification schemes may have the same name across the world, but different meanings depending on where the product is made. Organic agriculture is one of these, which has been defined by the International Federation of Organic Agriculture Movements (IFOAM) as

> "a production system that sustains the health of soils, ecosystems, and people. It relies on ecological processes, biodiversity and cycles adapted to local conditions, rather than the use of inputs with adverse effects [and] combines tradition, innovation, and science to benefit the shared environment and promote fair relationships and good quality of life for all involved." (https://www.ifoam.bio/why-organic/organic-landmarks/definition-organic)

In practice, organic production is defined by the use of "natural" processes and substances, versus use of "synthetic" substances, and a goal to reduce the use of inputs at all (Seufert *et al.*, 2017). For example, use of copper sulfate (a natural substance) rather than a fungicide like difenoconazole or planting resistant cultivars (a natural process) rather than persisting with those that are susceptible, so not as many applications of a control are needed. There is also a preference to use products produced on-site or locally to minimize transport costs (financial and environmental).

Organic viticulture is a significant player across the globe, with 562,000 hectares, or 8.3% of vineyard land managed organically in 2022 (Willer *et al.*, 2024). The trend has been an increase in area in the past decades, while noting that overall area under vines has been decreasing over that same time period.

Climate has a big impact on how much change is necessary when moving from conventional to organic production. In areas with high disease pressure, organic-approved pesticides must be applied often to maintain coverage and protection. This can be an issue if rain occurs so frequently that there are only small windows of dry weather to spray or the soil is too soft for machinery to pass through. The impacts are lower in drier climates with less disease pressure.

Generally speaking, there are more operations necessary in organic production, even in a dry climate. One study in the Languedoc, France found that on average, 15% more interventions were needed after conversion to organic, though the range was between 0% and 57% (Merot and Wery, 2017). Most had to do with weed control, and a lesser amount for disease control, resulting in a 13% increase in labour required. The effect of this is greater cost per hectare: in a more humid climate with a greater number of disease organisms to manage, the costs would be even higher.

On a carbon footprint basis, conventional versus organic viticulture is largely similar, with the increased fuel use in organics being offset by the greater carbon emissions associated with synthetic pesticides and fertilizers (Ghiglieno *et al.*, 2023). Ultimately, the decision to move to organic production will depend on the market, manager's ideals and cost issues.

There are a number of different regulations/regulatory agencies for organic production, including the EU's 2018/848 regulation on the production and labelling of organic products, the USDA's "National Organic Program", and Asia Regional Organic Standards (AROS) to name a few. Details in the regulations can be different between agencies (Seufert *et al.*, 2017), so it is important to find the appropriate set of rules for your area.

Notwithstanding all of this, there is much confusion among consumers as to what "organic" means (Yiridoe *et al.*, 2005), and their perception may not align with commercial practice. As an example, a survey of Romanian consumers found they perceived organic farming as protecting the natural environment and not using chemical fertilizers and pesticides (Oroian *et al.*, 2017), which may or may not be true. Some consumers would be shocked to find out, for example, that vineyards use broad spectrum insecticides like pyrethrins, or that organic farming requires many more applications of fungicides in more humid climates compared to a conventionally run vineyard. This mismatch between production and consumer only leads to greater confusion in the marketplace, which could hinder wider acceptance of the practice.

Biodynamics

Biodynamic viticulture has its roots in a series of lectures delivered to 111 people (Paull, 2011) by Austrian philosopher Rudolf Steiner in 1924. His vision was for an alternative form of agriculture, diverging from the then expanding practice of using synthetic fertilizers to boost plant productivity, with the idea that the farm is an organism and subject to cosmic forces (Steiner, 1924). The lectures were always intended to be about laying down the principles of his ideas around farming and that the "Experimental Circle" of attendees would carry out exploring and adapting these principles over time. Included in these sessions were the ideas for nine preparations, which would be essential to his principles.

The term biodynamics came along later and the ideas were adapted to viticulture at some point, with Alsace, France being noted as one of the first areas with biodynamic wines (Wallace, 2019). In 2024 25,000 hectares of vineyards were certified under biodynamic farming (Willer *et al.*, 2024). An advantage biodynamics has over organics is that there is the single international organisation, the Biodynamic Federation Demeter International, that oversees regulations and licences auditors, ensuring that the brand means largely the same thing to consumers around the globe.

The regulations for organic and biodynamic viticulture are effectively the same aside from the use of the nine preparations (Turinek *et al.*, 2009; Chalker-Scott, 2013) and adherence to lunar and astrological calendars for the timing of some operations (Chalker-Scott, 2013). These nine preparations, which are not considered to be fertilizers but rather vitalizers, are usually applied at the rate of grams per hectare or what can be considered homeopathic amounts (Steiner, 1924).

Sustainable Systems

After these more internationally recognized systems is a large number of those associated with sustainability in some form or another. Typically, these are more local in nature and more adapted to particular locations and situations. A few of many, many examples of these are

- OIV Guide for the implementation of principles of sustainable vitiviniculture (international) https://www.oiv.int/standards/oiv-guide-for-the-implementation-of-principles-of-sustainable-vitiviniculture-
- Vignerons en Développement Durable (France) https://sustainablewinegrowing.org/
- Programma VIVA (Italy) https://viticolturasostenibile.org/
- California Sustainable Winegrowing Alliance (California) https://sustainablewinegrowing.org/
- Sustainable Winegrowing New Zealand (New Zealand) https://www.nzwine.com/en/sustainability/swnz/

Historically, these types of programmes took into account vineyard production practices, focussing on encouraging use of those that had lesser impact on the environment. Today, more these programmes take into account the idea of the triple bottom line, which includes three factors: social, environmental and economic. This looks after sustainability, balancing your workforce and consumers, the ecosystem that you operate in, and the long-term viability of your business. Some have also included systems that try to drive members to actively improve their practices over time (New Zealand Winegrowers, 2025).

At its core, the goals of sustainability programmes are not different to those of organics or biodynamics: the main difference is in what materials can be used to meet those goals.

With regard to the consumer, of issue is the number and scope of all of these organisations: as there are many competing systems, it is no wonder that there is a lot of confusion in the marketplace (Yokessa and Marette, 2019). Despite this, these programmes have the possibility to engage with the consumer (Szolnoki 2013) and participation in some official form of environmentally conscious programme may be necessary for market access. Similarly as to what Fiore *et al.* (2016) found with ISO certification, Jourjon *et al.* (2016) reported that larger firms benefitted the most from Terra Vitis (a French sustainable programme for winegrowing) certification because of the advantages it held for operational efficiency or facilitating export.

Regenerative Viticulture

In recent years the concept of regenerative farming has increased in popularity. Originally the term was promoted by Robert Rodale who disliked the term "sustainable" and preferred "regenerative" as it connoted that things are "expanding and going better" (USDA, 1989). In practice this has evolved to a special focus on soil health, recognizing that this is the starting point to plant growth and animal development.

Guiding principles are to leave soil unworked, keep biodiversity at high levels and integrate the use of animals into the management system (https://www.regenerativeviticulture.org/faqs/). While organic farming practices are involved in the origins of regenerative farming, there was no explicit statement that materials considered non-organic could not be used because there was no formal definition of what regenerative farming was. IFOAM Organics Europe released a statement in 2023 that attempted to clarify that regenerative farming should be based on organic farming principles, and therefore not allow the use of synthetic materials: this statement was adopted and released two years later by the international organisation (IFOAM, 2025). Though certification for regenerative viticulture is available through third parties, there is less agreement about what specific rules should be part of the certification process.

Management Systems in General

As with any certification scheme there is the choice of choosing to become certified, or to practice a system and not bother with formal certification, perhaps because of its cost in terms of money and time. Accredited compliance with a management system can have benefits, for example marketing to consumers, or perhaps in some instances, access to the market, which could be a significant factor in a business sense. The downsides include data collection and reporting, and the fees associated with the scheme.

Data Management

The management systems described so far are limited to grape production, though some programmes may stretch to post-harvest processing, packaging and shipping, for example. Each has data collection, collation and reporting as part of its scheme, which is also true of the process of grape growing. As a viticulturist, you require information about each of your vineyard blocks, their characteristics, special needs or quirks that inform your management decisions. On top of this, you would historically gather seasonal information on a block-by-block basis, for example, time needed to prune, date of budburst/ flowering/ veraison, pesticide treatment dates, harvest dates, notes on fruit composition etc.

Developing and running a vineyard business has been getting more and more complicated as the years go on. In addition to increased competition in the marketplace, increased regulation and bureaucracy, results of climate change etc. our knowledge about managing vines, the factors that affect crop quantity and quality, and new technologies that allow for gathering increasing amounts of data mean that trying to make sense of it all is getting more, not less difficult. So, there is more and more opportunity to collect additional information, increasingly on a row by row or even vine by vine basis. These data could be used to help make better management decisions in the future, but given the magnitude of the dataset, how can this be done?

Vineyard Management Software/Services

This is not a new concept. A precursor to software used in viticulture was the heat accumulation model, which was used to predict when grapes would ripen in a certain area (Angot, 1883): an incredibly useful, but simple equation. Later on, disease forecasting models were developed that were slightly more complex, like that for powdery mildew, which required temperature and leaf wetness data (Gubler *et al.*, 1999), but they, too, were straightforward and it was easy to take information from the model and use it to make a management decision.

Today we have the goal of being able to enter data into a platforms that will take that information and allow us to use it to greater advantage. A spreadsheet is a generic, but powerful tool to organize data and manipulate it to gain additional knowledge. However, all too often data are collected and even entered, but never fully utilized afterward. Online services can aid in collation of data, through the camera on a smartphone, real-time data feed from sensors, or GPS information from tractors, and then manipulate that information into automated reports on disease incidence, yield estimates, soil water status, efficiency of spray applications etc. This can help make practical decisions about spray timings, irrigation scheduling, harvest planning and so on.

In fact, integrated pest management (IPM) programmes were one of the first holistic decision support systems (DSS). In the 1950s California researchers proposed the "integrated control concept" (Stern *et al.*, 1959) in order to give growers guidance on how to best manage an aphid pest after a biological control method had been introduced. The idea was to make sure all management methods worked in concert rather than the possibility they might be at odds. From these origins, more and more factors began to be incorporated into IPM, making it more about the production system as a whole.

This ties in nicely with the idea of precision viticulture, where information about the physical characteristics of a block are tied together with all the information needed to manage the individuals (vines in our case) in it. The goal is to optimize inputs by applying them only where they are needed, for example, directing a leaf removal crew to only areas of the block with vigorous growth, or applying compost to each vine according to its need.

There are many DSS available to the viticulturist now, all with the basic goal of collating information and distilling it down to suggestions on how best to manage the vines. Since data collection is increasingly automated, done remotely and collated automatically, it can be piped online, directly to very large, high-resolution datasets. As we understand more about the factors that go into making a decision, these databases and the algorithms applied to them should provide us with reliable information with which to make our choices (Mikram *et al.*, 2025).

This is an area of potentially huge growth and significance that has the possibility of increasing efficiency, reducing environmental impact and helping to increase the survivability of our business through the challenges that are to come.

8 Nutrition

Abstract: Maintaining vine nutrition is a balance between monitoring soil and plant nutrient status. Soil analysis provides valuable information about the presence of nutrients, but also their availability to the vine. Plant tissue analysis provides a more direct picture of nutrient status and how the vine might respond. The most important macronutrients are nitrogen, phosphorus, sulphur, potassium, calcium and magnesium and micronutrients iron, manganese, copper, zinc, molybdenum, boron and chlorine. Each of these plays an important role in the function of plants and their availability to vines should be managed between deficiency and toxicity. Commonly seen symptoms of unbalanced nutrient status are presented.

Grapevines have the same basic nutritional requirements as other higher plants, but what is a nutrient? In this chapter we will be talking mainly about mineral substances that the vines need to grow and produce fruit, but oxygen, hydrogen and carbon are also vital and are derived from water and carbon dioxide (CO_2) in the atmosphere.

Luckily, in most cases mineral nutrients are provided by the environment without our interference. However, best practice dictates that problems should be anticipated and corrected before they occur. The area of vine nutrition is an excellent example of why this is so, as deficiency in boron or zinc, which may not be apparent visually, can result in low percentage fruit set, leading to poor vine productivity. In contrast, vines that are deficient in nitrogen have low amounts of growth and usually have light green leaves, symptoms that are readily apparent. Vines that have been growing well for some years can, relatively quickly, develop a problem that can be traced back to nutrient imbalances that have come about from that nutrient having been mined from the soil by the roots, but not replaced by the manager.

Because the soil is a complex medium and the uptake of nutrients by plants is, in some cases, unregulated and in others carefully metered, determining when plants are deficient in, or suffer from excess amounts of, nutrients is not terribly straightforward. Compared with a simpler nutrient delivery system,

DOI: 10.1079/9781800627062.0008

hydroponics (where plant roots are grown in liquid where its composition can be easily altered), there is less certainty in knowing whether a nutrient that is applied to the medium will become immediately available to the plant. In dealing with soil nutrients it is important to remember that whatever is done to amend a nutrient may take a long time to show an effect and, if something has been done incorrectly, recovery time may be very long. Therefore, the process used to determine how much and of which nutrient is to be applied is very important.

Nutrient Analysis and Correction Strategies

In most cases the soil can provide sufficient amounts of nutrients for plant and fruit growth, at least in the shorter term. Since soils differ in their ability to hold needed nutrients (this can be thought of as a reservoir of nutrients), soil type and condition are important in site selection. Under the same cropping scheme, some soils will be exhausted of their nutrients before others. The question becomes: how can the soil be tested to determine when a nutrient threshold is approaching?

This, and the amount of nutrients needed by vines, is determined by soil analysis, knowing the capacity of the soil to supply the nutrients (for example, knowing soil depth and organic matter content) and by plant tissue analysis. Generally, plant nutrient deficiency symptoms occur after the most beneficial time for application of nutrients, e.g. after fruit set or near the end of the growing season. Thus, preventative monitoring and proactive application of nutrients is essential for continued vine health and productivity. The long-term goal is to maintain the nutrients at adequate levels despite changes in seasonal environment, crop loads or catastrophic weather events (such as heat, frost or hail).

What constitutes 'adequate' for producing grapevines is not necessarily the same in different vineyards. As soil characteristics, weather patterns, management systems and cultivars change so much from place to place, the values that are acceptable for production may change (Wolf *et al.*, 1983; Christensen, 1984; Schreiner, 2002). Referral to a local source of knowledge is helpful in interpreting soil and plant tissue test results.

Grape mineral nutrition management should be implemented in a long-term programme: one that reflects the ability of the soil to regularly provide optimum nutritional elements and considers the amount of nutrients that are removed by the crop. The latter can be significant, even when assuming that all vine prunings are left in the vineyard: up to 2.4 kg nitrogen (N), 3.2 kg potassium (K) and 0.4 kg phosphorus (P) are contained in each tonne of grapes. If prunings are removed (for example, due to concerns about them acting as a disease inoculum source), the amounts of these nutrients removed per tonne of prunings are 2.68, 2.76 and 0.46, respectively (Winkler *et al.*, 1974). New vine growth in the spring can acquire nutrients from either stored reserves or the soil. Up to 50% of the N and P, 15% of the K and 10% of the calcium (Ca) and magnesium (Mg) used may be from stored reserves each season (Schreiner *et al.*, 2006).

Testing and nutrient addition

Before vines are planted, there is little option but to test the soil as an indicator of what the nutrient status will be. It is still a useful exercise, however, as gross deficiencies, such as in pH or individual nutrients, are more easily corrected prior to the vineyard being established. Soil collection for testing should always be done as specified by the laboratory doing the analysis because results will vary according to the depth of soil taken (Jackson and Caldwell, 1993), and the time of year. Consistent collection methods ensure that the recommendations made from the results will be as useful as possible. Despite all care, however, because vines have such extensive and deep rooting systems, getting an accurate representation of the soil and its nutrients that are actually available to a vine is practically impossible (Robinson, 1992). Nevertheless, along with soil profile examination (Chapter 5), it remains as good a tool as we have for evaluating potential vineyard soils, and should not be overlooked.

Similar to the situation for measuring plant water status versus water in the soil, measuring the nutrients in the plant tissues is a better method of knowing what the vine situation is. However, there are potentially many different parts of the vine to sample, and each can contain a different level of nutrients (Christensen, 1984; Schreiner, 2002). Additionally, the time of the season also causes variation (Smith *et al.*, 1972; Christensen, 1984; Schreiner, 2002). Because of this, there has been some standardization in the procedure.

Most commonly used are leaf petioles because they are thought to represent a snapshot of plant nutrient status at the time of sampling. Leaf blades are also used as they represent the photosynthetic workhorse of the plant. Benito *et al.* (2013) did a comprehensive survey of nutrients in blades and petioles and found that they differed in their ability to accurately represent nutrients. For example, blades were better at indicating N, P, K, manganese (Mn) and zinc (Zn) and petioles were better for iron (Fe) and boron (B). They also reported that measuring the leaf blade at veraison was preferable over measuring at anthesis because nutrient concentrations were relatively stable during that time, making year to year comparisons more reliable. Schreiner and Scagel (2017) manipulated soil N, P and K and tested the use of blades and petioles at bloom and veraison to predict vine performance, yield and must composition. Leaf blade data were found to correlate best to these parameters, while timing of collection was not so important.

Results from frequent plant analysis (yearly is ideal, though in established vineyards with a management history it can be stretched out to once every

3 years) can be used to predict trends in plant nutrient status and then to make adjustments to the fertilizer programme before any symptoms appear. This, again, is the overall goal of plant nutrition management – to keep the vines from entering into a limiting situation.

Because of the complexity of the plant–soil system, there is not a blanket recommendation for application of fertilizers in vineyards. Some soils, such as those with a high sand content, are commonly N deficient, and therefore the application of N is essential for continued production of high-quality grapes (Spayd *et al.*, 1993). However, application of N to vines growing in heavier and younger soils is often not beneficial as it raises its level even higher, resulting in more vigorous growth and decreases in fruit quality. Fertilization programmes should be developed on a site-by-site (or even block-by-block) basis, using information about soil and plant nutrient content.

Foliar fertilization is useful in situations where the soil may not have the nutrients available, particularly for those needed in only small amounts. Where industrial fertilizers cannot be used, properly composted organic materials can be applied, though it may take longer for its effects to be seen.

The Nutrients

Important plant nutrients are often categorized into macronutrients and micronutrients, so called because of the rough amounts of them that are required for healthy growth. The macronutrients are carbon (C), hydrogen (H), oxygen (O), N, P, sulphur (S), K, Ca and Mg. The first three come from air and water and will not be discussed further. Micronutrients (just as important, but not used in high concentrations) are Fe, Mn, copper (Cu), Zinc (Zn), molybdenum (Mo), B and chlorine (Cl). A brief review of each of these follows.

With reference to pictorial symptoms of deficiency or excess, readers are referred to the excellent resources of Wilcox *et al.* (2015) and Nicholas (2004). Many times the term chlorosis (loss of chlorophyll, leading to yellowing of leaves) will occur in describing symptoms: care should be taken in interpreting chlorosis, as it can be caused by shade, ozone damage, sun scald, drought, wet soil, *Eutypa*, winter injury and some pesticides, as well as by nutrient deficiencies (Jordan *et al.*, 1981). Always take into account as much information as you can obtain about a problem before diagnosis.

Macronutrients

Nitrogen

This element is a key part of many plant cell functions because it is used in making the amino acids used to build proteins, which are used to make the enzymes responsible for much of the work done in plants as well as structural components, etc. (e.g. RuBiSCO and chlorophyll in photosynthesis). If it is lacking, vine leaves will become chlorotic, lower their rates of photosynthesis and cease to grow (Ryle and Hesketh, 1969; Keller *et al.*, 1998; Chen and Cheng, 2003). The root to shoot ratio will also change, with lower N levels resulting in more roots produced for the same amount of shoot (Marschner, 1986). Excess N leads to overly vigorous growth (significant lateral development, long internodes and thick canes) and large and dark green leaf blades. This can have both indirect and negative effects on vine fruitfulness (Winkler *et al.*, 1974).

Nitrogen is a mobile nutrient in the plant, so can be scavenged from one organ and transported to another that needs it more. For example, labelled N fed to vines at the beginning of growth was found in varying amounts between the various vine organs as the season progressed (see Fig. 8.1) (Conradie, 1990).

To correct deficiency, N should be applied before budbreak and possibly a second time at bloom, as these are the times when grapevines take up most N (Conradie, 1980; Schreiner *et al.*, 2006). However, the forms of N that are available to plant roots (nitrate, NO_3^- and ammonium, NO_3^+) are mobile

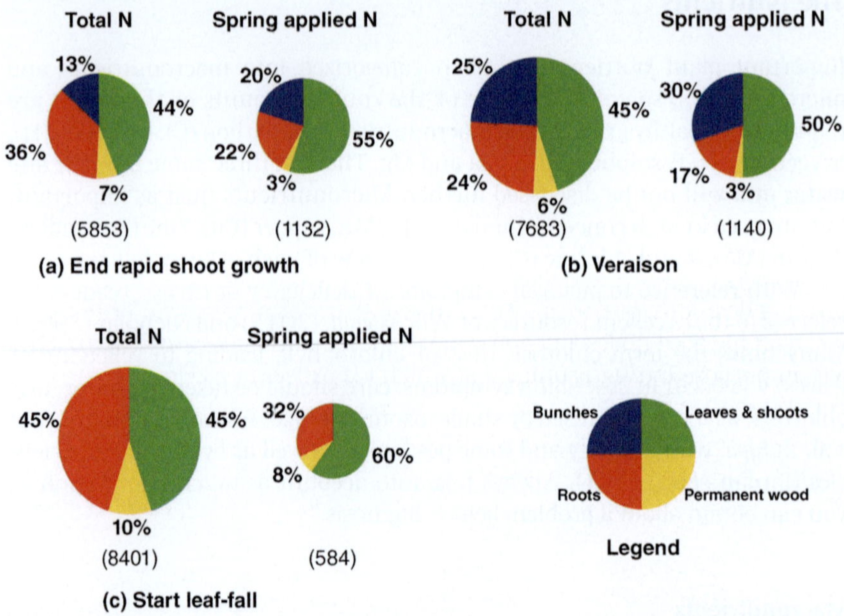

Fig. 8.1. Remobilization of spring-applied nitrogen in sand culture-grown grapevines. The charts indicate the relative proportioning of nitrogen between permanent wood, roots, shoots and leaves and fruit (when present) at end of rapid shoot growth (A), veraison (B) and start of leaf-fall (C). The numbers in parentheses indicate the total amount of nitrogen/vine (mg) (adapted, redrawn and used with permission from Conradie © 1990 American Society for Enology and Viticulture. AJEV 41:241-250).

in the soil, so if excessive water falls or is applied it can be washed (leached) from the soil profile (Bergström and Brink, 1986). Therefore, in areas with springtime rainfall, N is best applied in multiple smaller amounts of fertilizer rather than as one large dose, and starting in mid to late season rather than in early season, to maximize vine uptake (Peacock *et al.*, 1989). Application of compost or growing of leguminous plants inter-row can provide nitrogen in the longer term.

Phosphorus

This major plant nutrient is a constituent of phospholipids and nucleic acids used in cell building and is therefore abundant in meristematic regions such as actively growing shoot and root tips. It is also involved in sugar metabolism, respiration and photosynthesis. Vines are rarely deficient in this nutrient, possibly due to efficient capture from the soil and its remobilization within the plant. Deficiency symptoms appear as yellowing of interveinal areas of older leaves, which can turn red if the deficiency is severe (Nicholas, 2004). Unlike N, P is not mobile in the soil, so there must be exploration of new soil to secure it. Fortunately, soil is generally able to supply large quantities of P, which roots take up in the form of orthophosphate (PO_4). In those soils lower in P there are often mycorrhizal associations with the roots that assist in its uptake (Possingham and Groot Obbink, 1971, refer to Chapter 2). Application of P fertilizer can result in lower K levels in the vine (Conradie and Saayman, 1989) and so should not be applied without consideration of the available K supply (Haeseler *et al.*, 1980).

Sulphur

Sulphur is taken up by roots in the form of sulphate (SO_4), though it can also be taken up by leaves when it is in its gaseous form (De Kok, 1990), which is the active form of S that prevents powdery mildew infection. Many plant proteins contain S: for example, cystine is a major S-containing amino acid. Sulphur-containing compounds called thiols have been found to be important determinants of white wine aroma, lending box tree, citrus or cooked leek characteristics to the mix (Tominaga *et al.*, 2000), though these compounds are not found in the grapes themselves, but are products of the fermentation process. Sulphur is rarely deficient in grapevines as it is frequently applied in a spray for control of powdery mildew.

Potassium

Potassium is a very important element in protein and fat synthesis, enzyme activation and as an osmotic charge balancer. Its uptake form is the cation, K^+, and is used in large quantities by fruit crops, especially. The quantity removed with a moderate crop of grapes is frequently greater than that of N (Winkler *et al.*, 1974). Xylem exudate, sap that bleeds when vines are cut close to bud-break, is high in K (Glad *et al.*, 1992), which is evidence that the roots are actively absorbing soil nutrients.

Vines with K deficiency usually develop interveinal or marginal chlorosis, or both. At more severe deficiencies, marginal necrosis and leaf cupping develops (Nicholas, 2004). False K deficiency (also known as spring fever) is a temporary appearance of symptoms on basal leaves prior to bloom and is associated with cool and cloudy weather conditions, which cannot be corrected through the application of K fertilizers (Christensen *et al.*, 1990). It is thought that N reduction is lowered under these conditions and an accumulation of ammonium ions results, which interferes with potassium's role in protein synthesis (Maynard *et al.*, 1968). In sand culture, excess K results in interveinal chlorosis of basal leaves, slightly reduced shoot growth but unaffected root development (Li, 2004). There are also problems with high juice pH associated with excess K in grapevines due to it displacing H^+ in the solution (Mattick *et al.*, 1972), so K should be kept within moderate levels in the soil and plant.

Potassium deficiency symptoms are known to be corrected by potash (which is any of a number of K salts) application (Shaulis, 1961), and any deficiency should be corrected as soon as symptoms become visible. Potassium chloride (muriate of potash) is usually the most economical source of K, but in growing areas with high salt concentrations it can result in chloride toxicity, so potassium sulphate is the recommended alternative in this case. Deficiencies of K in low-pH soils may be caused by excess application of Mg (Bates *et al.*, 2002), as these two cations compete for the same uptake channels in root tissues.

Calcium

Known as the cement of plant structure, Ca is a component of the middle lamella that holds plant cells together and maintains cell integrity (Epstein, 1961). It also participates in both carbohydrate translocation and N utilization (Gossett *et al.*, 1977). Calcium is infrequently deficient in vineyards (Nicholas, 2004) unless the soil is acidic (Robinson, 1992), and applications of calcium sulphate (gypsum) can be used to increase soil pH and to treat high-sodium soils (Winkler *et al.*, 1974). Calcium deficiency in some fruit is associated with internal degradation disorders, like bitter pit in apple (Torres *et al.*, 2017), and in grape it can be associated with berry skin strength (Chardonnet and Donéche, 1995).

Magnesium

The central mineral element in the chlorophyll pigment is Mg, which assures the importance of this nutrient in plant growth. Magnesium also functions in chlorophyll synthesis, enzyme activation and can play a similar role to Ca in membrane stability. It also is important in the partitioning of carbohydrates from the leaves to the roots (Cakmak *et al.*, 1994). Deficiency is most severe on basal parts of the shoots due to its remobilization to newer parts of the vine. Symptoms of moderate deficiency are chlorosis between the large veins, which remain green, as does the tissue at the leaf margins (Fig. 8.2; Wilcox *et al.*, 2015; Nicholas, 2004). Severe deficiency results in chlorotic leaf margins and red coloration in red grape cultivars (Nicholas, 2004). Magnesium deficiency is most common on very low-pH soils and is increased by excessive K fertilization (Bates *et al.*, 2002). Regardless of pH, however, high soil K interferes with the uptake of Mg (Wolf *et al.*, 1983) and should be avoided. To correct deficiencies, soil pH can be raised (particularly with dolomite, also known as calcium magnesium carbonate – $CaMg(CO_3)_2$) or magnesium sulphate ripped into the soil.

Micronutrients

Iron

Although not present in the chlorophyll molecule, Fe is required for chlorophyll synthesis and is also involved with electron transport in respiration. Iron deficiency is observed first in early-season chlorosis in the youngest leaves, which causes reduced photosynthesis (Smith and Cheng, 2007). Iron shows very low mobility within the plant, so symptoms appear in developing plant tissues. Iron chlorosis is often ephemeral, appearing when growth is rapid and soils are cold and more likely to be wet, which limits the ability of the vine to take it up (Tagliavini and Rombolà, 2001). As the soils dry out and warm up, supply from the soil improves and chlorosis no longer occurs in the newly forming leaves. Chlorosis can occur even when the soils have adequate Fe, for different reasons, however (Mengel *et al.*, 1984; Davenport and Stevens,

Fig. 8.2. Examples of magnesium deficiency in leaves of 'Chardonnay' (left) and 'Pinot noir' (right). Content author's own.

2006). Chlorotic symptoms are greatly increased in high pH (alkaline)-soils (Fig 8.3), so plant nutrient analysis under these conditions will detect low concentrations of Fe. Iron chlorosis-resistant rootstocks are available for grapevines, but often result in excessive growth and reduced grape yield (Tagliavini and Rombolà, 2001). Therefore making sure soil pH allows for Fe uptake by the rootstock desired is important. Vesicular-arbuscular mycorrhizae can ameliorate the soil immediately surrounding the roots (the rhizosphere), altering its pH and increasing nutrient availability (Bates *et al.*, 2002), and thus providing another reason to encourage their presence. In most cases Fe deficiencies are more easily corrected by iron salt or iron chelate foliar sprays rather than by soil amendments.

Manganese

Manganese is essential in the synthesis of chlorophyll and plays a role in enzymatic reduction–oxidation reactions that power most functions in the plant tissues. Like for iron, manganese deficiencies are usually found on high-pH soils, with symptoms of Mn deficiency identified by interveinal chlorosis on shaded and mature leaves, as this mineral can be remobilized within the plant. Manganese

Fig. 8.3. Example of iron deficiency in a high-pH soil. The shoot tip is most affected because of iron's low mobility within the plant. Content author's own.

sulphate applied to the soil under the vine or foliar micronutrient sprays will correct Mn deficiency, although plant analyses should be conducted in following seasons to track changes in plant concentration. The use of mancozeb for disease control can maintain Mn levels in plants (Deckers *et al.*, 1997), as mancozeb is a coordinated product of Zn (1.9%), Mn (15%) and ethylene bisdithiocarbamate ion.

Copper

Copper is a non-protein component of several oxidizing enzymes such as ascorbate oxidase and tyrosinase, and forms an important part of the electron transport chain in photosynthesis. Deficiency symptoms for Cu in grapevines are similar to those of Fe-induced chlorosis and, like Fe, it is not moved from one place to another within the plant. Copper is generally very toxic to many organisms, and one reason that it is not frequently found to be deficient in vineyards is that some variation of Bordeaux mixture (copper sulphate and hydrated lime) is very often applied to vines in the control of downy mildew. It is possible for high Cu levels in the soil to reduce root growth and development (Wainwright and Woolhouse, 1977), but analysis of above-ground parts of plants does not seem to indicate high soil and root Cu status (Brun *et al.*, 2001). Therefore, if high Cu levels in soils are suspected, the nutritional status of the root tissues should be determined.

Zinc

Zinc is important in the synthesis of the growth regulator indole acetic acid, is a component of many different enzymes and plays an important role in pollen and fruit formation (Marschner, 1986). Zinc is not very mobile within the plant, so if the nutrient becomes deficient part-way through a season, younger rather than older plant tissues will show symptoms. Zinc deficiency is characterized by abnormal development of internodes ('zig-zag' growth pattern of shoots), interveinal chlorosis in early summer, small leaves (also known as little leaf) and a widened petiolar sinus (see Fig. 8.4). Straggly clusters with undeveloped shot berries and generally poor fruit set are also common, and can occur without the appearance of leaf or shoot deficiency symptoms (Nicholas, 2004), but fruit set like this can also be associated with poor weather during flowering. High pH soils have less available zinc, so deficiencies are more likely to appear in these situations. Foliar feed products containing zinc sulphate or zinc chelate can alleviate symptoms.

Molybdenum

Molybdenum is a component of the enzyme systems that carry out reduction–oxidation reactions. Enzymes that require Mo include nitrate reductase – which is essential for the incorporation of root-supplied N into cell components – and other enzymes involved in cell function (Kaiser *et al.*, 2005). It also appears to be important in pollen formation and fruit set (Williams *et al.*, 2004). Molybdenum is a mobile nutrient within plant tissues and required in only very small amounts, even in comparison with other micronutrients (Marschner, 1986). In grapes

Fig. 8.4. Left: Shoot symptoms of zinc deficiency. Note the short internodes, small, misshapen leaves that have an open sinus, and jagged leaf margins. Right: Poor fruit set often associated with zinc or boron deficiency - there are some large, seeded berries, but most are small and seedless. This is also called millerandage or hen and chicken. Content author's own.

the major economic symptoms of deficiency are shot berries (millerandage or hen and chicken) and leaf chlorosis, caused by a lack of usable N from nitrate reductase. Molybdenum deficiency has been identified as the cause of the 'Merlot problem' in South Australia, where foliar sprays of the nutrient applied at bloom greatly improved vine yields (Williams *et al.*, 2004).

Boron

In grapes, boron is important in cell division and in the normal development of pollen and pollen growth, so it follows that boron-deficient vines have poor fruit set. It is an important component of cell walls (Thellier *et al.*, 1979) and is not mobile within the plant. Boron is limiting for growth in many grape-growing regions of the world (Cook *et al.*, 1960), and symptoms of deficiency include zig-zag development of shoots (similar to those for Zn, Fig. 8.4), inter- node swellings (Cook *et al.*, 1960; Wilcox *et al.*, 2015) and leaf chlorosis, later turning into necrosis, though this tends to occur on older leaves as compared with Zn deficiency (Nicholas, 2004). Boron deficiency is also associated with dry soils and high pH soils, but is easily corrected by ground and/or foliar applications of soluble borates. Boron toxicity symptoms are misshapen and small leaves on new growth (see Fig. 8.5), brown spots on the margins of

Fig. 8.5. An example of altered leaf development on shoots of 'Italia Pirovana' in Chile caused by boron toxicity. Content author's own.

older leaves and sometimes death of the shoot tips as a result of movement of B from the roots upwards in the transpiration stream (Marschner, 1986). Boron toxicity may occur through the use of high-boron-containing potash as fertilizer or in high-boron irrigation water. Because B toxicity is as much of a concern as deficiency, plant analysis should be used to maintain moderate concentrations (Shorrocks, 1997).

Chlorine

Chlorine is abundant in nature (for example, between 4 kg and 8 kg of chloride/ha can be supplied through rainfall (Marschner, 1986)) and is important in photosynthetic reactions as an activator of enzyme reactions and as a charge-balancing ion (Marschner, 1986). Very little is required by plants, but it is usually found at much higher levels than this. Chlorine is very mobile within the plant and very few incidences of Cl deficiency are reported – the usual concern is with toxicity (see Fig. 8.6). Increasing levels of Cl supplied to vines, usually from increasing salinity of the soil, result in reduced shoot growth (Groot Obbink and Alexander, 1973), and the critical toxic concentration for grapevines has been determined to be 400 mg Cl/kg soil (Xu *et al.*, 1999), though this can vary depending on cultivar and environmental situation.

Fig. 8.6. An example of severe chlorine toxicity in an Australian vineyard. The marginal chlorosis is not diagnostic of this however, as these symptoms can appear through a number of different ways. Content author's own.

9 Mechanization

Abstract: Viticulture and use of machinery are inextricably linked in the modern vineyard due to the number of operations needed to manage vines each season. Mechanical harvesters have been used since the 1950s and additional machines have been produced to the point where it is possible to manage vines with almost no hand labour. For equipment to work in concert with vineyards uniformity is key, but the promise of practical robotics may mean that machines work to meet the requirements of the vineyard, not the other way around. At present, there are multiple options for weed and canopy management, crop adjustment, and sustained mechanical pruning is possible in some climates. Equipment to gather data are also accessible, such as weather stations that can provide data for disease forecasting, sensors to gather soil or plant water status or remote sensing to map vineyard variability.

The cultivation of grapes is a highly labour-intensive process. Unlike agronomic crops, the mechanization of horticultural crop management has been slower due to the more complex nature of the plants and production of their fruit. The proportion of harvest costs within the total cost of production for crops was higher for grapes in the 1960s (Shepardson *et al.*, 1970), and already by that time it was recognized that finding sufficient labour to harvest the fruit was problematic.

Practical research on harvest mechanization harvest started in the 1950s on the west coast of the USA, with researchers in California developing a device that would cut clusters hanging from an inverted-'L' trellis to assist with raisin production (Studer, 2000). It was an innovative thought, but of limited practical use because the clusters did not always hang in the right place to be cut off efficiently (Winkler *et al.*, 1957). In the late 1950s, researchers at Cornell University, USA, also started developing a mechanical harvester that used mechanical vertical shaking to dislodge the fruit from the vine (Shepardson *et al.*, 1962). This developed into a commercial device in 1967 that was particularly aimed at harvesting 'Concord'-type grapes grown on the Geneva Double Curtain trellising system (see Chapter 6; Shepardson *et al.*, 1970). The advent of the modern horizontal-type shaking system had its beginnings in

DOI: 10.1079/9781800627062.0009

the mid-1960s through the efforts of two 'Concord' grape growers. Their innovation was to use plywood boards to knock a vertical trellis, dislodging the berries, which were then collected by a conveyor belt (Morris, 2000).

From these beginnings have developed a wide range of machinery that help with not only harvest, but also shoot positioning and thinning, canopy trimming, leaf removal, crop reduction and pruning. Application of pesticides is also highly mechanized, but will be discussed in more detail in Chapter 10. The ultimate goal of an all-mechanized sustainable vineyard is currently not realized, but work is continuing in this area (Clingeleffer, 2013; Mazzetto *et al.*, 2013; Kurtural, and Fidelibus, 2021) and will surely succeed, especially with increasing difficulties of sourcing skilled labour.

Refinements to trellis design are leading to systems that lend themselves to greater degrees of mechanization, and therefore to labour and cost reduction. The Movable Free Cordon system, for example, is an adaptation of the single-wire trellising system that uses curved vine trunks and a cordon wire that is free to move up and down through slotted guides installed at every post. Harvest is facilitated by the ability to shake the vines up and down with minimal impact on the trellis system or direct contact with the grapes, which detach about 1 m from the impact head (Intrieri and Filippetti, 2000). Mechanized pruning is kept simple because there are no foliage wires. Curiously, this system is quite similar to the first mechanized harvesters developed in the late 1950s, which also used an up-and-down motion to dislodge the fruit (Shepardson *et al.*, 1962).

A key to mechanization in vineyards up to this point has been uniformity, as often machines are not very forgiving of driver error, vine parts or trellising structures that get in the way (see Fig. 6.12). It is the earlier effort that goes into developing as uniform a vineyard as possible that pays off in terms of greater ability to reach management goals, such as precise leaf removal in the fruiting zone or the ability to spray four or more rows at the same time. Lasers or GPS are now commonly used to make sure that rows are put in straight, with tolerances measured in millimetres so that machinery work can be more efficient.

Drone tractors, controlled by GPS, LIDAR and/or optical scanners are already in use in some agricultural cropping systems for planting, tilling, harvesting, etc. (Aravind *et al.*, 2017). In row-based vineyard systems adaptation of these technologies is relatively simple and can result in time savings and other benefits from greater levels of automation.

Currently, there are limitations to mechanization in terms of physically being able to traverse or access some vineyards. Most machinery developed for vineyards is not very amenable to steep slopes, for example, though gentler ones can be adjusted for with the use of hydraulics to keep the machine level while the wheels are at different heights on the incline.

All of these things are rapidly changing, however, with developments in robotics and artificial intelligence. Complete vineyard mechanization up to this point has largely revolved around adjusting the vineyard structures to be compatible with the machines. With robotics, we can make the machine work to our specifications, and with artificial intelligence, we should be confident that the machines will be able to cope with situations they were not specifically programmed for.

We are tantalizingly close to easily accessible and effective autonomous robotic workers in the vineyard. There are a number of products on the market, usually operated under a lease agreement because the technology is changing so rapidly. Virtually all vineyard tasks can be performed by robots now, with actions like undervine weeding, mowing, spraying bird scaring and data gathering most common, with trimming, leaf removal, disease scouting and others coming quickly (Boiling, 2025). Many of these are electric, some are diesel-electric hybrids. The ability to work on slopes as steep as 45 degrees is claimed, as is autonomy of eight hours for some electric models. True independent operation and speed/efficiency of work are still question-marks at this point, but surely this will change rapidly.

Another area to watch is the development of AI-aided pruning, which has the potential to quickly train new workers in how to prune vines to the viticulturist's specifications. Augmented-reality vision through a smartphone has been used to scan vines in the field and suggest pruning cuts (Häring *et al.*, 2024) and there is work to use virtual reality headsets to lead trainees through the process of pruning (O'Connor, 2023).

A side benefit to robots in the vineyard is the increased amount of information about each vine due to the constant monitoring of position they require. This could aid development of precision viticulture at a micro-scale. For example, in determining which cuts to make for vine pruning, a 3D model of each vine is created (Botterill *et al.*, 2017) that can be analysed from year to year to look for conformational changes, tracking the longer-term effects of different pruning practices, and monitoring growth in terms of trunk and cordon volume (which are related to the carbohydrate storage capacity of the vine) in response to seasonal trends.

This does highlight an issue with developing robotic/artificial intelligence: we have to tell the robots what to do and how to do it. At present, many viticultural operations are done by relying on the experience of the person doing the work – trying to quantify that experience is not a trivial pursuit. Again, using the pruning example, while a pruning program could make good decisions, if it came up against a situation that was not in the training set, it did not perform well (Corbett-Davies *et al.*, 2012). Additionally, there is more than one way to prune a vine and individuals approach this in different ways. Those questions led to a project that evaluated how pruning decisions are made and what constitutes 'quality' in pruning (Kirk, 2016). Clearly, developing automated systems for something as complex as pruning is not simple, and further research will be needed to achieve the goal of truly independent robot workers.

The following sections provide additional detail around mechanization of a variety of vineyard operations.

Weed Management

Between-row management of cover crop or weed growth in modern times is through the use of tractor-mounted mowers. These have the advantage of being much quicker than traditional methods (use of animals as grazers or powered cutting tools such as sickle-bars), which is especially helpful in areas where frost is an early-season occurrence as the vegetation needs to be kept as low as possible. However, machinery does also have disadvantages, such as fuel use and contribution to soil compaction, which can be mitigated by using smaller machines (such as quad bikes) or reducing the number of passes through the vineyard by performing more than one operation at once. There are mowers that deposit the clippings under the vine rows rather than leaving them between the rows, which can assist with weed control under the vines as well as increasing soil organic matter over the longer term.

As machinery use does have an impact on the soil as well as carrying a financial cost, the use of animals grazing to manage cover crop growth in vineyards has increased considerably in some areas. In New Zealand, for example, it is now common to see sheep in dormant vineyards. The desire to have cover crops eaten during the growing season has led to research into teaching the animals not to eat leaves (Manuelian Fusté *et al.*, 2014), but in some areas their predilection for leaves has been used to help with leaf removal. They are quite effective, though you need to keep them moving through the vineyard so that they do not start eating the fruit (J. Hickman, Taimate Vineyard, New Zealand, 2014, personal communication).

Flail mowers (see Fig. 9.1) are increasingly used to eliminate the need for pruners to remove brush from the vineyard manually or by tractor. As well as

Fig. 9.1. A narrow, PTO-driven flail mulcher suited to moving between vineyard rows. As opposed to rotary mowers, flail mowers are better suited to stony ground and can also break prunings up more fully to allow quick decomposition. Content author's own.

cutting cover crops, these attachments can break the prunings up into small pieces that have less potential to release disease spores (Seyb, 2004) because they are more quickly degraded and therefore do not carry disease inoculum sources in the following season. Dealing with the prunings in the vineyard also means that they do not have to be disposed of through burning or burial, and the nutrients contained are returned to the soil ecosystem.

In management systems where synthetic herbicides are not able to be used, weeds have traditionally been controlled through the use of hand labour. However, in large vineyard developments it is impractical to do this, so several tools have been developed to help mechanize the process of cultivation under the rows, which disrupts weed development. Examples of these include tools such as the mechanical tiller, hydraulically driven rotary heads mounted on an articulated arm (see Fig. 9.2). The rotor, which may have tines or blades, retracts automatically or semi-automatically using trunk sensors and fingertip controls that move the head away to prevent damage to the vines. A similar device consists of a single blade that is designed to undercut plants, again, mounted on a movable arm to allow control of plants between the vines.

These devices can provide good control of weeds providing they are used often enough. They are much faster than hand weeding, but do require tractor movement through the vineyard, and repeated cultivation causes a decrease in the number of roots in the surface soil layer if the soil disturbance is allowed to run too deeply (Saayman and Van Huyssteen, 1983). Additionally, if there are rocks under the row, they have a tendency to throw them out, potentially creating a hazard for vehicles and attachments like mowers and mulchers. Properly used, however, they can be a viable, and economical (Shrestha *et al.*, 2013), alternative to other methods, particularly in vineyards where the use of herbicides is to be avoided.

Fig. 9.2. Example of a scarifier, a mechanical weed management tool that uses pressure sensor controlled rotary heads with blades or tines to allow under-row cultivation. This one is front-mounted to make it easy to observe while moving through the vineyard. Tines are fitted to the machine, but blades can also be used (photographs courtesy Jonathan Hamlet, used with permission).

Similarly, devices that hill-up and take away soil from under the vines are used to bury and then unearth winter freeze-sensitive canes in those areas where vines are too sensitive to cold temperatures to survive in the open (see Fig. 4.4). This procedure can be quite effective when grapes are being grown in a region outside of the optimal climatic requirements, but there are limitations depending on soil type, where large stones make this process difficult.

Other options for weed control include: (i) repeated mowing under the vines (see Fig. 9.3); (ii) the use of directed heat in the form of flame (Ascard, 1998), steam (Melander *et al.*, 2002; Shrestha *et al.* 2013), hot foam (Collins *et al.*, 2002), microwaves (Sartorato *et al.*, 2006); (iii) electrocution (Vigneault *et al.*, 1990) and others (Bajwa *et al.*, 2015). These too can be effective in certain situations, but generally require frequent repeated use and are fairly resource intensive (Pinel *et al.*, 1999).

Canopy Management

The management of grapevine canopies can now be completely mechanized, but this is very dependent on the trellising/training system in use. Single fruiting wire trellising, using mechanical pruning, mechanical shoot trimming and mechanical harvesting, utilizes almost no manual labour, but in

Fig. 9.3. An under-vine mower where the mowing heads are pushed aside by vine trunks as they move down the row. As the mower heads are held out with springs, this sort of system works only with vines that have significant trunk development, so weak or young vines need to be staked. Content author's own.

the right climate can be a sustainable method of grape production. With the Geneva Double Curtain, devices have been developed that both thin the shoots and brush the shoots into the two distinct, downward, canopies the system requires (Pool *et al.*, 1990; Morris, 2007). Brush-type devices can also be used to rub off non-count shoots on vine trunks, which is otherwise a laborious undertaking.

During the growing season, for hedge-type systems like VSP tractor-mounted cutter bars are useful for trimming shoot tops and canopy sides, and though there are systems for raising foliage wires and 'binding' shoots so they stay more upright, precise positioning of shoots in a vertical shoot-positioned trellis, for example, is not yet viable. For leaf removal, a movable head that either sucks in leaves toward a cutter bar or blows high-pressure air out to rip leaf blades apart can be used to obtain more fruit exposure and a more open canopy (see Fig. 9.4). These have been found to be as effective as hand leaf removal in terms of reducing botrytis incidence (Gubler *et al.*, 1991).

The systems that use pulsed air to remove leaf blades do have one disadvantage in that the cut plant material often stays in the canopy or can damage the fruit, creating potential hot spots for disease development later in the season, though in warmer and drier climates the risk of this appears low (Gubler *et al.*, 1991). In fact, the air movement can dislodge flower caps and abscised berries within the clusters, leading to a decrease in bunch rot (Wolf *et al.*, 1997; Molitor *et al.*, 2015).

Fig. 9.4. Mechanical leaf removal equipment. Left: a system that uses a vacuum to suck leaves into the reciprocating blades. Right: a system that used pulsed, high-velocity air to remove the leaf blades. Content author's own.

Crop Management and Harvesting

In addition to taking ripe fruit off of vines, mechanical harvesters can also remove it earlier in the season, as a form of crop reduction (Pool, 1987; Clingeleffer, 1993a; Pool *et al.*, 1993). Originally, the machines were tested in the same configuration as used at harvest, which resulted in fruit being removed, but remaining fruit being damaged, which subsequently did not ripen properly (Pool *et al.*, 1993; Petrie and Clingeleffer, 2006). More recent suggestions have been to remove the crop from only a section of the canopy, accomplished by removal of some of the harvester's beater bars, which has given good results compared with hand thinning (Petrie and Clingeleffer, 2006). Best results come from running the machine through after berries reach about 4mm through to pea-sized (Kurtural and Fidelibus, 2021; Petrie, and Clingeleffer, 2006).

Raisins have been mechanically harvested for many years (May *et al.*, 1974), and special training systems have been developed to increase efficiency (Christensen, 2000) and allow automated distribution of harvested grapes onto paper trays between the vineyard rows (see Fig. 1.5). More recent research has refined systems for producing dried-on-vine (DoV) raisins using conventional trellises and mechanical wine grape harvesters (Fig. 9.5; Peacock and Swanson, 2005), which makes the practice much more grower-acceptable as less capital expenditure is needed for specialized equipment and trellising. Improvements on this can be made if the vineyards are designed for DoV from establishment, where trellis systems are larger, there is less need for space between rows and earlier-ripening cultivars are used (Fidelibus, 2014).

Fig. 9.5. Dried-on-vine raisins being harvested and loaded directly into bins in California (photograph courtesy of Matthew Fidelibus, used with permission).

Mechanical harvesting of wine grapes (see Fig. 6.66) has been commercially practised for some time, and is now an accepted way of obtaining quality grapes at low cost and at any time of day or night. Most reports agree that a well-set-up mechanical harvester will provide fruit quality that is close to that of hand-harvested fruit (Johnson and Grgich, 1975; Clary *et al.*, 1990). Modern equipment with further refinements (see Fig. 9.6) are now approaching the

Fig. 9.6. The interior of a mechanical harvester. Note that the bars are removable, curved and flexible, which allows for adjusting to fruiting wire height and lessens the physical impact on the vines. Content author's own.

ability to harvest berries with little or no skin breakage, which may lead to further changes in how the industry perceives the practice. Research indicates that there is little difference in sensory assessment ratings of wine caused by the practice (Hendrickson *et al.*, 2016), especially if the crop is well looked after in the vineyard. Other advances on recent harvesters are in the areas of removing material other than grapes, optical berry sorting and even crushing and juicing in the vineyard.

However, this labour-saving device is not without its challenges. Because machines are not discriminating when it comes to harvest, a second set crop (fruit arising from lateral growth, which is far behind in terms of development) and anything that is dislodged from the canopy can be caught and end up in with the target grapes. Therefore, for mechanical harvesting to be effective, viticultural management up to this point must ensure that fruit that is not ripe will not be harvested, as well as any other items that may be collected during the process. Though material other than grapes (MOG) is found in hand-harvested fruit as well, the size of these items tends to be smaller in mechanically harvested fruit (Petrucci and Siegfried, 1976), which makes it harder to remove and can lead to greater extraction of unwanted substances during winemaking. Noble *et al.* (1975) reported that the inclusion of leaves as part of MOG did not significantly affect the acceptability of wines, but others (Wildenradt *et al.*, 1975; Huang *et al.*, 1988) have reported deleterious effects. It is likely that wine style and winemaking practice both have a significant effect on this, which is why varying results have been reported.

One advantage of mechanized harvest is that cluster size and position within the canopy is no longer as great a concern, which can lead to greater freedom in developing trellising/training systems. Hand harvesting is most efficient when clusters are large and within a distinct zone in the canopy, but with machine harvesting this does not matter (Petrie and Clingeleffer, 2006). This lends itself to the adoption of mechanical pruning (Clingeleffer, 1993b), where mechanization is a great cost and time-saver, and allows for more flexibility in getting the fruit in when it is ready.

Robotic picking machines, which are able to cut clusters at the peduncle and place them into a tray are being developed for use in the table grape industry (Boiling, 2025), although if successful they could also be used for high-quality sparkling or still wine production.

Pruning

Once mechanical harvesting had become common, the most time-consuming aspect of grape production became pruning (Morris *et al.*, 1975; Pollock *et al.*, 1977). Mechanizing this completely was accomplished through hedging the dormant growth (mechanical pruning, Shaulis *et al.*, 1973; Pollock *et al.*, 1977) or cutting only the hanging canes back (minimal pruning, Morris *et al.*, 1975).

Machinery can be used to make some hand-pruning jobs easier. A barrel pruner (see Fig. 6.9) is a set of overlapping discs that cut canes into short sections while avoiding damaging the foliage wires. Cane pieces fall from the wires or can be easily removed by hand, avoiding the tiring task of stripping canes out of the trellis. A variation of this has a cutting disc optically guided to maintain a certain distance above the cordon, allowing for variations in vine height, increasing precision and reducing the amount of hand follow-up needed. These systems can be used for spur pruning, where workers trim out the unwanted growth, or cane pruning if someone goes through prior and ties down those canes needed for wrapping to the fruiting wire.

Another device calls for some modifications to the trellising system, but performs two important jobs at once. The Klima vine pruner system, and others like it, requires fruiting and foliage wires to be secured to the posts with special clips and someone to go through the vineyard beforehand to make the major pruning cuts to the vines. Once this is done the machine can lift the wires and attached canes and cordons earlier cut from the vine, up and away from the trellis. As it moves down the row the plant material is fed into what is effectively a wood chipper, rendering the canes into small pieces that are scattered onto the vineyard floor. After this the wires need to be re-clipped and the canes cut to length and wrapped to the wire.

These machines do have limitations, but both can reduce labour inputs significantly, as well as making the pruning process less of a physical task for the workers.

The practice of hedging, which is relatively easy to mechanize and practise on simple trellising systems, can result in good fruit quality and reduced production costs. However, there can be minuses due to increased levels of botrytis in cooler growing areas caused by more shoot and lateral growth and fruit shading (Clingeleffer, 1993a) if vine vigour is not managed correctly. The persistence of older wood inside the trimmed box can create a place for disease inoculum to develop, which is a problem in cooler-climate areas (Kurtural and Fidelibus, 2021). In a non-irrigated vineyard in Spain, an 11-year study using simulated hedge pruning found little effect of the practice on fruit quality, but a significant increase in yield as well as in vine capacity (de Toda and Sancha, 1999). For 'Concord' grapes destined for juice, the practice of mechanical pruning was found to be a commercially sustainable practice after a 6-year study in Michigan, USA (Zabadal *et al.*, 2002), where there are short summers and cold winters, though additional hand work was necessary to maintain good canopy structure. Reynolds (1988) reported similar results in Canada using simulated mechanical pruning. In many areas, hedging, as a way of mechanizing pruning, has worked very successfully and has been adopted widely, particularly in Australia (Clingeleffer, 1999).

The promise of viable robotic pruning (and its potential to cope with current and varied grape production systems) has been around for some time (Ochs *et al.*, 1992; Gunkle and Throop, 1993; Monta *et al.*, 1995). Challenges are many, as the natural variation found between vines is high and the environment itself is not necessarily very robot friendly (Kondo and Ting, 1998). However, research into this area has made advances, where there are now viable systems for 3D sensing and for the fine control of cutting implements (Botterill *et al.*, 2017), which are able to make vine maps accurate within millimeters (Williams *et al.*, 2023). We will no doubt see this emerge as a commercial option in the coming decades, and it promises to have a large impact on reducing dormant-season labour inputs.

Environmental Monitoring and Remote Sensing

Mechanization carries through to other tasks in the vineyard as well. Remote and real-time monitoring of many environmental and vine characteristics is now possible thanks to the use of wireless mesh technology and drones. Weather stations, monitoring combinations of air and soil temperatures, relative humidity, solar radiation, wind speed and direction, rainfall, etc. provide information used for predicting disease infection periods, insect stages, projecting growth of vines and timing applications of weed sprays. Similarly, monitoring of plant and soil water status is possible, and can be used in the determination of irrigation frequency.

If a multi-sensor weather station is not within your budget, then purchase of a temperature logger and Stevenson-type screen is still worthwhile (Fig. 9.7). Regularly downloading and plotting the data allows the heat accumulation of the season to be followed, which can be helpful in planning vineyard operations (refer to Chapter 3). This data can be complemented by information from a nearby online weather station, if available.

Fig. 9.7. A Stevenson-type screen containing a logger, mounted near fruiting wire height. Content author's own.

Advances in miniature sensors (for example, that could be embedded in plants to monitor water status) will result in a greater ability to monitor the vine and its environment, both in terms of the number of factors and also the intensity or frequency with which they can be determined in a given area. This kind of information can be supplemented with GPS and proximal sensing data from vehicle traffic in the vineyard, and a large range of data types gathered via smartphones. Drones or satellite imagery can be used to monitor canopy density, health or nutritional status of vines through normal photography, infrared or multispectral scans (Dobrowski *et al.*, 2002; Giovos *et al.*, 2021)

With increased mechanization comes the opportunity to gather more information. Clearly, the problem is not be getting information, but developing methods to be able to collate and analyse it all, culminating with useful recommendations for management. Complex data sets anchored with GPS coordinates are manipulated using geographic information systems, or GIS. The spatial database created can overlay a great amount of information on a vine-by-vine basis if necessary, creating the possibility of truly precise viticulture. Incorporation of variable rate technology to currently available equipment, or use of the information for robotic workers, would allow for operations to be individualized for each vine and could mean optimal use of resources (e.g. highly targeted disease control, nutrient addition, selective harvesting) for the many operations performed each season.

10

Grapevine Pests, Diseases and Disorders

Abstract: Grapevines are susceptible to a wide range of diseases and pests and express symptoms in various ways. The principal fungal diseases are botrytis, powdery mildew, downy mildew and phomopsis, though black rot, anthracnose, and trunk diseases like eutypa, esca and black foot are found in many areas. Bacterial diseases of import include Pierce's disease, the various grapevine yellows, crown gall and bacterial blight. Viral disease like grapevine fanleaf, leafroll, red blotch and Pinot gris virus are present in many areas and their spread likely to increase with climate change affecting their vectors' ranges. Other damage in the field may be caused by off-target herbicides, insects such as phylloxera, leafhoppers, mealybugs, thrips, beetles and mites. Animal pests include millipedes, nematodes, various species that eat foliage or fruit, and the last pest type is weeds. Successful management depends on knowledge of the pest's lifecycle and accurate delivery of the control method.

There are a wide range of situations where grapevines and their fruit can be put in jeopardy. This is important not only for vine health reasons, but also for the economic health of the business. Therefore, minimizing vine exposure to the adverse effects of biotic and abiotic origin is essential.

Pests cause or have a significant potential to cause loss. There are many, many different organisms in the environment, but none are considered pests unless they cause a problem. Despite this, grapevines have many pests and thus problems. Depending on the climate, a large proportion of ongoing vineyard costs can be devoted to managing diseases and pests, and with good reason. Many pests are difficult to suppress and in most cases it is best to adopt a tolerance approach and accept some damage. The point where the cost in money or time is greater than the economic return from the practice is called the economic threshold. The economic threshold, however, varies with pest, location and end use of the grapes. For example, if the consumer sees the actual fruit, then its appearance is very important, whereas this is not so important if the grape is to be processed.

Damage to grapevines can occur in the roots and in the above-ground parts, the trunk, arms, cordons, canes, shoots, leaves, rachis and the berries. Pests of grapes include diseases (bacterial, fungal and viral), insects, mammals

DOI: 10.1079/9781800627062.0010

(including humans) and birds. Disease organisms generally have ecological niches: some flourish under wet, damp conditions (e.g. downy mildew), some under dry conditions (e.g. powdery mildew), some in heavy soil (e.g. phylloxera), and some in sandy soil (e.g. nematodes). Grapes are grown in most of these environmental conditions. Pests can cause acute damage to vines or cause symptoms that are frequently described as 'decline' where vines become progressively weaker and ultimately stop producing. Examples of classic decline pests are phylloxera, nematodes and eutypa.

Some grape pests are problems everywhere grapes are grown, while the significance of others, although profound in their localized areas, is not nearly as high on a global scale. Many of the regionally limited pests are contained because of climatic limitations on the disease, pest or vector of the disease. For example, Pierce's disease (of bacterial origin), does not proliferate within the vine in cool conditions (Feil and Purcell, 2001), and the geographical location of the vector (such as the glassy-winged sharpshooter) is also limited by temperature (Hoddle, 2004). Climate change is modifying the areas where pests can survive and become an issue. In the case of Pierce's disease, which is a problem in a wide range of horticultural crops, Giménez-Romero *et al.* (2024) predict the Mediterranean region is at high risk for the spread of the vector and disease from a 2°C rise in global average temperature. This is especially worrying because the bacteria causing the disease has been found (and subsequently eradicated) in Italy, France and Germany in the past, but is well established on Mallorca and other Balearic Islands (Olmo *et al.*, 2017). However, this chapter will concentrate on the more ubiquitous, global pests, but be mindful that localized pests can be the major problems in any given region.

Many diseases have been introduced from distant areas; for example, major diseases of grapes were introduced into Europe from North America where the wild grapes were tolerant; examples of this follow. The potential exists, and is greater than ever, for more pests to follow grapes to new growing areas, so the lessons of history should be heeded.

Powdery mildew is a fungal disease originating in eastern North America and first described in the US in 1834 and introduced to Britain, where it was first reported by a gardener named Tucker, which led to the European name of *Oidium tuckerii*. From there it made its way to France in about 1847, causing widespread damage: production losses of up to 80% due to the berry splitting and the degradation of grape flavours it caused. In 1853 sulphur was reported to control powdery mildew was and this technique widely spread with the 1856 publication of La Maladie de la Vigne (Perold, 1927). Sulphur is still the main fungicide used for powdery mildew control today.

Phylloxera, the root louse native to eastern North America, was probably first introduced into Europe in the 1850s when American vine material was brought in to help find a way of controlling powdery mildew (Campbell, 2004), and although the cause of grapevine death associated with the pest was unknown until the late 1860s, it radically reduced grape production in a relatively short time span. The French government offered a reward of 300,000 francs in 1873 to anyone who could find a way to stop phylloxera.

Prize-induced treatments ranged from the ridiculous to the universally lethal, but that prize was never claimed (Campbell, 2004). The failure to come up with a workable solution led to another startling decrease in wine production in France: from 83 million hectolitres (hl) in 1875 to 23 million hl in 1889. The best solution so far has been the grafting of European *V. vinifera* grapes to American rootstocks (initially) or to hybrid rootstocks. Phylloxera has spread to almost all grape-growing areas of the world (Chile remains one of the exceptions for the moment), causing the replacement of millions of vines as the own-rooted plants succumbed to the pest.

With the introduction of phylloxera-resistant rootstocks, other North American native pests were introduced as well, leading to a third wave of disaster for the French vignerons. Downy mildew, imported with the American grapevine material that was used to combat phylloxera, was first reported in France in 1878 and caused another downturn in French grape production until a control method was found. The story goes that a grower just outside the town of Bordeaux was having trouble with townspeople stealing grapes as they walked by his vineyard (Prial, 1987). He decided to paint clusters near the road with a greenish blue paste made by mixing copper sulphate and lime, which gave the leaves and fruit an unappetizing, splotchy, green–blue colour. It is not reported whether this deterred the pilfering, but a passing scientist observed that the grapes did not develop downy mildew. A few enquiries and experiments later, in 1885, he published results that showed the effectiveness of this *bouilli bordelaise* (Bordeaux mixture), and it is still used around the world for control not only of downy mildew but also other disease organisms.

Diseases

Disease in grapevines can be caused by a wide range of organisms, including those of fungal, bacterial and viral origin. The most common diseases of grapes will be discussed in this section.

Fungal diseases

On a global scale the diseases caused by fungi that are most of a concern are botrytis (bunch rot), powdery mildew, downy mildew and phomopsis. A discussion of each follows, but an expanded view of botrytis control is taken. Many aspects of managing botrytis can be applied to other grapevine diseases, so when considering a disease management strategy look for areas where these ideas have multiple uses.

Botrytis and fungal diseases in general

OVERVIEW Botrytis bunch rot or grey mould (caused by *Botrytis cinerea*) (see Fig. 10.1) is ubiquitous and exists in all vineyards of the world. Its genus

Fig. 10.1. Badly affected botrytis cluster, demonstrating how it acquired its alternative name of grey mould. Content author's own.

name was coined in the early 18th century and taken from the Greek term for a cluster of grape berries (*botrus*), so it is likely to have been associated with grapes for some time (Rosslenbroich and Stuebler, 2000).

The fungus favours cooler temperatures than many others, but warm and damp conditions are best for rapid development (Nair and Allen, 1993). Although generally detrimental to grape quality by causing rotten berries in table grapes and off-flavours in wine (Loinger *et al.*, 1977), in specific cultivars and climatic conditions it is known as 'noble rot' (Müller-Thurgau (1888), cited in Rosslenbroich and Stuebler, 2000), where the berry skins retain their integrity and shrivel, resulting in a concentration of sugars and flavours inside (see Fig. 10.2). The rot contributes to the production of the exceptional sweet wines of Hungary (Tokay), France (Sauterne) and Germany (Auslese, Beerenauslese and Trockenbeerenauslese). *B. cinerea* is not restricted to grapevines: it attacks many cultivated and wild plants and can live on dead tissue as a saprophyte (see Fig. 10.3) or on living tissue. Because of its wide host range and adaptability to different situations, it is hard to avoid in a typical vineyard situation.

In addition, just as there are different clones of 'Pinot noir' or 'Gewürztraminer', botrytis is made up of a population of similar, but slightly different organisms (van der Vlugt-Bergmans *et al.*, 1993), each of which may have a different capacity to grow, reproduce and survive attempts at eradication. Because there is a potentially diverse population of individuals rather than millions of identical clones, attempts to control 'botrytis' through a single method are usually met with only limited success.

Fig. 10.2. Botrytis infection contributing to 'noble rot' of some Riesling grapes. Note the shrivelled, but whole berries. Content author's own.

Fig. 10.3. Botrytis has infected frost damaged shoot tissue, creating a source of inoculum for later in the season. Content author's own.

Just as there is a choice between single-site (those that act on a particular point in the target's physiology) and multiple-site (those that act on more than one particular point) fungicides, there is also the choice between single-idea and multiple-idea strategies to manage botrytis. At every possible step, practices in the vineyard and how they may influence fungal progression must be considered. In order for this to be possible, a thorough knowledge of the life cycle of the disease organism is necessary.

Under harsh conditions botrytis can overwinter in hard, dark-coloured structures called sclerotia, which can resist extreme drought, heat and cold, and can be found on canes, rachis material or mummified grapes. In mild climactic areas, however, the fungus can easily survive between growing seasons as dormant hyphae inside buds, under bark etc., and resume growing when environmental conditions improve (Wilcox *et al.*, 2015).

Hyphae of botrytis grow under a wide range of conditions (even quite cool ones) and produce conidia-bearing structures called conidiophores, which are spores that can be spread by water, wind, and mechanical means (see Fig. 10.4). Ideal conditions for infection are 23.7°C and 20.8°C for leaves and berries, respectively, with free water or a high relative humidity (Nair and Allen, 1993). The time required for infection (the infection period) is from as little as 1 hour, depending on the tissue type, temperature and wetness, with drier and cooler conditions requiring longer amounts of time (Nair and Allen, 1993).

Fig. 10.4. Close-up of an infected berry, showing the conidiophore structures that release spores. Content author's own.

Various models have been developed to predict infection periods based on temperature and leaf wetness data (e.g. Nair and Allen, 1993; Broome *et al.*, 1995) or aerial spore counts in the vineyard (Rodríguez-Rajo *et al.*, 2010). Some of these have been used commercially, but more comprehensive models, which take into account growth stage of the fruit (González-Domínguez *et al.*, 2015) may be able to improve forecasting and further reduce chemical use.

Berries appear to be more resistant to infection than leaves due to the protection of the intact skin, although infection can happen through natural openings, such as stomata (in young berries these transform into less open, but still important for gas exchange, lenticels a few weeks after fruit set), and even through the cuticle itself (Northover, 1987), but any breach of its integrity results in rapid colonization (see Fig. 10.5; Coertze *et al.*, 2001). Mechanical injury is the most frequent origin of botrytis infection, and can be caused by hail, insects, birds or even powdery mildew infections. Mechanical leaf removers may also cause physical damage to berries in the process of removing the leaves.

Spring infections of leaves and young clusters are probably caused by spores produced from sclerotia in debris on the vineyard floor or in the vine (Nair *et al.*, 1995; Seyb, 2004). Under certain conditions, infections during bloom remain latent until veraison or later (Nair *et al.*, 1995), causing damage that appears shortly before harvest.

Fig. 10.5. Point infections of botrytis on 'Sauvignon blanc' berries. It is likely that the entry point was a defect or a stomatal pore in the berry skin. Content author's own.

Grape cultivars differ greatly in their susceptibility to botrytis bunch rot. Cluster architecture, microclimate around the berry, chemical composition (Hill *et al.*, 1981), including tannin concentration in the skin (Deytieux-Belleau *et al.*, 2009) and the possible intervention of phytoalexins (Langcake and McCarthy, 1979; Creasy and Coffee, 1988) are involved in cultivar differences.

Cultural practices – primarily promoting air movement and exposure of the clusters (leaf removal) – can reduce the severity of the disease, although chemical protection is generally required on susceptible cultivars. Chemical treatments frequently include applications at particularly important stages of development, such as bloom, before berry touch (closure of the cluster, to allow the fungicide to get inside the cluster) and just before harvest.

A coordinated approach is necessary to minimize the risk of infection by botrytis or any other disease organism. The goals for this are to: (i) minimize potential infection periods; (ii) achieve high vine health; (iii) reduce sources of inocula; (iv) monitor weather and the vines; and (v) use pesticides intelligently.

MINIMIZING POTENTIAL INFECTION PERIODS Site selection is the first point of call for many important management concerns, botrytis included. Consideration of the macro or mesoclimatic conditions and how they would favour botrytis growth may alter plans, including the possibility of planting elsewhere rather than have ongoing costs and increased risk of infection. Cultivar and rootstock choice is important to the potential growth of the vines and thus density of the canopies, and should be matched to the site, e.g. choosing a shallow-rooting rootstock for a deep soil. Loose-clustered cultivars or clones are desirable over tight-clustered types, as the fruit dries out more quickly, leading to fewer infection periods. If the weather consistently turns wet in the harvest period, an earlier-maturing cultivar or cultivar/rootstock combination should be considered.

Trellising and vine spacing should also be matched to the site, as spacing that is too close or trellising that is inadequate for the amount of growth at the site can result in vigorous and crowded canopies and, potentially, a greater number of infection periods. If vigour is expected to be high, a divided canopy should be considered, which can reduce individual shoot growth and open up the fruiting zone.

Row orientation has an impact as well. North–south oriented rows have east- and west-facing sides, and therefore the sun will spend half the day shining directly on one side and half the day shining on the other, leading to even light distribution and early drying of the eastern side in the mornings. East– west-oriented rows have a north and a south side to the canopy and only the one side (which one depends on which hemisphere the vineyard is in) gets direct sunlight. Fruit from the continually shaded side of canopies may be more susceptible to disease because of the lack of sunlight and greater fruit wetness that results.

Prevailing winds during the cropping season, particularly when the grapes are at their most susceptible (pre-bloom up to fruit set and again after veraison), can also be an important consideration in determining row orientation. Breezes down the row can aid in air circulation within the fruiting zone, reducing fruit wetness, so one consideration for row orientation would be the prevailing wind direction during the season.

Inter-row crop choice and irrigation and fertilization regimes influence vegetative growth of the vine. Vines will grow most with luxurious levels of water and fertilizer and with bare soils, so limiting the amount of water to which the vines have access is a good way of managing this. Managing water inputs can be done directly if irrigation is used and/or indirectly, e.g. by altering between-row cover crops. Soil moisture levels can be controlled to change vine growth (see Chapter 6). In soils with high water availability, cover crops with deep rooting systems, such as chicory, can be very effective in drawing water away from vines and thus reducing vegetative growth (Caspari *et al.*, 1997), with concomitant effects on reducing the frequency of botrytis infection periods.

One major goal of leaf removal is to reduce the incidence of botrytis. As such, there can often be significant benefit from removing leaves from just the east side of a N–S-oriented row. This exposes the fruit to the morning sun, helping to dry out the area more quickly. On an E–W-oriented row it is usually best to take leaves from the shaded side of the canopy, to improve air movement where the sun has little influence and allow some protection from exposure for the vines on the sun-side. Be careful where the discarded leaves end up – if left within the canopy, they can cover clusters and promote humidity underneath.

Shoot thinning to remove non-count shoots and open up the head area is invaluable for ensuring the canopy you planned on at pruning actually develops in the summer. Shoot positioning is equally important to prevent crowding in the fruit zone. Additionally, if your shoots are tending to run horizontally along the row (or there are excessive numbers of laterals developing in the fruiting zone), the leaf density in the fruiting zone will be greatly increased, which will enhance conditions conducive to botrytis development.

Higher yield is also associated with greater risk of botrytis, as a fruiting zone heavily laden with clusters provides many places for free water to accumulate and persist. Overlapping clusters should be avoided, even to the point of instructing hand crews to reposition clusters laying on other clusters or remove fruit in areas where there is too much. These clumps of clusters are hot spots for the development of botrytis and should be avoided.

As mentioned earlier, cluster morphology can have an impact on botrytis infection, where open and loose clusters allow for better spray penetration and drying. High percent fruit set contributes to tight clusters, so one way of opening the cluster, aside from flower or berry thinning, which is a laborious task, is to reduce the percent fruit set. This can be accomplished with pre-bloom leaf removal, where the reduced carbohydrate availability results in fewer berries being set (Poni *et al.*, 2006). This reduces yield but has been found to be effective in lowering losses to bunch rot and has positive effects on fruit composition. The reduction in botrytis from using mechanical harvesters to crop thin after fruit set has already been mentioned (Chapter 6)

A wide, dense canopy presents a problem in disease control as it decreases the amount of light reaching inside, reduces air movement, increases relative humidity inside and decreases spray penetration. All of these contribute to increased risk of botrytis incidence via a greater likelihood of infection periods occurring. It is therefore important to monitor vine vigour and canopy density as part of a disease management programme.

ACHIEVING HIGH VINE HEALTH Other ways to combat botrytis infection are to take advantage of what nature has to offer. Aside from the physical properties that vines have to repel infection attempts (such as the cuticle (Kolattukudy, 1985)), vines possess chemical mechanisms to slow or stop botrytis development. There are also other microorganisms that can prevent botrytis from becoming established in the grapes, but which do not interfere with production.

If the fungal spore's hypha has successfully overcome the physical barriers or has found damaged tissue, chemical barriers then come into play. Plants – particularly woody ones like grapevines – produce a number of secondary metabolites that may contribute to a passive resistance capability. These preformed phenolic compounds may have a significant influence on the success or failure of an infection attempt (Jersch *et al.*, 1989). For example, tannins found in grapes may bind to enzymes produced by the fungi during the infection process (Goetz *et al.* 1999), thus preventing them from acting on the plant cells.

Compared to fruit, grapevine leaves generally are more resistant to infection as they age – perhaps because they accumulate relatively high concentrations of lignin and other polyphenolic compounds (Prins *et al.*, 2000). Grape berry resistance varies according to developmental stage, the type of pathogen attempting infection and the grape cultivar.

Part of the reason for this could be the active responses of the vines to attempted disease infection. There are a number of these, and it must be emphasized that one or more may be occurring at the same time. One visible example of an active response is hypersensitivity (Kombrink and Schmelzer, 2001). In this case the plant isolates the infected tissue by producing impermeable substances, like lignin, around the area and/or by accumulating toxic substances that kill the affected tissue, thereby containing the infection.

Pathogenesis-related proteins (PRPs) are induced upon infection to a variety of pathogens including viruses, bacteria and fungi. In the grapevine, *B. cinerea* induces the production of a number of different PRPs (Renault *et al.*, 1996) that appear to act upon fungal cells. Chitin and glucans are structural components of most fungal cells, and chitin in particular is not found in plant tissues. The production of PRPs that have enzymatic activity and break down fungal cell walls may be an important defence response by the plant (Robinson *et al.*, 1997). It appears that fungal hyphal growth is inhibited by the enzymes, supporting this theory (Giannakis *et al.*, 1998).

Developmental stage of the tissue may also influence its capacity to produce PRPs. For instance, their production in grape berries was greater after veraison once the berries had begun the ripening process (Robinson *et al.*, 1997). This may change the susceptibility of berries to botrytis during development.

While the presence of pre-formed phenolics may be a barrier to infection, the synthesis of more phenolic compounds following infection can also slow or stop the pathogen. One advantage to this class of response is that it can happen very quickly, as the compounds do not need to be evolved from scratch, but rather can be modified versions of molecules that are being made already, such as from primary metabolites. The effect of this is that cell walls near the infection site are rapidly 'reinforced' against fungal penetration and able to wait for the more permanent formation of lignin, which can take in the order of days, rather than hours, to occur.

Phytoalexins are plant-produced compounds that inhibit microbial growth and are produced rapidly upon attempted infection by a pathogen or exposure to UV radiation. Many phytoalexins are phenolic in nature, and the known examples in grapevines belong to a class called stilbenes. Resveratrol is the first stilbene to be synthesized by grapevines and the first to be discovered as a fluorescing compound made along the margins of botrytis infections on leaves (Langcake and Pryce, 1976). Modifications to this molecule produce a variety of other compounds such as piceid, pterostilbene and ε-viniferin, which accumulate more slowly than resveratrol. The speed at which the vine can accumulate antifungal substances like resveratrol may influence the establishment of the fungus (Creasy and Coffee, 1988), and therefore be important in determining the success or failure of the infection attempt. The vine's environment also has an influence, as it has been suggested that the cyclic content of resveratrol in berries mirrors rainfall events (Creasy and Creasy, 2016).

Leaves and newly formed berries are able to produce resveratrol quite effectively, but for fruit this ability drops as the berries enter the ripening phase (see Fig. 10.6) (Creasy and Coffee, 1988). UV radiation trailers are available for pulling through the vineyard, partly to induce a phytoalexin response but also to kill disease organisms directly. Note, however, that the vine response is ephemeral, lasting up to a week in leaves and half that time in green fruit (Creasy and Qi, 2005) and localised to the tissue exposed to the radiation, so its practical effect is probably not worth the cost of taking the equipment into the vineyard. Gadoury *et al.* (2023) seemed to confirm this, finding that UV radiation was able to control powdery mildew to the same extent as fungicide, but it took two passes per week, and at night, to do so.

Coincidentally, the stage of development that berries lose their ability to manufacture phytoalexins is the same phase at which PRPs start to be manufactured (Robinson *et al.*, 1997), which means that grapes have some sort of defence throughout their development. However, if there is a breach of the berry skin, phytoalexins and PRPs will have little effect as their production is primarily in the skin (Grimplet *et al.*, 2007).

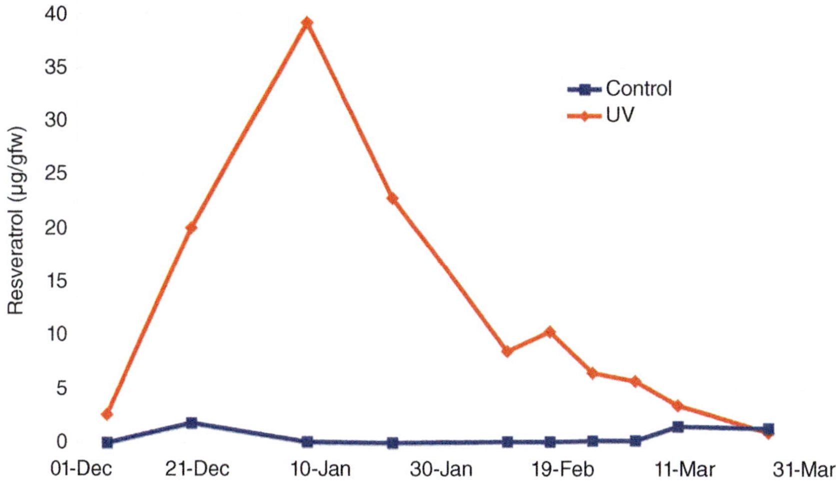

Fig. 10.6. Potential of grape berries to produce resveratrol in response to UV irradiation through the (southern hemisphere) season (the difference between the UV and control lines), which is taken to be indicative of the berry's response to botrytis infection (GL Creasy, unpublished data).

However, botrytis may be able to combat the effect of stilbene production by the vine. One enzyme secreted by the fungus is stilbene oxidase, which can cause the degradation of resveratrol, pterostilbene and perhaps other stilbenes such that they lose their antifungal properties (Hoos and Blaich, 1988; Adrian *et al.*, 1998).

Induced resistance and systemic acquired resistance are concepts based on the notion that plants can be 'immunized' against future infection by pathogens (Hunt *et al.*, 1996). This is similar in concept to human immunization, where we can be inoculated in the arm and then, for some time afterward, be resistant throughout our bodies. This response can occur in plants as a result of an infection, where an active defence response occurs remote from the infection site, or it can be induced by application of a chemical or other signal. For example, treating vines with fosetyl-aluminum increases the ability of those grapevines to produce resveratrol if an elicitation event occurs after the application of fosetyl-Al (Creasy and Qi, 2005). There is much interest in the area of induced resistance, as environmentally benign substances could be substituted for fungicides that have a direct action on botrytis.

Development of a fungus may also be inhibited by organisms that occupy the same environmental niche as botrytis, such as *Trichoderma* spp. The presence of these non-pathogenic fungi living on plant surfaces inhibits the growth of a number of plant pathogens by taking up the space that botrytis would use, or possibly interacting directly to prevent it from getting established or even inducing a systemic response from the plant tissue (Bélanger *et al.*, 1995; De Meyer *et al.*, 1998; Elad and Kapat, 1999). Advantages to using *Trichoderma* are their short withholding periods and their eco-friendly nature (being able to be used in organic production in most cases). Disadvantages are in finding a strain of the fungus that will survive on grapes in the field, getting high enough populations of them on the berries to compete with botryis and possible interactions with use of fungicide. Efficacy of this type of control is also highly dependent on environmental factors, such as temperature and relative humidity, and so what works in the laboratory often does not translate well to commercial situations (Hjeljord *et al.*, 2000). Finding the right strain for the job is extremely important.

Despite potential problems, there remains a lot of interest in biocontrol strategies. Many trials indicate that these methods can be just as, or even more, effective than a regular spray regime, bearing in mind that they must also be used in combination with cultural practices.

REDUCING SOURCES OF INOCULA Cleanliness in the vineyard can contribute to reducing the chance of infection. Up to 90% of leaf trash found in some New Zealand vineyards was found to be infested with botrytis (Seyb, 2004). Old tendrils, petioles, berries, canes, etc. have been found to harbour botrytis, which sporulates in the following growing season. Therefore, machine-harvested vines can carry over a large amount of botrytis if cluster remnants are not removed at pruning (Fig. 10.7). Material that may contain botrytis can be taken away from the vineyard and buried or burned, or it can be mulched into small pieces (<2 cm lengths of cane) and left to decay on the vineyard floor. In drier climates, a cover crop will aid in this process due to increasing the amount of morning moisture on the vine tissues (Jaspers *et al.*, 2013).

Research in New Zealand has highlighted the effectiveness of mechanical thinning to reduce botrytis incidence, even if this thinning is done shortly before veraison. The mechanism for this is more open clusters and therefore reduced chance of infection periods, but it also appears the remaining fruit is more resistant to infection for other reasons (Mundy *et al.*, 2021). This work has also led to the practice of shaking the vines shortly after fruit set where to goal is not to reduce crop load, but remove bunch trash (calyptra, unfertilised ovaries, etc.). This has resulted in a consistent and significant reduction in botrytis at harvest, to the order of 50% in both low- and high-pressure years (Nicholson, 2015).

MONITOR WEATHER AND THE VINES A weather station can provide more than just a record of rainfall events, wind direction, etc. The data collected can be used in botrytis infection forecasting models, which take into account the weather conditions that are needed for botrytis spores to germinate and note when an infection period has occurred (Broome *et al.*, 1995). Rather than using a preventative spray programme, fungicides are applied only when needed, thus saving money on purchases of the chemicals as well as their application, especially in low-disease pressure seasons.

However, the best way to stay on top of what is happening in a vineyard is to spend time in it! There is no substitute for getting out there and getting into the canopy. Weather stations and disease forecasting models are wonderful

Fig. 10.7. A cluster from the previous season acts as a source of disease inoculum. If present in the canopy as in this example, it is in a prime location to spread spores onto sensitive tissues. Content author's own.

innovations, but they cannot actually look through the vines and see that botrytis is developing in the clusters. They also cannot tell how effective the spray coverage has been, so remaining vigilant and proactive is still a must.

Keeping records of the location of trouble spots in the vineyard will also pay off. If there is one area that is particularly prone to botrytis infection (a hot spot), it can be used as an indicator. If disease starts developing there, then the rest of the vineyard is probably at risk.

USE PESTICIDES INTELLIGENTLY Use of chemical controls should not be the first line of defense: the goal should be to minimize the impact on the environment and the costs associated with purchasing and applying products. Many synthetic pesticides have a high per-hectare cost of application, and even under organic regulations, the cost of copper-based products is rising as governments try to reduce their use (Anon., 2020). Organic-approved insecticides, for example costs (at time of writing) over 70€ per hectare per application, with three applications mandated per season (re: vector control for flavescence dorée in France).

Once that becomes necessary, however, deciding when and how to apply them is a crucial point. For efficient use, details on the life cycle of each target organism should be studied and the information applied when making a pesticide programme for the season. In the case of botrytis, this means applying chemicals at the key points of control (flowering and pre-bunch closure), keeping track of weather than induces infection periods, being ready to apply when needed, and verifying that the chemical has reached its target. Further details on pesticide application are given towards the end of this chapter.

Powdery mildew

This disease is also called oidium and is caused by *Erysiphe necator* (alternate name *Uncinula necator*) (see Fig. 10.8). If inoculum is present in the vineyard

Fig. 10.8. Powdery mildew. Top: infection on overwintering canes (dark areas). Bottom: a flag shoot, which occurs when the fungus infects the pre-formed shoot inside the compound bud. Notice the dusted-with-flour appearance, stunted growth (resulting in the zig-zag pattern of the shoot axis) and leaf cupping. Content author's own.

it will develop at the first instance that conditions are favourable during the season. Initial spring infections are due to spores produced by overwintering fungal bodies called cleistothecia in the buds or on the surface of the vine. The fungus spreads to new green tissue after the production of asexual spores, which can be wind-blown or splash transported. Powdery mildew is a fungus that grows on the epidermis of green vine tissues. Special fungal organs called haustoria enter epidermal cells to nourish the fungus (Wilcox *et al.*, 2015), and the cells must be living for the fungus to survive and reproduce (i.e. it is an obligate parasite). Temperatures of 20–27°C and dry conditions are optimal for infection and disease development, unlike those for some other common fungal diseases. Infection forecasting models based on temperature and leaf wetness have been developed for powdery mildew (e.g. Thomas *et al.*, 1994; Gadoury *et al.*, 1997) and with relatively minor tweaking (Gubler *et al.*, 1999), they have been successfully used in many areas around the world. Use of these models avoids a calendar-based spraying programme, instead, spraying only after there has been an infection period, thus reducing chemical use and cost.

Powdery mildew looks like a white powder on the surface of the plant. Leaf powdery mildew is usually seen on the upper surface, but can appear on both abaxial (lower) and adaxial (upper) sides (Fig. 10.9). All green parts of the vine are susceptible. Infected leaves often appear less green and suffer from reduced photosynthesis (Lakso *et al.*, 1982), which can debilitate the whole vine. Infected areas can turn dark, for example, on shoots (see Fig. 10.10), which acts as an inoculum source for the next season if retained after pruning.

The most economic damage, however, is from cluster infections, where infected berries fail to ripen, are prone to splitting and the infected rachis becomes brittle. Once berries approach veraison, however, they are not susceptible to new infections. *Vitis* species vary in their susceptibility to powdery mildew, and choosing resistant cultivars might be part of a management scheme if the grower is willing to grow these types of vines. The desirable *V. vinifera* grapes are susceptible, but most North American and Asian species are resistant (Zhang *et al.*, 2014). Hoffmann *et al.* (2008) found that 'Kishmish vatkana', a table grape grown in Uzbekistan, was extremely resistant to infection, tracing it to highly localized responses at the point of infection. These could be due to cell death (which, as powdery mildew is an obligate parasite, would stop the infection) or production of callose or oxidised phenolic compounds. A California study estimated that using powdery mildew resistant grapevines could save growers from 2% to 6% of their total production costs, depending on the type of grape being grown (Fuller *et al.*, 2014), so there is considerable value in investigating this area further. Taking advantage of some *Vitis* species' natural resistance to powdery mildew, some breeding programmes have developed "mostly" vinifera vines that reduce the number of sulphur treatments needed per season. Some of these are branded as PIWIs (based on a German word for "fungus resistant": pilzwiderstandsfähiger), developed in Europe and now being planted, and even acceptable as part of a blend in some appellations' wines. From an environmental and economics of production point of view, adoption of these resistant cultivars would be logical,

Fig. 10.9. Top left: powdery mildew infection at bloom. The leaves are slightly yellow and cupping, and the flowers/ developing berries have been infected already. Top right: powdery mildew cleistothecia, the tiny black specks, on the underside of a 'Sultana' grapevine leaf. Bottom: late season infection of second set fruit. Berries have split due to the infected skin not being able to expand to accommodate the growing mesocarp. Content author's own.

Fig. 10.10. Mid-season powdery mildew infection on shoots. Content author's own.

but consumer acceptance remains the barrier to widespread planting, though with increasing awareness of climate change and environmental issues, there is hope (Vecchio *et al.*, 2022).

Cultural practices can control the powdery mildew under low-disease pressure and improve the control success when chemicals are used. These strategies rely on keeping the canopy open so that there is good air movement; however, if spores are present in the area (for example, if a neighbouring vineyard has powdery mildew), some form of chemical control will be necessary due to the continued exposure to inoculum. UV-B radiation has been shown to inhibit spore germination and mycelial growth of powdery mildew (Willocquet *et al.*, 1996), so open canopies may be beneficial for more than just better air movement and spray penetration.

Powdery mildew is typically controlled by either a programme that includes one or more modern fungicides or by elemental sulphur. Sulphur is active both as a vapour and through direct contact with the fungus. If spray coverage is not complete, control is not as good at lower temperatures due to lower volatilization of the sulphur (Wicks *et al.*, 2003). Sulphur causes phytotoxicity in some cultivars of grapes (e.g. those of *V. labrusca* origin) and is thought to be phytotoxic to *V. vinifera* vines at high (>32°C) temperatures, particularly if applied in dusting form. Some research suggests that there is little danger when using wettable sulphur, though the influence of relative humidity on toxicity needs to be investigated (Magarey *et al.*, 2003). In any case, leaf damage and berry scarring can still occur under extreme circumstances

(Fig. 10.11). For effective control, any tissue to be protected must be covered with sulphur, so thorough spray coverage is necessary (hence the importance of canopy management for spray penetration), as well as reapplication to newly emerging leaves or surfaces where the sulphur has been washed off. Sulphur can also disrupt the life cycles of beneficial insects and arthropods (spiders, mites, etc.) in the vineyard, and sulphur residues on berries are associated with off-flavours in wines (Thomas *et al.*, 1993). Fungicide-grade sulphur can persist on the berry surface, so to minimize the risk of it interfering with wine quality, it is recommended that no sulphur be applied to grapes intended for red wine production within 8 weeks of harvest (Kwasniewski *et al.*, 2014).

Modern synthetic fungicides (such as demethylation inhibitors (DMIs) or quinone outside inhibitors) have the advantage of being functional over a wider range of temperatures and with less phytotoxicity (Wilcox *et al.*, 2015), and of course must be used on sulphur-sensitive cultivars and/or when high temperatures are expected. However, the development of resistance to both of these chemical families has already occurred (likely due to their specificity in action (Köller and Scheinpflug, 1987), which has resulted in recommendations to limiting their use in spray programmes (Beresford *et al.*, 2016).

Fig. 10.11. Leaf necrosis and berry scarring caused by high heat (~40 °C) after application of wettable sulphur. Content author's own.

Potassium and sodium bicarbonates have been used in viticulture and reported to control powdery mildew as effectively as sulphur (Kauer *et al.*, 2001). Mono-potassium phosphate fertilizer has been used successfully to manage powdery mildew, and its alternation with DMI fungicides has resulted in control equivalent to the use of DMIs alone (Reuveni and Reuveni, 2002), which thus reduced the amount of DMI used. Phosphates have also been trialled (Reuveni and Reuveni, 1995) and found to be effective, perhaps more so than bicarbonates. Other alternatives to sulphur and synthetic fungicides include milk or whey (Crisp *et al.*, 2002, 2006), which were effective at controlling powdery mildew, possibly in part due to antimicrobial activity in the milk products.

Certain oils of low phytotoxicity (such as JMS Stylet-Oil® (JMS Flowers Farms, Inc., Florida), which is a refined form of paraffin) can provide adequate control (Dell *et al.*, 1998). However, these oils inhibit vine photosynthesis in the order of 20% or more (see Fig. 10.12; Finger *et al.*, 2002) which is severe enough to affect grape composition even if used only a few times in a season. There appear to be cultivar differences of this effect, with a *V. labrusca* grape ('Catawba') being much more affected than the *V. vinifera* grapevine 'Chardonnay' (Baudoin *et al.*, 2006) and the latter more seriously affected than 'Cabernet Sauvignon' (Finger *et al.*, 2002). The presence of trichomes (fine hairs) on the underside of leaves appears to be related to the severity of photosynthetic decrease (Baudoin *et al.*, 2006). Due to toxicity and possible plant damage, oil and sulphur must not be used in close succession: at least 10 days should separate their application.

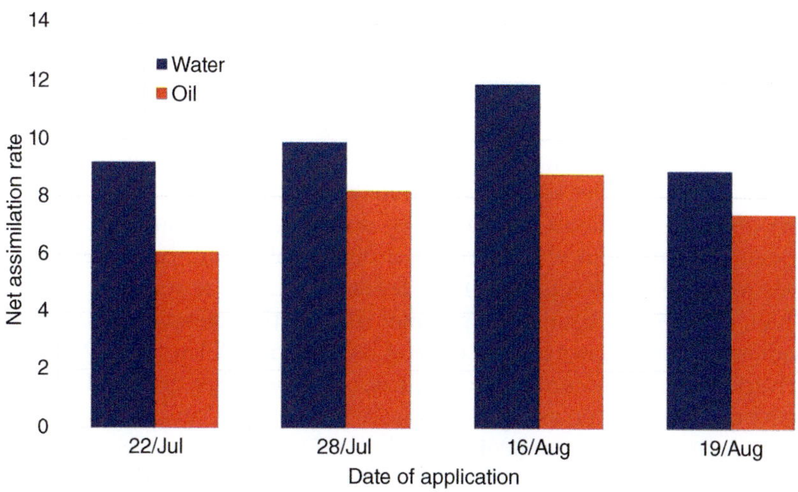

Fig. 10.12. The effect of application of a horticultural oil on 'Chardonnay' vine net assimilation rate at four different dates during one season. Differences are statistically significant (p-values 0.02 or less in all cases) (adapted from Finger *et al.* © 2002 American Society for Enology and Viticulture. AJEV 53:116-124).

Chitosan, a derivative of a fungal cell wall component, has been applied to grapevines and found to decrease powdery mildew incidence almost as well as synthetic fungicides (Iriti *et al.*, 2011). Brulé *et al.* (2024) confirmed that the mode of action was through stimulation of resveratrol and chitinase production, and there was also an increase of fruit phenolics and antioxidant capacity (Iriti *et al.*, 2011).

Biocontrol of the fungus is another possibility A parasitic fungi, *Ampelomyces quisqualis*, has been trialled and found to have promise, though how well it copes with a typical vineyard fungicide programme is still uncertain (Hofstein *et al.*, 1996; Németh *et al.*, 2023). *Trichoderma spp.* are also a possibility (Sawant *et al.*, 2017) and have the added benefit of being more adapted to real-world vineyard situations. Mycophagous mites that feed on the fungus may also have promise (Falks *et al.*, 1995; English-Loeb *et al.*, 1999a, 2007). Interestingly, one study reported that these mites were present on wild examples of grapes, but not commercially grown ones, which was thought to be due to the use of sulphur and other fungicides (English-Loeb *et al.*, 1999a). This provides food for thought when it comes to rethinking disease control programmes, where beneficial fungal or insect populations can be reduced as a result of controlling the target organism, increasing the reliance on the control substance.

Nitrogen and UV radiation were also found to modify grapevine susceptibility to powdery mildew, with higher UV and lower N resulting in less disease (Keller *et al.*, 2003). This is thought to be due to the higher levels of phenolics, particularly flavonols and hydroxy-cinnamic acid derivatives, synthesized in the epidermal cells under these conditions, as well as the direct effect of UV radiation on spore germination and hyphal growth (Willocquet *et al.*, 1996).

Downy mildew

Downy mildew (*Plasmopara viticola*) causes leaf, shoot and cluster damage. The upper surface of the leaf will have a characteristic 'oil spot' appearance and sporulation on the lower surface, which looks downy white (see Fig. 10.13). Berries and infected shoots have a similar downy white appearance and discoloration. Downy mildew is primarily a disease of warm, humid growing regions as these conditions favour its development. The absence of rainfall and high humidity in spring and summer reduces the spread of this disease in Mediterranean climates such as Chile and California (Wilcox *et al.*, 2015).

Overwintering spores of downy mildew, called oospores, germinate in the presence of water as soon as the temperature reaches 11°C and, shortly thereafter, produce a sporangium (Wilcox *et al.*, 2015). This asexual reproduction organ produces clonal spores dispersed, usually, by rain splash. The fungus enters the plant through the stomates, its hyphae developing inside the leaf, and eventually a spore-producing body emerging through the stomatal pore to disperse propagules by wind and water splash (Langcake and Lovell, 1980). Interestingly, such is the importance of this opening being the entry point

Fig. 10.13. Downy mildew fungus on the upper and lower surfaces of grapevine leaves and on mid-season fruit. Content author's own.

for the fungus, once the berries reach the point at which its stomates become lenticels (about 3 weeks after fruit set), they are no longer susceptible.

Because free water is so crucial to spore germination, rain and high humidity are the most important factors in promoting epidemics of downy mildew. The sporangia are sensitive to desiccation and sunlight (Kast and Stark-Urnau, 1999; Rumbolz *et al.*, 2002), so the first point of control is canopy management to hasten the drying-out of the leaves and fruit, along with ensuring there is as little shade within the vine as possible.

There are a number of downy mildew forecasting models that work on the basis of temperature, moisture and other factors (Magarey *et al.*, 1991; Orlandini *et al.*, 1993; Dubuis *et al.*, 2012; Reis *et al.*, 2013), but these are mostly useful in those areas where the mildew is not very prevalent. If you are in a humid and warm region, the disease pressure can be such that regular spraying is unavoidable.

Control of downy mildew is primarily by fungicides containing copper (e.g. Bordeaux mixture, copper oxychloride). Copper is phytotoxic to plants, but the addition of lime reduces this, hence the creation of Bordeaux mixture. Some cultivars are more sensitive to copper than others, and application at high temperatures is more likely to cause phytotoxicity in any grapevine (Pertot *et al.*, 2006). Leaf damage through this (as with sulphur) results in reduced photosynthesis and low vine and berry quality. Some fungicides with curative (ability to kill fungi after they have infected the plant tissue) and systemic (taken up into the tissue and transported for some distance into other, non-treated cells) ability specific for this family of fungi, such as phenylamides (e.g. metalaxyl or banalxyl) and carboxylic acid amides (dimethomorph or mandipropamid), have been developed specifically for downy mildew. As with any chemical control, use should be managed to minimize the development of resistance within the fungal population: the story with chemicals for downy mildew control is strikingly similar to that of powdery mildew.

Copper-containing pesticides have been used for many decades in some vine-yards, resulting in high copper concentration in the surface soil layer (Parat *et al.*, 2002; Pietrzak and McPhail, 2004). In a French study, for example, some vineyard soils were found to have up to 1500 mg Cu/kg soil (Flores-Vélez *et al.*, 1996), which is many, many times higher than that considered normal (Pietrzak and McPhail, 2004). Recognizing the environmental impacts of this, in 2018 the European Union put a limit on the application of copper in vineyards, with a maximum of 28 kg over the course of seven years (EU Commission Implementing Regulation EU 2018/1981), which allows for higher rates of use in high disease pressure seasons, as long as this is made up with lower use in other seasons. In addition, taxes on copper-containing fungicides have increased, encouraging efficient use and exploration of other means of controlling downy mildew.

Alternatives to copper or synthetics include phosphonic acid and its derivatives, which were recognized to have activity against downy mildew by 1980 (Raynal *et al.* 1980). While they have no pre-infection activity, some of these have considerable ability to kill the fungus post-infection (17 days according to Magarey *et al.* (1990)) and may stimulate the production of stilbenes (Raynal *et al.* 1980). There is also a non-protein amino acid that triggers a stilbene response in the vine (Reuveni *et al.*, 2001; Slaughter *et al.*, 2008), as well as chitinases (Brulé *et al.*, 2024). Treated areas with infection showed localized death of cells, a reduced rate of development and sporulation compared to the controls. As with powdery mildew, it seems stimulating the vines' innate antifungal capacity may be a useful tool in disease management.

Biocontrol through fungi such as *Trichoderma spp.* and bacteria in the *Streptomyces* (El-Sharkawy *et al.*, 2018) and *Bacillus* genera (Zhang *et al.*, 2017) were found to provide at least some control of downy mildew, even under field conditions.

Downy mildew resistant cultivars have been discovered or developed from the 1800s (Gessler *et al.*, 2011) and are quite successful at fending off infection, either completely or with less assistance from chemical control. American species and some crosses with *V. vinifera* can grow quite happily without chemical control, such as 'Catawba', 'Marechal Foch' and 'Chambourcin', though the quality of the wine is not sufficient for many consumers (Gessler *et al.*, 2011). Delmas *et al.* (2016) reported that three more recently developed resistant cultivars, 'Regent', 'Prior' and 'Bronner', showed lesser infections than 'Cabernet Sauvignon'; however, while this is very promising, there are signs that the fungus could overcome this resistance. The PIWI cultivars mentioned earlier also have downy mildew resistance, and like those, there are the Bouquet varieties developed specifically for southern France, the latest generation of which will incorporate multiple genes for more durable resistance to downy and powdery mildews (Bouquet *et al.*, 2000).

Phomopsis

Phomopsis cane and leaf spot (*Phomopsis viticola*) is found in all major grape-growing regions of the world (Wilcox *et al.*, 2015). It was formerly thought to be the sole cause of a disorder called dead-arm of grape, but later this was confirmed to be due to two separate organisms that occurred at the same time, phomopsis and eutypa dieback (see below). Phomopsis prefers dampness for infection and, if this occurs during budbreak and early shoot development, there is a greater chance of developing serious problems in the vineyard. Although the fungus does not often cause damage to berries or clusters, it is able to if conditions are wet enough for it to become established early on (Nicholas, 2004; Wilcox *et al.*, 2015).

Symptoms include scarring and dark, narrow lesions in dormant canes and, in the growing season, black longitudinal cracks developing light-coloured centres (Fig. 10.14). Petioles may develop similar lesions. Affected leaf tissue develops brown, roughly circular necrotic spots surrounded by a yellow margin.

As for the other diseases of note, phomopsis has been carried into new growing areas by propagation material. The fungus overwinters in the shoot periderm (which is frequently described as having a bleached appearance, though DNA examination of non-bleached canes also found phomopsis present (Melanson *et al.*, 2002)) and dormant buds. Spores are spread by rain splash, and any green tissue can be infected (Cucuzza and Sall, 1982; Pscheidt and Pearson, 1989a). Historically, younger tissue was thought to be more susceptible

Fig. 10.14. Phomopsis symptoms on: top: late season canes; bottom left: mid-season shoot; bottom right: leaf. Content author's own.

than older, but some work suggests that tissue of any age is equally susceptible (Erincik *et al.*, 2001), while other research on berries shows that, indeed, younger fruit is more susceptible than older (Pscheidt and Pearson, 1989b).

Dead, infected wood is a rich source of spores, and therefore any infected material should be removed from the vineyard. For example, the pruning stubs at the base of the vine are a source of spores that can infect the basal shoots needed for trunk renewal. Machine or minimal pruning leaves lots of dead material in close proximity to susceptible tissue, which can be a major source of infection. Infected shoots are weak and can be more easily broken, but the major loss is caused by infection of the rachis, which can kill part of the cluster and prevent further berry development. If weather conditions are sufficiently wet, phomopsis can infect the fruit and contribute toward significant rot before harvest. The rot may be due to latent infections established shortly after bloom but, although latent infections have been identified in other phomopsis species (Williamson *et al.*, 1991; Velicheti *et al.*, 1993), it has not yet been reported as occurring in grapevine.

Control primarily involves keeping the canopy open so it can dry out as quickly as possible, and removing infected material, prior to the times of peak sporulation and when the vine tissue is at its most susceptible (spring). Infected tissue can be burned, buried or mulched onto the ground. There are some chemical sprays available, such as those containing benzimidazoles, dithiocarbamates, or captan, but eradicant activity of any fungicides does not seem likely (Nita *et al.*, 2007). In organic production, a dormant season spray of copper or lime sulphur has been recommended, and sulphur is sometimes registered for use shortly after budbreak. A *Bacillus subtilis* product has also been registered for use against phomopsis, and there is promise of the use of *Trichoderma spp.* (Sofia *et al.*, 2024), and myricetin (a flavonol, see Chapter 3, Yuan *et al.*, 2024). Weather-based modelling systems are now accurate enough to use for determining when to apply fungicides (Gonzalez-Dominguez *et al.*, 2022).

Other fungal diseases of importance

BLACK ROT (*GUIGNARDIA BIDWELLII*) Native to north-eastern North America, this disease is also an introduced pest problem in most other grape-growing areas of the world aside from Chile and Australia (Wilcox *et al.*, 2015). Although shoot and leaf symptoms occur (Fig 10.15), they are not usually an economically significant problem and appear not to be correlated with fruit infection (Jermini and Gessler, 1996). Severe economic damage occurs from infection and death of rachis tissue or fruit rot. Rot in the latter case progresses to shrunken, black, mummified berries, unsuitable for any use. Fortunately, fruit becomes resistant to new infections about 5 weeks after bloom (Hoffman *et al.*, 2002), though this may vary with species and cultivar. As with many other diseases, wet conditions favour infection, with warmer temperatures needing less time for this to occur. Control is through keeping a canopy that dries out quickly, removal of infected material (e.g. infected cane or cluster material from last season) or the application of synthetic fungicides with protectant activity or myclobutanil, which has some eradicant capability (Hoffman *et al.*, 2004).Under organics, lime sulphur application pre-budbreak is recommended, but not entirely effective, and after that, copper can control the disease, but is not as effective as synthetic fungicides and coverage must be maintained in high-pressure areas (Wilcox *et al.*, 2015).

ANTHRACNOSE (*ELSINOË AMPELINA*) Also known as black spot, this disease is of European origin and a particular problem in rainy, humid regions. After the introduction of Bordeaux mixture in 1885 and its extensive use to control

Fig. 10.15. Black rot symptoms on leaves and shoots. In the bottom right you can see the dark pycnidia, which can release spores later in the season. Content author's own.

downy mildew, the severity of the disease was greatly reduced (Wilcox *et al.*, 2015). It infects *V. vinifera*, *V. labrusca* and muscadine grapes (Schilder *et al.*, 2005; Yun *et al.*, 2007). Symptoms include dark necrotic leaf spots, and if this occurs early in leaf development, the leaves will develop a 'crumpled up' appearance with holes in areas earliest affected (Fig. 10.16; Nicholas *et al.*, 1994; Wilcox *et al.* 2015). Shoots, petioles, rachis and tendrils can develop roundish patchy, light-coloured lesions with dark margins that may coalesce as infections progress. Serious infections result in raised lesions that are not as long and thin as phompsis (Fig. 10.17). Fruit can also be affected, developing circular spots with a light centre and dark margins, sometimes described as 'birds' eyes'.

Sanitation is a primary management measure, with (as for most fungal infections) failure to removed infected berries or stems on the previous season's canes being a high-risk activity. Modern synthetic fungicides are capable of controlling anthracnose in a way similar to phomopsis and black rot,

Fig. 10.16. Anthracnose-derived necrotic spots on a leaf and an early-season shoot and petioles. Content author's own.

although, as usual, there is concern about resistance. The latter is why DMIs are not recommended for control even though anthracnose may be on the label. Copper-containing chemicals are effective, and since they are used for control of other fungi that are common in humid growing regions, specific management of this disease is usually not needed. In problem areas, pre-budbreak application of lime suphur has also been recommended.

Fungal trunk diseases

EUTYPA DIEBACK (*EUTYPA LATA*) This is a disease that infects the woody tissues of grapevines and many other woody plants, and which is found in areas with rainfall greater than 350 mm per year (Wilcox *et al.*, 2015). The symptoms include weak and stunted shoot growth (Fig. 10.18), particularly noticeable

Fig. 10.17. Anthracnose on berries and the base of a cane. Content author's own.

early in the season. Shoots can grow in a zig-zag pattern if badly affected, with small, cupped and frequently yellow leaves. The organism affects the plant's vascular system and causes a slow vine decline, where the vigour of the plants reduces each year. In some cases, this temporarily results in greater fruit quality, as the amount of vegetative growth is lessened leading to more fruit exposure.

Cross-sections of infected trunks will have a pie slice-shaped brown area that indicates the wood is infected, dead and no longer functioning (Fig. 10.18). Spores of the fungus are released under wet conditions in the winter and spring and generally infect through injured areas, such as pruning cuts less than 4 or 5 weeks old (Wilcox *et al.*, 2015). Hence, reducing the chance of infection can be accomplished by pruning in dry conditions when

Fig. 10.18. Top: early-season shoots, showing stunted growth and leaf cupping. Bottom: cross-section of vine trunks, showing the characteristic wedge-shape of dead tissue. Content author's own.

spores are not being released and by pruning closer to spring and budbreak, after the reserve of spores has been exhausted and the vine cuts are not as susceptible (Munkvold and Marois, 1995).

Pruning wounds can also be protected using a physical barrier, an antagonistic fungus or treatment with a fungicide, though in some areas there are limited options available due to regulatory concerns. Acrylic paint was found to be an effective barrier (Sosnowski *et al.*, 2008) although if pruning is close to budbreak and sap is bleeding from the cuts, it may not adhere to the cut surface. Colonizing the cut surface with another organism, such as *Trichoderma harzianum* (John *et al.*, 2005) or *Bacillus subtilus* (Ferreira *et al.*, 1991) may also assist, though there is a delay in protection as the organisms become established on the cut and there is the same issue as with paint when pruning close to budbreak. In many areas, there are no fungicides registered for this use, so it is best to check with your local officials as to what may be available. Application of these materials is usually by hand, and normally only to cuts that are on second-year old wood or older due to the small surface area of cane cuts (Fig. 10.19). This is a laborious process, so it is possible to coat cuts with fungicide applied by mechanical sprayer (Sosnowski and Mundy, 2019), though as for anything regarding chemical control, complete coverage of the cut surfaces is essential.

Some fungicides have been found to have curative effects, such as tebuconazole, for infections six days old, and protective effects for 14 days (Ayres *et al.*, 2022), but under organic regulations, investigation of which physical barriers and preparations, such as those with *Trichoderma*, are permitted is needed.

If infected material is found, it should be removed and disposed of by burning or burial. Since the same organism also infects many stone and pip fruit crops, it is important to remember that inoculum may come from outside the vineyard.

Fig. 10.19. Left: a large pruning wound that was painted with a *Trichoderma* product. Right: a eutypa-affected vine that has had a disease-free shoot near the base of the trunk tagged for use as a replacement. Content author's own.

Since the fungus moves from cuts downward, the more basal parts of the vine may be free of the disease. If the vine is cut back beyond affected wood, it is possible to develop a new trunk from below that point. This can be a successful way of rejuvenating a vine (see Fig. 10.19).

OTHER TRUNK DISEASE ORGANISMS There are a range of other fungal organisms that can inhabit the woody tissue of the vines and for the most part their symptoms arise from the same causes – restriction of xylem capacity, resulting in water or nutrient-type symptoms, and through the presence of toxins that the fungi may make.

Examples of fungi, other than eutypa, that have been associated with trunk diseases are *Phaeomoniella chlamydospora*, *Phaeoacremonium aleophilum* and a range of species from the *Botryosphaeria* genus. Often more than one of these species will infect vines that are showing symptoms, and the term esca complex may be the most appropriate to describe this situation (Mondello *et al.*, 2018). In terms of symptoms, however, there is little to distinguish them. There may be reference to Petri disease in the literature as well, which was an early name for *Phaeomoniella*, particularly when it was found in young vines (Whiteman *et al.*, 2007).

Externally, there may be stunted growth, shoot dieback and chlorotic leaves (the yellowing being in a distinct pattern than has been called tiger stripe or grapevine leaf stripe) (Fig. 10.20). In some cases, the vine may suddenly and catastrophically fail (called apoplexy), usually in response to increased water demand during early-season growth outstripping the ability of the compromised vascular system to deliver water. Fruit may develop blotchy darker patches, which earned it the name of 'black measles' in California. Internally, there would be darkening of the wood where the xylem has been affected by fungal development, and sometimes exudation of 'black goo' (an early name for the disease) from individual xylem vessels (Wilcox *et al.*, 2015). In the dormant season affected canes will appear black and have little substance (Fig 10.20).

Curetage (sometimes referred to as curettage), or removal of previously infected and dead wood in the vine, is an old technique (Ravaz, cited in Lafon, 1921) that has been popularised in recent years to try to manage esca. The goal is to remove active infected material and potential inoculum source, open up the interior of the trunk to drying, and potentially induce natural defenses, such as phenolic or phytoalexin production. Traditionally, this was done with a billhook or similar curved blade, followed by application of copper sulfate (Lafon, 1921), but now is done with chainsaws of varying sizes. The operation is difficult and slow, taking 10–15 minutes per vine depending on the amount of wood to be removed (Lecomte *et al.*, 2022), but could be useful to extend the life of particularly valuable vines.

In fact, for trunk disease organisms it is likely to be the norm, rather than the exception, to find them in what look like healthy and productive plants (Halleen *et al.*, 2003; Aroca *et al.*, 2006; Bruez *et al.*, 2016). This state, where the fungi are present, but quiescent, remains until the vine becomes stressed, whereupon they proliferate and become pathogenic. This highlights the importance of maintaining appropriate balance in vine growth and production.

Fig. 10.20. Esca symptoms. Top: Late-season leaf on Sauvignon blanc; Bottom left: Mid-season Syrah leaf; Bottom right: Badly affected shoots will appear as black canes in the dormant season. Content author's own.

BLACKFOOT DISEASE (*CYLINDROCARPON* SPP.) This affects roots and trunks and is caused by fungi in the genus *Cylindrocarpon*. It becomes apparent in young vines by a loss of vigour and symptoms resembling water stress (see Fig. 10.21). If the lower section of the vine looks as if part of it is growing and part is not, then uprooting the vine may reveal death of the root system and new roots growing above that as a way of the vine compensating (see Figs. 10.21, 10.22). The dead roots are sometimes in a 'J' shape from when the vine was planted,

Fig. 10.21. A vine with red foliage and failing early in the season. Closer examination (inset) shows that part of the trunk leading to the roots is dead while roots have emerged from the scion. Content author's own.

the roots having been turned upward from being placed into a hole that was too small. This weakens root growth and allows for infection by the fungus. Infection is almost always fatal.

Blackfoot and other trunk pathogens can cause significant losses (Scheck *et al.*, 1998), but these can be almost eliminated by using clean planting stock, making sure the holes for planting vines in are sufficiently large that their roots are free to point downward when planted, ensuring vines have access to adequate nutrients and water and avoiding early cropping (Halleen *et al.*, 2006).

Bacterial diseases

*Pierce's disease (*Xylella fastidiosa*)*

Originating from southern North America, this disease struck California grapevines in the Napa Valley in 1887 which, by 1906, had destroyed 14,000 ha

Fig. 10.22. Blackfoot disease-affected part of the vine showing 'J'-shaped rooting habit of the original roots and the new root growth from scion tissue above. Content author's own.

(35,000 acres) of grapes: Perold (1927) even referred to it as "California disease" or "Anaheim" disease. Named after the Californian plant pathologist who worked with it in the late 1800s, it was not until the 1970s that its cause was established as a bacterium (Mollenhauer and Hopkins, 1974). It has been identified as occurring in Europe, apparently introduced with nursery stock (Boubals, 1989; Berisha *et al.*, 1998). Pierce's disease is transmitted by some xylem-feeding insects: leafhoppers (various types of sharpshooter in California) and spittlebugs, which can feed on many alternate hosts that may carry the bacteria (Severin, 1949; Raju *et al.*, 1980; Varela *et al.*, 2001; Costa, *et al.*, 2004), making it difficult to control.

The bacteria live in the xylem, blocking flow, which results in some of the symptoms associated with the disease: marginal chlorosis, followed by death of leaf blades and the fruit may raisin. Symptoms appear when vines are under higher levels of water stress, such as during the later stages of ripening. However, some of the more diagnostic symptoms are different, such as the abscission of the leaf blades from the petioles, leaving a 'matchstick' appearance, and development of 'green islands', regions of green in late season shoot tissue that is otherwise brown from maturation of the periderm (Thorne *et al.*, 2006; Wilcox *et al.*, 2015). Symptoms may differ between cultivars, however (Borkar and Yumlembam, 2016).

All *V. vinifera* and *V. labrusca* cultivars appear to be susceptible, but there is variation. For example, 'Chardonnay', 'Pinot noir', 'Fiesta' and 'Red Globe' are examples of susceptible *V. vinifera* types, whereas 'Chenin blanc', 'Sultana' and 'Zinfandel' appear to tolerate it more (Varela *et al.*, 2001). This may be due to the structure of the vascular system (Baccari and Lindow, 2010). *Vitis* species that evolved with the threat have an ability to grow and fruit normally despite its presence, such as *V. rotundifolia* and *V. arizonica*, and these were used by Andy Walker at UC Davis to develop resistant vines that have since produced marketable fruit (Svyantek *et al.*, 2020), though commercial plantings are still young. This may be the best hope for dealing with the effects of the disease.

Avoidance is another approach, as once vines are infected they decline over a number of years until death. The vectors do not live in areas with cold winters, and freezing temperatures have been shown to kill the bacteria inside vine tissues (Purcell, 1980), so site selection is a valuable tool. The effects of climate change could be taken into account using systems that model the effects of cold temperatures on the viability of the bacteria (Lieth *et al.*, 2011). Management consists of planting clean material and preventing potential vectors from feeding on the vines. If a vector is present where grapes are being grown, aside from the use of insecticides, which should be avoided due to their non-selective activity, there are few options. Kaolinite (a form of clay) was found to be effective in reducing the transmission of the disease into vines (Tubajika *et al.*, 2007), and is an organic-friendly option. There may be possibilities with parasitic wasps that predate the vector eggs, but this will not eliminate the vectors, only limit their population. Alternate hosts may be removed from the area, but this may not be a viable option due to the large range of plant hosts that the vectors can feed on.

Several promising control methods may help in the future. One involves inoculating vines with a benign strain of *X. fastidiosa*, which has led to 'Cabernet Sauvignon' vines that live for years after neighbouring, unprotected, vines have died (Hopkins, 2005). However, a commercial product based on this technique is not yet available as testing continues (Hao *et al.*, 2017). Topical application of another bacterium, *Paraburkholderia phytofirmans*, was shown to reduce incidence and severity of Pierce's Disease symptoms in field trials and may be a viable control method (Lindow *et al.*, 2024). Another tested and promising product is derived from a bacteriophage, a virus) that targets and destroys bacteria like *X. fastidiosa* (Das *et al.*, 2015), and researchers in Italy have developed a zinc and copper-containing solution, along with an injection system to introduce it into the wood, reduced symptoms and reinvigorated production in olive trees (Scortichini *et al.*, 2021).

*Grapevine yellow diseases (*Candidatus *Phytoplasma)*

There are a number of grapevine yellows diseases found globally, all caused by phloem-dwelling phytoplasma (a type of bacteria that lacks a cell wall) and transmitted via insect vectors or infected plant material. Symptoms include leaf rolling, yellow (or red depending on grape colour) and necrotic leaf patches and uneven periderm formation on canes (Fig. 10.23) (Wilcox *et al.*,

Fig. 10.23. Flavescence dorée symptoms on (left) Syrah, (centre) Syrah late season cane and (right) Merlot.

2015; Pierro *et al.*, 2024). Flower clusters may abscise prior to bloom, and if fruit does set, it might shrivel before veraison.

Known types of grapevine yellows diseases include flavescence dorée, Australian grapevine yellows, bois noir (also known as vergilbungskrankheit), palatinate grapevine yellows and North American grapevine yellows. All of these are caused by closely-related phytoplasmas that are transmitted by phloem-feeding insects, such as leafhoppers (in the case of flavescence dorée, *Scaphoideus titanus*). Vector insects often feed on a variety of plants, and there are multiple alternate hosts for the phytoplasma (Chuche and Thiéry, 2014). A four-year study in Tuscany found flavescence dorée widespread in vineyards and secondary hosts and vectors (Pedrelli *et al.*, 2024), and there are aggressive European campaigns to slow its spread (e.g. Jermini *et al.*, 2007).

Control is difficult and limited to regular scouting for symptoms and trying to manage vector populations. Insecticides for this can be targeted to the susceptible life-stages with temperature-based models to limit effects on other insects. Infected vines cannot be cured, so removal as soon as symptoms appear is the recommended procedure. Resistance to the phytoplasma exists, for example, in some traditional Georgian cultivars (Quaglino *et al.*, 2016), so there is something for grapevine breeders to work with, and intense treatment of vines with substances that elicit vine defense responses were found to reduce symptoms over the course of a four-year study (Romanazzi *et al.*, 2013). Further work in this area, however, is needed.

*Crown gall (*Allorhizobium vitis, *formerly* Agrobacterium vitis*)*

Crown gall is found wherever grapes are grown and often at the limits of their cold tolerance. This bacterium is closely aligned with grape and infects cells, altering their DNA so that they produce plant growth regulators and special amino acids called opines. The former cause the occurrence of masses of undifferentiated cells (callus) that look like a gall (see Fig. 10.24). The disease proliferates on trunks damaged by winter freezes or through physical means, and can eventually girdle and kill the vine. It is a ubiquitous soil organism (along with the closely related and also pathogenic *A. tumefaciens*) and cannot be eradicated from the environment.

If galls go unnoticed, serious infection will result in water-stress-like symptoms due to the disturbance of the vascular tissue. The starting point for gall formation is the cambial layer, so both phloem and xylem can be disrupted. Because of the former, there could be stunted shoot development after budbreak, due to less carbohydrate being able to be transported from the roots.

V. vinifera is highly susceptible to crown gall infections, while American species and many French hybrids are as well. It also appears that resistant rootstock grafted onto a susceptible scion reduces galling and vine death due to the disease, and after a number of years, the pathogenic bacteria population in the scion is greatly reduced (Süle, and Burr, 1998).

Though crown gall-free plants can be produced, they are easily infected in the field. Control best relies on trying to prevent the kind of damage that encourages infection. Failing this, removing the affected plants is the only option but, as the disease is so prevalent it is likely that, if a clean vine is planted

Fig. 10.24. Crown gall disease symptoms on grapevine cane (left) and trunk (right) growing in New York State, USA caused by cold damage during the dormant season. Content author's own.

as a replacement, it, too, will become infected. Biological control methods are being investigated (Habbadi *et al.*, 2023), but a commercial means to manage crown gall in this way is not yet available.

*Bacterial blight (*Xylophilus ampelinus*)*

This is another bacterial pathogen that colonizes vascular tissue of the vine. As such, symptoms can be water-stress-like, but there is usually the development of cracks on shoot bases and along petioles that progress toward the tip, eventually becoming dark cankers, anthracnose-like spots on leaves and staining of the xylem tissue (Wilcox *et al.*, 2015; Borkar and Yumlembam, 2016). However, the bacteria may be present in otherwise healthy-looking vines. The only known host is grapevine and it is presently found in Europe, South Africa, Asia and South America. Spread is through infected nursery material, wind, water splash and contaminated tools. As infected vines cannot be cured, control is through avoidance and preventing spread through vineyard sanitation and sterilizing pruning tools. Applying foliar copper fungicides may help if applied from completion of pruning through just after budbreak.

Viral diseases

Viral diseases are transmissible diseases for which (at least originally) no obvious pathogen could be found. These could be spread by propagation, by (in)vertebrate vectors (generally man, insects or nematodes) or by sap inoculation. The particles identified in diseased plants by electron microscopy were determined to be the disease agents and were called viruses. These microscopic particles are composed of genetic material inside a protein coat. Viruses are much smaller than bacteria and fungi and, once inside a living plant, they have the ability to reprogramme plant cells into making more virus particles. Once a plant is infected, the virus will spread to all parts of the plant and the plant will remain infected for life. Depending on the virus involved, the vine may or may not show visible symptoms, but general loss of vine health and reduced fruit yield and quality are associated with the most significant virus diseases. Detection methods initially used symptom expression on the plant, then indicator plants, which were more sensitive and showed symptoms with lesser titres of virus. The development of the enzyme-linked immunosorbent assay was a major step forward in terms of postive and accurate identification, but has been supplanted by DNA analysis through polymerase chain reaction, which is quicker and more precise.

Grapevine fanleaf virus (GFLV, or fanleaf degeneration)

This is one of the oldest described grapevine virus diseases, causing a variety of symptoms (Wilcox *et al.*, 2015) and also known as Roncet or court noué. One, caused by the distorting strain, is a characteristic leaf shape, which has an open petiolar sinus, more closely spaced radial veins and exaggerated marginal points, somewhat like glyphosate damage. The chromogenic strain results in bright yellow shoot parts, including whole or parts of leaves or clusters (Figure 10.25). Additionally, there may be a proliferation of shoot fasciation.

Spread of the virus is most rapid by transfer of infected propagation material. In the field, the virus is transmitted from grapevine to grapevine by *Xiphinema index*, the dagger nematode. Their control in infested soils is virtually impossible so there was interest in developing rootstocks resistant to their feeding. One was developed and released (e.g. O39-16), but probing by the nematodes was enough to infect the plants, and it had *V. vinifera* in its parentage, so new crosses were made. Information about five of these crosses was published in a thesis in 2021 (Nguyen, 2021) and one in particular performed well - testing continues.

Modern molecular approaches may be a more viable option, however, as it has been shown that rootstocks with a gene silencer (production of an anti-gene that cancels out an existing one) have been able to prevent proliferation of the virus in (non-genetically altered) scion material grafted to it (Torres *et al.*, 2014). This technique could potentially be used for control of any virus and may yield huge benefits for any industry that grafts its production plants.

Grapevine leafroll-associated viruses (GLRaV)

This is a family of viruses found in all countries where grapes are grown, and which probably originated in the Near East with *V. vinifera* (Wilcox *et al.*, 2015). There are various virus types within the leafroll family that have been identified, but GLRaV-3 is most debilitating and therefore the most discussed (Charles

Fig. 10.25. Examples of the chromogenic strain of Grapevine fanleaf virus, where symptoms can appear on individual leaves, shoots or entire vines. Content author's own.

et al., 2006). Leafroll has serious impacts on both vine health and grape quality, with growth and yield significantly reduced in the majority of cases. Ripening is affected by delayed or increased variation in maturity, resulting in reduced sugar content and inhibited fruit coloration. Grapevine leafroll can change fruit composition of French hybrid cultivars without acute leaf symptoms (Kovacs *et al.*, 2001), which makes identification of the disease more challenging. Leaf symptoms include rolling of the leaf margins, reddening between major veins in red cultivars and yellowing of leaves between the veins in white cultivars (Fig. 10.26; Goheen and Cook, 1959; Cabaleiro *et al.*, 1999; Borgo and Angelini, 2002). However, symptoms vary with cultivar, location and titre of virus(es).

As an example of its effects, comparing GLRaV-3 virused vines of Crimson Seedless to virus-free ones showed a 50% increase in berry weight, with elongation of its shape, more than 50% loss of berry color and a slight decrease in Brix and increase in titratable acidity (Salo *et al.*, 2024). While increasing berry weight is desirable, the loss in color would downgrade the value of the fruit in the marketplace. Notably, the results of this study differ from many others, which typically report a decline in berry size and vine yield (unfortunately, yield was not reported in the Salo *et al.* 2024 study). Further research to figure out under which conditions berry size is affected is needed.

Rapid movement of the virus is caused by infected propagation material, which may not show symptoms. Several mealybugs have been shown to vector grapevine leafroll (Cabaleiro and Segura, 1997; Golino *et al.*, 2002) as well as soft scale insects (Mahfoudhi *et al.*, 2009). Controlling vectors is problematic due to where they tend to live on the vine (e.g. under bark), so getting chemical controls to them is best through the use of systemic insecticides. There is the possibility of biocontrol through parasitic wasps, beetles or even moth larvae (Rakimov *et al.*, 2015), and promise in creating virus resistant

Fig. 10.26. Leaves showing grapevine leafroll virus symptoms in mid-season. Left: a green grape cultivar. Right: a red grape cultivar. Content author's own.

vines using CRISPR (Jiao *et al.*, 2022), though these still permit some viral replication in the tissue. However, unless vectors are kept out of the vineyard altogether there is still the chance of virus spread, though it can be slowed. Management of the virus is through control of vectors, identification of affected vines late in the growing season (easier for red-fruited grapevines as they show symptoms at lower titres), followed by removal when they are dormant. In New Zealand it was not necessary to remove adjacent vines in the row as few of these were infected (Bell, 2015), though it was beneficial to remove as many of the infected vine's roots, as these remained sources of the virus for years (Bell *et al.*, 2009).

Grapevine red blotch

Grapevine red blotch disease was first noted in California and New York vineyards in 2008 due to the similarity of its symptoms and their timing to GLRaV, but the tissue tested negative for leafroll (Al Rwahnih *et al.*, 2013; Krenz *et al.*, 2014). The virus associated with the symptoms was definitively identified in 2018 (Yepes *et al.*, 2018). In red cultivars it is distinguished by the more blocky nature of the leaf reddening, the veins of the affected areas also turn red and that leaves do not roll down as in GLRaV. White fruited cultivars may show leaf chlorosis similar to GLRaV or magnesium deficiency. A particularly detrimental effect of this virus is that fruit ripening is severely affected, with fruit sugar being lowered by up to 4 Brix and colour production being affected as well (Sudarshana *et al.*, 2015; Blanco-Ulate *et al.*, 2017). It appears to be widely spread through different cultivars and regions around the world (Krenz *et al.*, 2023).

Although primarily spread through infected plant material, the most significant vector appears to be the three cornered alfalfa leafhopper, *Spissistilus festinus* (Cieniewicz *et al.*, 2018), though it does not feed on grapevines preferentially and cannot complete its lifecycle on vines, either. Management methods are the same as for pretty much all the viruses - scout for disease symptoms every season and remove infected vines, and try to control the vectors directly or through removal of alternate insect hosts (Preto *et al.*, 2018) and virus hosts. Although no alternate virus hosts are known at this time, grapevine red blotch has been detected in wild vines growing near established vineyards (Perry *et al.*, 2016) and it spreads to previously uninfected vines. Clearly, more information about how it spreads is needed to be able to manage it effectively.

Grapevine Pinot gris virus

First noted in an Italian vineyard in 2003, this virus was described and named in 2012 (Malossini *et al.*, 2012). It results in chlorotic mottling of misshapen leaves, stunted shoot growth and lowered yields and fruit quality (Giampetruzzi *et al.*, 2012). The virus may be present in symptom-less vines and titre seems to be positively associated with visible signs of infection (Saldarelli *et al.*, 2014; Bertazzon *et al.*, 2017), but the yield and fruit ripening of infected, but symptomless, vines is not affected (Messmer *et al.*, 2024). Alternate hosts have been reported (Gualandri *et al.*, 2017; Demian *et al.*, 2022), as well as certain mites that appear to be able to act as vectors (Malagnini *et al.*, 2016). As the scope and seriousness of this virus, its hosts and vectors is not yet fully known, we should expect to see further research in this area.

Control of Grape Diseases

Grapes are, in general, susceptible to many diseases. If a specific disease does not yet occur in a growing region it is possible that it will at some later date. The grape grower is therefore advised to be aware of many vine disease symptoms and be alert to these appearing in the surrounding region. Most countries have rigorous controls over the movement of plant materials across their borders, to deter the introduction of unwanted pests. However, even the most diligent of border precautions cannot prevent the entry of an organism that is not yet known (recall the introduction of phylloxera and various North American diseases to Europe). All growers of grapevines should be aware of diseases that are already present in their region as well as those that are potentially a threat, in order to minimize transfer of unwanted organisms. History has shown that man is the most effective of vectors when it comes to pests and diseases!

Despite the multitude of potential diseases, grapes are resistant to most microorganisms and, therefore, disease is the exception rather than the rule. Therefore the defence mechanisms of grapevines are effective in preventing many potential diseases (refer to the earlier discussion about phytoalexins).

Many cultural practices are available that can aid in disease control. Most disease organisms favour moist conditions for spore germination and infection, so therefore leaf removal, which increases air movement through the vine and the drying rate of wet leaves and fruit, is a great help. Excessive fertilization (particularly nitrogen), irrigation or an over-vigorous rootstock can result in luxuriant growth that is conducive to conditions favouring disease. Overly compact clusters (which can be due to clonal, varietal or seasonal factors), for similar reasons, are more susceptible to fungal diseases as well. It is also worth emphasizing the role that vine propagators play: starting with clean material is a must, so when purchasing plants ensure that there are guarantees for their health status.

Chemical control of diseases is the norm in most grape-growing areas. As an example, we will discuss those used against fungal diseases. Fungicides can be regarded as organic or synthetic, and their modes of action include protectant, post-infection, antisporulant and eradicant. Protectant fungicides must be in place before the fungi attack the plant part. Since the vine grows and produces new tissues throughout the season, this new growth must be covered with the protectant fungicide in order to prevent infection. Fungicides may also wash off with rain, and therefore previously covered surfaces must be recoated.

Post-infection fungicides can be applied after the infection has initiated, as they kill the fungus after infection has started and will prevent further progression of the infection to the disease stage. The time following infection when these types of fungicides can be applied varies, but may be only a few days. In the case of myclobutanil, application up to 6 days post-infection provided complete control of black rot, whereas the efficacy of azoxystrobin decreased linearly after 4 days post-infection (Hoffman and Wilcox, 2003).

Antisporulants decrease spore production in infected tissue, reducing the spread of the disease. Hoffman and Wilcox (2003) also found that myclobutanil was effective at reducing spore production of the black rot fungus providing it was applied before symptoms appeared.

Eradicants kill all or most of the fungal colony when applied after symptoms have appeared. They are rare, but an example is the use of lime sulphur to kill overwintering powdery mildew inoculum (Gadoury *et al.*, 1994).

Other Disorders

There are many causes of unusual appearances of grapevine parts that are not due to pathogens or nutritional deficiencies/toxicities. Shoots and leaves are well exposed to the potentially damaging effects of environment and manmade hazards. Wayward herbicides are one cause of many leaf problems, as in some cases only very minute amounts of these can make a significant difference to the vine's appearance.

Contact herbicides, which only affect tissue they fall upon, can create localized leaf symptoms ranging from leaf discoloration (see Fig. 10.27) to

Fig. 10.27. Left: example of contact herbicide damage on grapevine leaves; paraquat and diquat mixture. Right: effect of a pre-emergent herbicide spray of flumioxizen on grape shoots that had been splashed with soil particles after rainfall. Content author's own.

necrotic regions on leaves and shoots. These types of herbicides are safer to use around actively growing vines since, if some chemical contacts the plant tissue, the damage will not spread.

In some cases, pre-emergent herbicides are used to discourage weed growth for longer periods of time. As the chemical persists in the soil, it is important not to let it contact the vine tissue, or damage can ensue (see Fig. 10.28).

Systemic herbicides move through the plant tissues and usually result in leaf and shoot distortions (see Fig. 10.28.), which can appear well beyond the

Fig. 10.28. Examples of systemic herbicide damage on grapevine shoots and leaves. A diuron and bromacil mixture; B 2,4-dichlorophenoxyacetic acid (2,4-D) on a shoot tip; C glyphosate on a shoot tip; D mild glyphosate on leaves. Content author's own.

point of contact. Therefore, their use around vines should be quite cautious, in order to prevent potential catastrophe.

Volatilization of herbicides is an important consideration in their use – this can result in vine damage many kilometres from the point of application. Do not apply under conditions where this is likely to happen, and encourage your neighbours to do the same.

Environmental hazards include hail, which can break off buds, damage or strip leaves, break shoots and flower clusters and wound canes (see Fig. 10.29).

Vines brought from a nursery shade house can be damaged by heat stress. This comes about from a combination of higher temperatures, increased sunlight intensity and small root systems that cannot supply enough water to the leaves, and results in necrosis of the leaf tissue (see Fig. 10.30). Leaves that develop in low intensity sunlight, such as those inside a canopy, are similarly ill-suited to full sun.

Insect Pests

In all grape-growing regions, roots, wood, shoots, leaves and fruits are subject to potential insect attack. Although the specific species of insect causing the damage may change in different grape regions, there is considerable similarity in their form. Some major insect pests for grapevines include phylloxera, leafhoppers, borers, mealybugs, leaf folders, cutworms and aphids. Grapes are also susceptible to infestation by many species of mites, which are not insects, but are included in the spider (arachnid) family.

In some cases, it is not that a pest causes damage to the vine or fruit, but contaminates it. As mentioned in Chapter 3, there are beetles and millipedes that can move into the clusters near harvest and if processed with the grapes, cause a noticeable taint in a wine even with less than one beetle per cluster on average. There can be similar problems with table grapes, where insects hitching a ride in the clusters would be an unwelcome surprise for the consumer.

Phylloxera

Phylloxera (*Daktulosphaira vitifoliae*) is an aphid-like insect that feeds on roots (see Fig. 10.31) that has worldwide distribution. It is only known to attack grapes. When there is a point origin for phylloxera entering a susceptible vineyard, only isolated vines may be affected at first, with the area of dead and declining vines enlarging in a lens shape (long axis down the rows, Fig. 10.32), sometimes quite rapidly (King and Buchanan, 1986) in subsequent years as the insect spreads. In the case of vines infested as planting stock, the entire vineyard may show decline after a few years of growth, though soil type was found to have an influence on phylloxera populations (Chitkowski and Fisher, 2005). High-sand content soils support smaller populations of phylloxera (Nougaret and Lapham, 1928;

Fig. 10.29. From top to bottom: hail damage to leaves on early-season shoots, with one that was on the leeward side protected; fresh damage to mid-season fruit; recent damage near the base of a shoot; old wound caused by hail on the cane below the shoot. Such wounds can be entry points for disease organisms. Content author's own.

Fig. 10.30. Top: necrosis of leaves on a recently transplanted vine, caused by heat, intense sunlight and lack of water supply. Bottom: damage to leaves resulting from shoot trimming, where shade leaves are not adapted to full sun. Content author's own.

Galet, 1982; Reisenzein *et al.*, 2007; Wilcox *et al.*, 2015), which may explain why some vineyards in infested areas have not succumbed to its effects.

The foliar form of phylloxera causes spherical galls on the underside of the leaf, as well as distortions and an exit hole on its upper side (see Fig. 10.32; Granett *et al.*, 2001).

The most serious damage is caused by the insect feeding on grape roots, the materials injected into the tissue causing the formation of nodosities

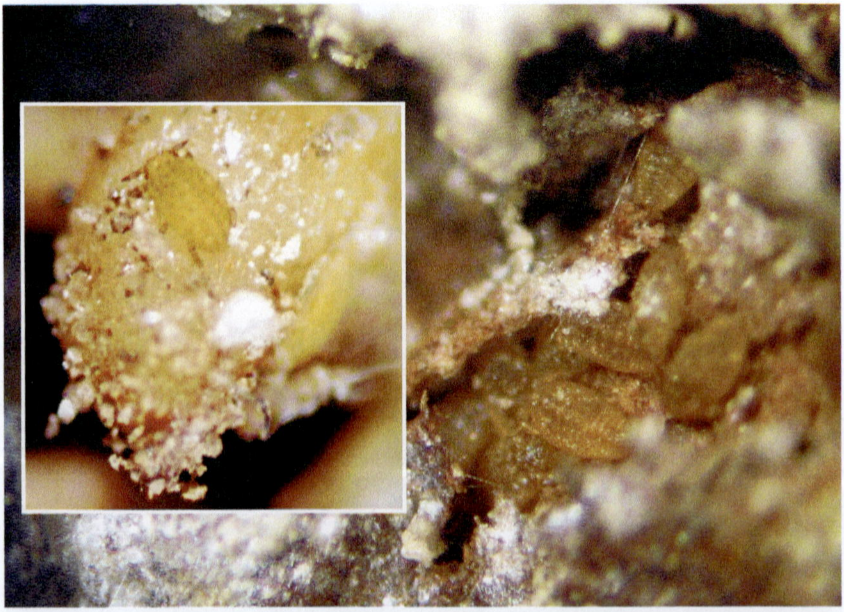

Fig. 10.31. Phylloxera nymphs on grapevine roots. The inset shows one on a root tip, the main photo shows a group of them nestled into the bark of the root. Content author's own.

Fig. 10.32. Left: weak vines in a row, the first potential signs of phylloxera infestation. Right top: leaf galls on the adaxial side; right bottom: on the abaxial side. Content author's own.

(hook-shaped galls on younger roots) and tuberosities (more generalized swelling) (see Fig. 10.33). Feeding by large numbers of phylloxera will kill a vine through the reduced capacity of its root system and potential infection of root wounds by pathogens (Omer *et al.*, 1995).

The presence of the foliar form of phylloxera has been reported in many areas around the globe and effect on the leaves has been severe enough to warrant research into its effect on vine performance and fruit quality (Wilmink *et al.*, 2022). Though leaf photosynthesis can suffer with galling, resulting in decreases in yield and grape Brix, the response was not uniform (Yin *et al.*, 2021), so further research is needed.

Phylloxera spread is primarily through infested nursery material, but the crawler form of phylloxera can be carried on vineyard machinery, and carried some distance by wind (King and Buchanan, 1986). The winged form of the insect is rarely found, but it is capable of movement over much greater distances (Granett *et al.*, 2001).

Control of phylloxera in infested and susceptible vines is not yet feasible, though some soil drenches have been tried. One problem with this approach is that the insect can live on roots far down into the soil profile and out of reach of applied chemicals, which are also difficult to apply in the clay soils that phylloxera prefer (Granett *et al.*, 2001). Eggs are fairly hardy and long lasting as well. Affected vines can be managed, through additional water and fertilizer inputs, to continue producing for some additional years but, in most cases, they will eventually succumb to the pest and need to be replaced. Some biocontrol agents are being investigated, such as an insect pathogenic fungus (Kirchmair

Fig. 10.33. Nodosities in grapevine roots, which are the result of phylloxera feeding. The club root or shepherd's crook shape is typical of phylloxera-infected roots, but tuberosities (inset) are also the result of their feeding. Content author's own.

et al., 2009; Ficiu *et al.*, 2015; Korosi, 2020) and nematodes (English-Loeb *et al.*, 1999b), but it is unlikely that these methods will work in more than very specific situations.

Successful management is therefore accomplished by avoiding the pest in the first place or grafting susceptible vines to resistant rootstocks, which have both been discussed earlier.

Leafhoppers

Leafhoppers are pests of many cultivated (and non-cultivated) crops and, since they have many alternate hosts, eradication of these from the vineyard is difficult. Leafhoppers feed through piercing the epidermis and taking up the cell sap. During this process there is exchange of material between the bug and the plant, which contributes to these insects acting as vectors for some disease organisms, most notably Pierce's disease.

Symptoms of leafhopper feeding include fine chlorotic spots on leaves (see Fig. 10.34), leading to more general chlorosis or virus-like symptoms such as leaf curling and angular reddening on leaves (Winkler *et al.*, 1974; Wilcox *et al.*, 2015). If the honeydew excreted from the insects falls on fruit, it can cause an unsightly sooty mould appearance, which reduces the quality of table grapes (Settle *et al.*, 1986). As for most pests and diseases, a zero tolerance to

Fig. 10.34. Leaf symptoms typical of those caused by leafhopper feeding, in this case, on 'Sauvignon blanc' vines. Content author's own.

their presence is not feasible, and one study has indicated that, at least from the standpoint of impact on vine health and productivity (crop loads and vine photosynthesis), a small population of leafhoppers can be tolerated (Candolfi *et al.*, 1993).

Removing basal leaves on shoots coming from the cordon or cane is a good first line of defence against leafhopper infestation in vineyards (Stapleton *et al.*, 1990; Prazaru *et al.*, 2023) as is removing watershoots and their leaves at the base of the vine, as this is the location where eggs are laid. Control of leafhoppers can also be through the use of insecticides, with pyrethroids being an option that has been in use for many years (Perring *et al.*, 1999), as well as *Bacillus thuringiensis*, which is a bacterium that produces a chemical that disrupts cells in the insect gut. Synthetic insecticides such as organophosphates and carbaryl have been used, but there are concerns about their environmental toxicity (Jenkins and Isaacs, 2007; Ore *et al.*, 2023). There has also been research into successful biocontrol agents, such as wasps (Kido *et al.*, 1984; Williams and Martinson, 2000), which may form a component of integrated pest management systems.

Borer insects

There are various types of borer insects that attack grapevines. In general, the adult forms are either beetles or moths that lay eggs in the soil near the vine or on the vine itself. Root and cane borer larvae hatch from the eggs and find roots or shoots, respectively, to infest. The larvae eat through the centre of the roots/canes/shoots, damaging the vascular tissue and causing water stress-type symptoms, or killing the vines outright (see Fig. 10.35). Control is best accomplished by removing alternate plant hosts near the vineyard and

Fig. 10.35. Left: the base of the trunk of a young vine killed by a lepidopteran root borer. Note aerial rooting. Right: the larva becomes visible when the base of the trunk is cut open. Content author's own.

sanitation of infested material. Mating of the insects can be disrupted with pheromone release into affected areas (Johnson *et al.*, 1991), which can be effective enough to negate the requirement for insecticides. At least two fungal pathogens of the larvae have also been reported (Dutcher and All, 1978), and the use of entomopathogenic nematodes also looks promising (Blaauw and Shapiro-Ilan, 2023).

Mealybugs

Mealybugs (species in the genus *Pseudococcus)* are a problem in vineyards because they are a vector for diseases but, also, through their feeding, increase botrytis infection and physically infest the clusters, causing the appearance of sooty mould, which is unacceptable for table grapes (see Fig. 10.36). In many areas economic control can be accomplished through the use of natural predators, such as parasitic wasps (Berlinger, 1977; Trjapitzin and Trjapitzin, 1999; Rakimov *et al.*, 2015), though there are many other types of potential control organisms (Walton and Pringle, 2004). Control is difficult because the insect often lives under bark or deep inside the clusters.

Moths

The grape berry moth of the eastern USA (*Paralobesia viteana*, formerly *Endopiza viteana*) is a serious pest in smaller vineyards with wooded boundaries

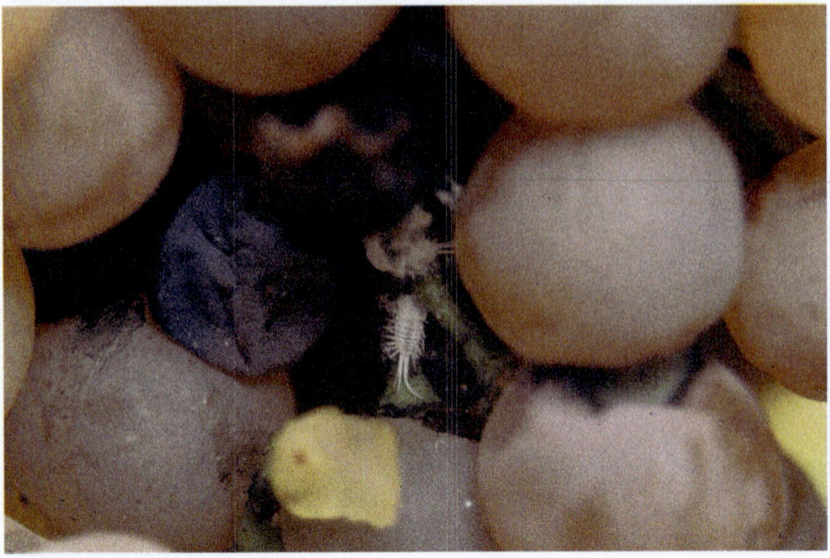

Fig. 10.36. Mealybugs nestled inside a late harvest 'Semillon' cluster. Content author's own.

harbouring many wild grapevines, as moths move from the wooded areas to the vineyard (Botero-Garcés and Isaacs, 2004). Larval development has been successfully modelled using a heat accumulation formula, allowing prediction of when control methods should be used (Tobin *et al.*, 2001). Although cultural practices that alternately bury and expose the overwintering stage of the insect are sometimes effective, chemical treatment is usually required. The sex pheromone of the grape berry moth is commercially available and has been successfully used for control (Martinson, 1995; Witzgall *et al.*, 2000), as has *B. thuringiensis* (Ifoulis and Savopoulou-Soultani, 2004). In larger vineyards with less boundary per area, chemical control can be limited to the outside rows. In other vineyards damage is rare and no chemical control is required. The European grape berry moth is a different species (*Lobesia botrana*), causing similar grape damage and requiring equivalent methods of control, though some aspects may differ (Gabel and Renczés, 1985; Witzgall *et al.*, 2000).

The light brown apple moth (LBAM; *Epiphyas postvittana*) can also be a problem in some areas. The adults lay eggs on leaves or fruit and the larvae feed on the tissues, causing direct damage, but also by creating wounds in the plant tissue and spreading around fungal spores, such as botrytis (Fig. 10.37).

Fig. 10.37. Top: light brown apple moth larva cocoon in a grape leaf before and after unrolling the leaf. Bottom left: larval webbing has floral debris stuck to it. Bottom middle: later season cluster infestation results in death of berry tissue where the larva had been feeding. Bottom right: active larva extracted from a cluster. Content author's own.

Control methods for the LBAM are similar to those of other moth pests, and include encouraging parasitic wasps to move into vineyards (see section on Pest and Disease Control in Organic Grape Production).

Thrips

Thrips are very small insects that cause damage when the eggs are laid in fruit (sometimes called halo-spotting), or when they feed by rasping on the berry surface, which causes scarring (see Fig. 10.38). Damage is primarily visual, which is important in the table grape industry, but berry scarring can also be an origin of berry splitting (Yokoyama, 1979). Enhancing plant diversity in vineyards can enhance natural control of thrips (Nicholls *et al.*, 2000), but these may also act as thrip sources as populations of adults. Schmidt *et al.* (2006) found high populations of thrips in flowering vineyards, but few larvae, suggesting that they were moving into the vineyard from other places.

Beetles

Beetles can also cause damage, either through the larvae feeding on vine roots or from adults feeding on the leaves (see Fig. 10.39). The bark of the grapevine

Fig. 10.38. Scarring on 'Pinot noir' berries as a result of thrip feeding earlier in berry development. Most damage is usually near the stylar end of the berry as the insects feed on pollen and tender ovary tissue. Content author's own.

Fig. 10.39. Left: vine weevil damage on leaves of a mature grapevine, giving a notched appearance to the leaf margins. Right: A Japanese beetle feeding primarily on the interveinal areas of leaves. Content author's own.

trunk is often host to a variety of insects, some of whom only come out to feed after dark. In New Zealand, the grass grub, or bronze beetle (*Costelytra zealandica*), has become a significant pest in the spring, as adults emerge from the ground, fly into vineyards and feed on the young shoots, leaves and flowers after mating (Fig. 10.40). A possible non-chemical means of management could be spreading of mussel shells under the vines, as this appears to alter flight patterns and potentially deter emergence from the soil (González-Chang *et al.*, 2017). The larva of the same beetle has also been found to feed on vine roots and been suspected of causing newly-planted vine death (Mundy *et al.*, 2005).

The Japanese beetle (*Popillia japonica*, Fig 10.39) is a significant pest in North America but tends to be found at the margins of vineyards as the preferred reproductive sites are meadows or woodlands, so control can concentrate on eliminating adults on foliage rather than larvae in the soil (Lessio *et al.*, 2025). Worryingly, this beetle was discovered in Portugal in the 1970s, Italy in 2014 and has already spread to neighboring countries. Unfortunately, with climate change its habitable range is predicted to increase significantly (Pulighe *et al.*, 2025).

The multicoloured Asian ladybird beetle can be a significant problem even though it doesn't damage leaves or fruit, but because, as mentioned in Chapter 3, it contains a potent methoxypyrazine that can taint wine made from it (Galvan *et al.*, 2007). Keeping the beetles from entering clusters centers around minimizing damaged fruit, as it attracts them, insecticides, chemical attractants or repellents, but at present there are no completely satisfactory methods (Pickering and Botezatu, 2021).

Fig. 10.40. Bronze beetle damage to spring-time shoots in New Zealand. Virtually all of the green tissue can be eaten in one evening. Content author's own.

Mites

Spider mites, grape erineum mites (and the closely related bud mites), grape rust mites and false spider mites are examples of those that can be found in most grape-growing areas. Mites cause leaf damage through their feeding (see Fig. 10.41). The leaves may brown or drop and reduce vine photosynthesis or buds can also become infested (see Fig. 10.42). Many problem mites are controlled naturally by predatory mites (Karban *et al.*, 1995), so chemical applications to control mites must be carefully selected to prevent decimating beneficial mite populations (English-Loeb *et al.*, 1986). For example, sulphur applications for disease control can disrupt the life cycles of both damaging and beneficial predatory mites (Hanna *et al.*, 1997). Alternate food sources for the predatory mites may also be beneficial for those times when pest mite populations are low.

Chemical control

In general, chemical controls should not be applied unless the insect is present and in a form that is susceptible to the chemical. Life cycles vary for different insects, and the appearance of the vulnerable form may depend on macro- or microclimate factors. Scouting, or monitoring through the use of sticky-traps,

Fig. 10.41. Erineum (or blister) mite colonies on grapevine leaves cause these deformations on their adaxial side (left) and light-coloured downy masses on the abaxial side (middle). Later in the season (right) some colour may appear on the upper sides of the blister. Content author's own.

Fig. 10.42. Example of shoot growth (left) and leaf (right) affected by bud mite damage. Note the zig-zag growth pattern of the shoot and open sinus of the leaf blade. Content author's own.

can be used to determine the presence and population of relevant insects, and in some cases, weather-based modelling systems can predict the emergence of the target form. Vines should be examined in detail for the number and type of insects and, if the number seen is known to result in economic damage (through experience or published information), a chemical or other control should be applied to reduce the risk. As stated earlier, the use of insecticides should be carefully considered due to their non-specific action.

Animal Pests

Millipedes

These are occasionally an issue in vineyards because they can climb into the grapevine canopy and end up being harvested with the grapes. This has been noted to happen when there is rainfall close to harvest, probably because they are escaping the flooded soil. They can be found on leaves, shoots and in the clusters (Fig. 10.43), and when disturbed, coil around whatever they are near, making them impossible to dislodge. Unfortunately, they secrete a chemical defense that contains mostly benzoquinones (Kania *et al.*, 2016), which have a sharp, offensive odor and only very small amounts are needed to taint wine.

With a life cycle spanning years, control of the pest is difficult. Soil cultivation was found to reduce their numbers in vineyards (Sharley *et al.*, 2008), and there has been research on the ability of a parasitoid fly to limit populations (Baily, 1989) and a parasitic nematode (Schulte, 1989), but this is an area that requires further research. Constructing light-traps could be a way of reducing populations leading up to harvest (Bailey, 2016).

Nematodes

Nematodes are generalized feeders present in most soils and many are serious pests of grapes. They insert mouth parts into the roots, removing sap, but also

Fig. 10.43. Millipedes in the leaves and clusters of Vermentino vines, just before harvest. Content author's own.

resulting in abnormal root growth and function, resulting in water-stress-like symptoms in the above-ground parts of the vine and reducing productivity. Root-knot nematodes (*Meloidogyne* spp.) cause general decline in grapevines, not usually killing them, but rendering them more susceptible to other diseases or environmental stresses. Dagger nematodes (*Xiphinema* spp.) also cause vine decline but are also a vector of GFLV. Lesion nematodes (*Pratylenchus* spp.) also cause vine decline by root death (Nicol *et al.*, 1999). Nematode populations tend to increase with time in vineyards (Ferris and McKenry, 1974), and this has been associated with replant disorders. If a vineyard was to be planted into a soil with high nematode populations, this was traditionally corrected through soil fumigation (McKenry, 1999).

With recent and strict regulations regarding the use of methyl bromide, a commonly used pre-planting soil fumigant, alternatives such as the planting of crops that discourage nematode infestations (e.g. marigolds) can be tried (Nicol *et al.*, 1999; Reynolds *et al.*, 2000). The use of soil drenches to control nematode populations suffers the same problems as for phylloxera – the organism can live quite deep into the soil profile and persist in the soil long after vines have been pulled out (Raski *et al.*, 1965). Residual roots in the soil can carry the pests, so killing roots with systemic herbicide and removal of as many roots as possible is helpful. Soil-borne bacteria have been found to have activity approaching that of organophosphate nematicides against viticulturally important nematodes (Aballay *et al.*, 2020), pre-infection of roots by mycorrhizae also reduced nematode feeding and improved vine growth (D'Addabbo *et al.*, 2023), and garlic extracts applied as a soil drench was successful in killing nematodes in a potted vine experiment (Hao *et al.*, 2012). There is hope that a more environmentally friendly means of managing nematodes will be found in the future.

Vine-grazing pests

Vine-grazing pests include many species of deer and rabbits. They do not usually consume significant amounts in a healthy vineyard, but are particularly damaging in newly-planted vineyards. They can defoliate vines in a new planting, eating the tender leaf blades while leaving the petioles still attached to the shoots (see Fig. 10.44). This can greatly delay the establishment of young vines through the loss of photosynthetic capacity. Rabbits, turkeys and other animals can also eat the bark from the young trunks, damaging or even killing vines (see Fig. 10.45).

Berry-eating pests

Berry-eating pests include many mammals such as boar, raccoons, skunks, opossums, rodents and others. Birds, including turkeys, can also be voracious berry consumers and can cause severe economic damage to grapes (see

Fig. 10.44. A young vine that has been defoliated by deer, who prefer to eat the leaf blades, not the petioles. This issue can be more of a problem in dry areas or seasons, where irrigated grapevines have the lushest growth available. Content author's own.

Fig. 10.46). Birds can nest in or near the vineyard, arrive in flocks (such as starlings) or be on a migratory path. Birds travelling in large flocks are particularly dangerous, as they can consume a large amount of fruit in a short amount of time.

Although there are many possibilities for managing animal damage in vineyards, exclusion methods are most dependable for both mammalian and

Fig. 10.45. Rabbit (left) and goat (right) damage on vine trunks. The latter have completely girdled the vine trunk, meaning the vine will need to be replaced unless a new shoot can be trained from above the graft union. Content author's own.

avian predation. Electric fences (permanent or temporary) or barrier fences are both effective for mammals, though costly. Netting is effective for birds (see Fig. 10.47), although stationary noise (see Fig. 10.48) and visual scare methods may suffice temporarily for moderate bird pressure (Aubin, 1990). Drones are also of interest to viticulturists, though like other scaring methods, tend to lose effectiveness as birds habituate to them. Many commercial operators in high-pressure areas use a combination of nets, propane cannons and frequent movement/noise in the vineyard (e.g. ATVs), with actual shooting of birds (if permitted by local regulations), the latter of which they find necessary for devices such as propane cannons to be effective.

Fig. 10.46. Left and centre: Bird damage to grapes. Particularly in dry seasons, birds can take fruit quite early, even just as it starts to colour. Some bird species tend to peck at, rather than take, whole berries, which can result in secondary infections (centre). Boar (right) take groups of berries, whole or part clusters, and if it has been raining, often leave muddy nose prints behind. Content author's own.

Bird netting can be applied directly on the vines and tied under to prevent entry, draped over several rows at a time, or applied on supports over the vines. It must be applied before birds become habituated to feeding from the grapevines, and removed just before or soon after harvest. The latter is to prevent shoot growth through the nets and to prolong their life, which can last up to 10 years with care in application and storage. If nets are left in the vineyard, UV radiation continues to degrade the plastic, and the weight of snow can cause irreparable tears. Though effective, nets require significant labour to apply and recover (especially if shoots have grown through), as well as space to store when not on the vines. In areas with lower bird pressure, netting may only be needed on the outside rows, or those closest to trees and other places where birds can perch and find shelter.

Due to differences in the behaviour of types of birds, research on learning more about the behaviour of specific species of birds in vineyards may yield more effective and holistic management methods (Saxton *et al.*, 2004a,b). For example, blackbirds and silvereyes prefer fruit of different sugar concentrations, and their feeding behavior is very different: blackbirds will select a whole berry and then leave, while silveryes spend much longer in the canopy, pecking at many different grapes, thereby causing greater damage (Saxton *et al.*, 2004b)

Birds of prey have been used successfully to deter bird activity in vineyards, while at the same time helping to preserve and grow threatened falcon populations (Kross *et al.*, 2012). Having these falcons nesting in vineyard areas resulted in significant drops in crop loss. This is a reminder that bird activity in and around vineyards can have beneficial effects - many are insect eaters and

Fig. 10.47. Examples of bird netting used in vineyards. Overhead nets (upper), over-the-row nets, which can cover multiple or single rows (left) and side-netting (right). Content author's own.

may help to reduce pest species, particularly in areas that have more bio- and eco-diversity (Barbaro *et al.*, 2017).

Fences are expensive to erect, but can last a long time with continuing maintenance. Fences for berry-eating mammals must be close enough to, or even extend into, the ground so the animal cannot crawl or dig under, and

Fig. 10.48. A stationary propane cannon situated in a vineyard; these are usually set to go off at variable intervals, sometimes with two reports in close succession. Content author's own.

must extend high enough to discourage those animals that can jump. Fences to keep out wild deer-like animals should be at least 3 m high if they are erected on uneven terrain (VerCauteren *et al.*, 2006).

An issue worthy of serious consideration when it comes to disease and pest control is the potential effects on neighbouring properties. Spray drift and noise used to manage these problems can often be objectionable to others, so it is best to communicate well with neighbours to ensure minimal disturbance. In some cases there are local regulations covering these types of activities, which should be taken into account when developing a vineyard.

Weeds

A definition of a weed is a plant growing where it is not wanted. Weeds in vineyards use both nutrients and water needed for vine growth and can block light and hinder airflow as well. In some floor management systems, nothing is allowed to grow under the vines or between the rows. More machine travel in the vineyards has increased the need for a planted middle, which allows for earlier entry into vineyards following rain events and lessens soil compaction. Native cover (also called a volunteer crop) can be used, and may be suppressed during the highest water use part of the season for vines. Rather than use volunteer plants as a cover crop, sowing a specific grass or other plant seed that has attributes better suited to vineyards is also practised. Several strategies for cover crop management are presented in Chapter 6.

Weeds can be prevented from growing, or be destroyed at some growth stage, by either mechanical (Chapter 9) or chemical methods. Chemical herbicides are used to kill established weeds, suppress weed growth in row middles and inhibit germination of weed seeds. It is possible completely to control weeds with chemical applications. Herbicides can be sprayed, wiped or broadcast as granules. Some herbicides (contact) kill only exposed green tissue and can be used near the base of growing vines, although because only the part of the weed that was in contact with the herbicide is killed, it usually recovers. Other herbicides are translocated (systemic) by green tissues to roots and kill the whole plant. They must be applied when the vines are dormant (green tissue is much more susceptible to absorbing chemicals) or physically protected from contact, as they can also be taken up by the vine. Germination inhibitors are applied in conjunction with other herbicides to prevent weeds from growing rapidly in the space created by dying weeds. Weed management under organic systems are discussed further on.

Regardless of method, weed control strategies must be developed that do not reduce grapevine health or productivity to below economic targets.

Pesticide Resistance

Disease-causing organisms, weeds and insects frequently become resistant to specific pesticides. Resistance to traditional insecticides in insect control has been shown with grape berry moth (Nagarkatti *et al.*, 2002), suggesting the need for alternative methods. The first documented resistance to the herbicide glyphosate was reported in 1996 and currently totals 61 species (www.weedscience.org). Many fungicides have become useless because of resistance development by their target pest organisms. Resistance develops because of the biology of the pesticide – usually its specificity to the pest target and how frequently it is used. The greater the specificity, generally the greater the chance that a mutation will develop that overcomes it. It is very important not to overuse new and effective pesticides, which will encourage resistance development and possibly render them ineffective. Several strategies are involved, including limiting the number of applications per season, alternating use with pesticides of different modes of action or use in combination with a second pesticide of a different mode of action (Staub, 1991; Clarke *et al.*, 1997). Despite these precautions, disease management using chemicals will be dependent on new pesticides being developed as either resistance develops or regulations force their withdrawal from the market.

Pesticide chemical classification

Pests develop resistance to chemicals such that a material that was effective when introduced may no longer be effective in controlling that pest. Hence there is the need for continuous development of new pesticides. Some chemical materials are less prone to these problems, usually the ones that have been in use for a long time. This is because their modes of action (MOA) are so broad that it is difficult to overcome the multiple ways the pesticide combats the pest. Newer pesticides are more specific to each pest, which is good from the standpoint of minimizing impacts on non-target organisms, but this makes them more vulnerable to resistance development.

Some pesticides have closely related chemistry (MOA), so if resistance to one develops it is likely that others that are similar will lose effectiveness. The Fungicide Resistance Action Committee (www.frac.info) was formed to provide information to maximize the life of fungicides by developing strategies around their use. It has categorized fungicide chemicals into groups based on their MOA, standardizing nomenclature and simplifying communication of best-use scenarios for chemicals. The first letter of the code signifies the MOA of the chemical: what part of the physiology it works on. For example, 'A' represents those chemicals that target nucleic acid synthesis, 'C' those that target respiration and 'G' membrane-based sterol biosynthesis. The number that follows represents a target site within the MOA.

As examples, the DMI fungicides (e.g. tebuconazole, myclobutanil, cyproconazole) are in Group G1 and strobilurin fungicides (e.g. azoxystrobin, pyraclostrobin, trifloxystrobin) in Group C11. Pristine, a fungicide sold in some markets, contains two active chemicals, boscalid and pyraclostrobin, that are in Groups C3 and C11, respectively.

The codes make it straightforward to rotate or combine chemicals with different MOA and minimize the risk of resistance building up in the pest population.

Even if chemical companies have different formulations within the same group, if there are poor results controlling a disease with, for example, one fungicide, it cannot be assured that application of another material in the same group will provide control. The rate of development of resistance is largely determined by the frequency of exposure to that chemical and how effective the chemical (and its application) is in killing the pest. Growers should always limit use of single-site MOA chemicals and alternate between pesticides with different MOA.

Pest and Disease Control in Organic Grape Production

The organic grape producer is faced with the same mix of pests as the non-organic producer. Organic production does not mean no-spray, and organic pesticides often do not provide as long-lasting protection as non-organic alternatives and must be applied more frequently. Diseases, insects and weeds can be controlled by organic methods, but with some limitations, particularly with regard to climate or the amount of pest/disease pressure. The more suitable the climate is for a pest, the more difficult it will be to control through organic means. If organic production is one of the goals of a vineyard, then site selection is the most important pest control activity.

Insect damage can be tolerated to some degree, particularly for juice or wine grape production, though less so for other end uses where the fruit is seen by the consumer. Insecticides for organic production must be approved by an accrediting agency, and could include toxic botanicals extracted from plants, such as rotenone, pyrethrum, ryania and sabadilla. Some oils and soaps are effective insecticides (Sams and Deyton, 2002), such as those made from the neem or tea tree. Biologically derived products like those from the bacterium *B. thuringiensis* can be as effective as synthetic insecticides (and their use should be treated with the same caution). Behavioural control agents such as repellants, antifeedants, attractants and sex pheromones can be used for insect control (Martinson, 1995), if registered. For example, biological control of lepidopteran insects (such as the grape berry moth) is possible with sex pheromones (Ifoulis and Savopoulou-Soultani, 2004). Release, or attracting to the vineyard, of parasitic wasps, for example *Trichogramma minutum*, which target moth eggs, has resulted in acceptable (i.e. below economic threshold) control of the North American grape berry moth (Nagarkatti *et al.*, 2003) and grapevine leaf roller larvae (Scarratt *et al.*, 2008).

Organic disease control depends on these major factors: climate, location, cultivar and pesticides. For example, in dry climates, downy mildew, phomopsis and anthracnose are usually not significant problems; however, they can present considerable control problems in humid growing regions. Locations away from wild grapes or abandoned vineyards reduce pressure from powdery mildew. Exposure to wind for rapid drying reduces pressure from many fungal diseases as well. Cultivar selection is important in avoiding disease damage, but frequently the cultivars required by the market are susceptible to those diseases most difficult to control, e.g. all *V. vinifera* are susceptible to powdery mildew, downy mildew and black rot. There is an available range of resistance to these diseases among PIWIs and hybrid crosses, but frequently these are not (currently) marketable cultivars, or are not officially recognized by an accrediting agency. Organic pesticides available for disease management include sulphur and copper and their use may be necessary for control of diseases, but some cultivars have a phytotoxic reaction to them, which creates an issue.

Weed control in organic vineyards is more difficult than insect or disease control. Under-row weed management depends primarily on mechanical cultivation. Propane burning, steam or electricity under row is permitted under some organic regulations, but is fuel/energy intensive and difficult to carry out safely (Pinel *et al.*, 1999). These methods must be performed frequently when weeds are small, to prevent potential damage to lower grapevine parts. Those plants with growing tips near or under the soil surface are not well controlled as heat-based methods cannot damage to these protected tissues.

Mulching to prevent or slow weed growth under the row shows promise. Plastic mulches are effective but expensive, may not be considered suitable for organic production and create disposal problems. More sustainable possibilities are the use of waste products, such as mussel shells or crushed glass. The former has shown promise both as a way of reducing early-season damage from grass grub beetle (González-Chang *et al.*, 2017), and as discouraging weed growth and in having a positive effect on the quality of wine made from the 'Pinot noir' grapes growing over it (Creasy *et al.*, 2006).

Removal of between-row weeds can be done mechanically, although it is easy to damage vine roots if done too deeply (Saayman and Van Huyssteen, 1983). The maintenance of continuous grass covering the between-row area is now common, although in dry years or climates this can provide excessive competition and should be eliminated.

Among the goals of pesticide application are the deposition of the recommended quantity of pesticides at the sites needed for effective pest control, and minimization of waste and escape of pesticides into the environment. Applications should also be rapid, and low cost in terms of both the equipment needed and the chemical itself.

Most pesticides are applied by spraying, but in specialized situations solid pesticides can be broadcast as granules, blown as dust, or solutions wiped on with damp rollers or flat pads. Wipers have the advantage of direct contact with weeds away from the crop plant, and avoid the dangers of spray drift to non-target locations.

Maximizing Spraying Efficiency

Vineyard sprayers, wipers or spreaders can be hand carried, tractor mounted or on a trailer: the difference is only one of scale (see Fig. 10.49). A vineyard needs to be covered within a time period that includes the interval when the pest is susceptible and troublesome. It would not be satisfactory to start an application at the right time, but be unable to finish before the disease progressed beyond control, the pest was no longer vulnerable to control or the weather no longer favourable. Most vineyards need application equipment capacity to cover the entire planting within a few days, for reasons such as the traditional sprays of elemental sulphur weathering rapidly, requiring repeat applications, and short windows between favourable weather.

For a fungicide to be effective a specified dose must reach the tissue it is meant to protect. To achieve this, the grapevine canopy must be open enough for the spray to penetrate, the sprayer must be properly calibrated and maintained and the spray applied at the appropriate time. Water-sensitive papers or clay suspensions that are visible when dried are available that allow evaluation of how well the spray is covering the target (Nicholas *et al.*, 1994). This testing should be done whenever a change is made to the spraying setup, and through the growing season as the canopy develops and thickens.

Pesticide applications use quantity–time calculations. The target area must be known, as well as the amount of pesticide required to control the pest. Good coverage is much easier to achieve early in the season, when the canopy is smaller and less dense, as opposed to during the ripening period (bear in mind that the quantity of chemical needed will also change with canopy leaf surface area). Herbicides used under the row require only a percentage of the

Fig. 10.49. Examples of vineyard sprayers (clockwise, from top left): airblast sprayer, fan-assisted sprayer, ATV-mounted, controlled-droplet applicator for herbicides and fan-assisted hydraulic sprayer. Note the safety gear, which is essential when working with most vineyard chemicals, including those registered as organic. Content author's own.

total ground surface area to be sprayed, which must be used in rate calculations to determine the amount of pesticide to apply per unit area.

Pesticides are usually applied in motion, so the speed of application is important in determining the dose of pesticide. A speedometer (if equipped), a tachometer (and a corresponding ground speed table) or a smartphone with GPS can be used to determine speed. Once the rate of application is determined and the total amount is calculated, the carrier is selected – commonly water or possibly a solid. In most cases water and sprayers are used, which is the basis of the example presented here, but application of solids or liquids with wipers use the same types of calculations.

Sprayers pump liquids containing suspended or dissolved pesticides through nozzles that disperse the liquid into small droplets. The size of the droplet is important with regard to how far they travel and how well they impinge and stick to the surface of the plant (Matthews, 2001). Maximum coverage (a high percentage of the surface area of the plant covered by pesticide) is a primary goal of application, though this may not be quite so important with systemic pesticides. Coverage depends greatly on the design of the sprayer and the positioning and density of the vine canopy. One problem with sprayers includes fine mists that can float away from the vine and result in the loss of (often expensive) material; which can create hazards for neighbouring properties or crops. Modern sprayer technology uses different approaches to minimize this problem.

- Droplet size (larger droplets will travel shorter distances) is adjustable by pressure changes and by nozzle design and type.
- Droplets can be given an electric charge to improve their attachment to leaf surfaces (Hislop, 1988).
- Hoods cover the part of the row being sprayed and reduce spray drift (Ozkana *et al.*, 1997; Derksen *et al.*, 1999).
- Hoods can be designed to collect and filter spray not deposited on the plant and return it to the tank (Panneton *et al.*, 2001).

The time required to spray a vineyard depends on the efficiency with which the tractor and sprayer can move between rows (i.e. turning space) and the size of the sprayer tank, because a significant amount of time is spent refilling with water and pesticides and in travel to and from the filling station. The speed of vineyard travel and the number of rows sprayed at a time are also very important. Travel speed can be limited by the condition of the row middles, with high speeds requiring smooth row middles. Multi-row sprayers generally move more slowly than single-row sprayers, but cover more area per unit time (see Fig. 10.50).

Application is also influenced by environmental conditions. The temperature should not be excessively hot, because the phytotoxicity of some pesticides is temperature dependent (i.e. copper and sulphur) (Pertot *et al.*, 2006). Hot and dry conditions can also result in the spray droplet evaporating before it reaches the target, and the pesticide is free to travel in the wind and is thus lost.

Windy conditions favour drifting as well, particularly for droplets that have not hit the target (Reichard *et al.*, 1992). Gusty winds during application may result in later patches of disease development where the fungicide was carried away from the canopy momentarily. The most favourable time for spray applications is in the evening or early morning, when wind is usually at its most calm. Note that insecticides should be applied when activity of non-target organisms is at their lowest, which is often at night. Adequate lighting for night-time spraying is needed so the operator can confirm the operation of all nozzles. The effect of wind can be ameliorated by sprayer designs such as hooded over-the-row sprayers or sensors that automatically shut off spray when passing a gap in the canopy.

Fig. 10.50. Multiple-row spray unit. For effective use, the vineyard trellising and vines must have been installed with high precision, and a skilled, alert, tractor operator is needed. Content author's own.

Pesticides are toxic; they are selected to kill some organism or group of organisms. Therefore, pesticide applicators and all those working within range should be protected from exposure and have proper and relevant training. Personal protection apparatus (PPA) is available for all pesticides, and these vary for the different levels of toxicity. The recommended (or required) PPAs will be listed on the pesticide label, which also incorporates application recommendations, safety requirements and recommended uses (this may vary with country and regulatory agency). It is also good practice to develop a working relationship with neighbors, notifying them of what chemicals are being used, upcoming applications, and precautions you are taking to minimize exposure outside of your property.

Pesticide Application Equipment

Pesticide application equipment should be carefully selected to fulfil the needs of the grower, and be calibrated each season (at least) for the delivery of known amounts of pesticides to the vines (see Fig. 10.51). Calibration is not difficult and should not be neglected. All sprayers have the ability to adjust the rate of pesticide application, at least within its design limits. This is necessary because nozzles and other parts of the spray delivery system wear out, changing the

Fig. 10.51. Equipment made to measure the output of various types of sprayers: each of the upright panels collects the spray that hits it and feeds it to containers at the bottom. The amount of liquid in each is then measured to see how evenly the spray is being applied. Here, an airblast sprayer is being evaluated. Content author's own.

rate at which they deliver material, thus eventually requiring replacement. Nozzles are available that provide varying delivery rates, which is one method of adjusting the rate of spray application and may be necessary for different sprays. Sprayers also have the ability to change the pump pressure, with an increase in pressure increasing the flow through a given nozzle, though most nozzles are designed to operate within a specific range of pressures. Spray patterns and droplet sizes may also change with alterations to pressure, so the effects of any changes should be verified before use in the field. Another control point is the speed at which the sprayer moves through the vineyard. Because of time limitations and surface conditions, changing the nozzles is usually a better way to make significant changes in spray volume.

Calibrating a sprayer requires knowledge of how much liquid is sprayed per minute and the speed at which it travels. A wiper is calibrated the same way, but a duster or spreader of solid uses the weight of material and ground area covered per minute. Note that in most cases pesticides are applied on a unit weight/volume per unit area basis, but for some products it is on a concentration basis (e.g. kg/hl). For this, and many other reasons, it is vital to read the label carefully.

Harvest and Postharvest Processing

11

Abstract: How and when grapes are harvested varies depending on what the target market is. Post-harvest processing and storage procedures can mean the difference between a marketable product and food waste. Global table grape production continues to increase. Because table grapes are seen by the consumer, their appearance is an important quality factor, as is sugar/acid balance, texture and colour. Fruit quality revolves around careful handling and cooling the fruit quickly. Dried grape production has been slowly increasing, supported by a large amount of international trade. "Dried on vine" production methods and increased mechanization have been areas of improvement in recent decades. Grape musts and juices are a lesser produced product, but are important as natural sweeteners, and increasingly, as sources of phenolics. Wine requires a lot of post-harvest processing and the nuances in the final product come from that as well as the grapes themselves.

Unlike many other fruit crops, grapes do not ripen after they have been taken off the vine. There is no conversion of starch into sugar, as the amount of starch in berries is very low: none if you remove the brush from the fruit (Amerine and Root, 1960). Sugar produced by photosynthesis in the leaves is translocated from there to the grapes only until they are harvested or until maximum Brix is reached (around 23.5°, though the concentration of sugar can rise further than this with dehydration of the fruit). Though the concentration of berry solutes can change after harvest, and some other quality-related compounds can be evolved or degraded, in general, grapes must be harvested when the target quality parameters have been reached. Ethylene, considered to be a plant growth regulator, has been implicated in the development of grape berries, particularly for the development of colour (Hale *et al.*, 1970; Weaver and Montgomery, 1974), but it does not play the same role as it does in other, climacteric fruits, such as bananas and apples (Mailhac and Chervin, 2006).

After harvest, grapes may be consumed fresh, crushed or dehydrated. Fresh table grapes are handled, stored and marketed much like other fruits consumed fresh. Crushed grapes may be directly fermented into wine, pressed into juice that may be fermented into wine, consumed fresh or concentrated for later use. Dehydrated grapes or raisins are also a major product of the vine.

DOI: 10.1079/9781800627062.0011

The largest use of grapes in the world is still for making wine, although this is declining. In 2000, 57% was used for wine production, but this has fallen to 45% in 2023, while the amount for fresh consumption increased from 24% to 45% over that time period (OIV, 2024).

Table Grapes

Table grape production has been increasing in recent years, rising from 16 to 32 m tonnes between 2000 and 2023 (OIV, 2024), largely due to planting in China, which is the largest producer and consumer of table grapes: in 2023 they produced 13.5 million tonnes. In production, India is second and Egypt third. Figure 11.1 shows how global production has increased since 2000 in the top six countries, which highlights how much investment China has made in this area.

World trade in table grapes stood at 3.7 million tonnes in 2022, with Peru (which has increased production by more than 1300% since 2000) exporting the largest amount, closely followed by Chile and China: together these three countries contribute 41% of the total trade (USDA, 2025b). Peru, Chile, and South Africa are highly dependent on selling their grapes overseas, shipping approximately 75% of their total production. Although China has invested heavily in table grape vineyards (Fig. 11.1), almost all of its production stays within the country.

Quality parameters

The quality of table grapes depends on sugar concentration, acidity, colour, texture and aroma and flavour characteristics of the cultivar. Though sugar

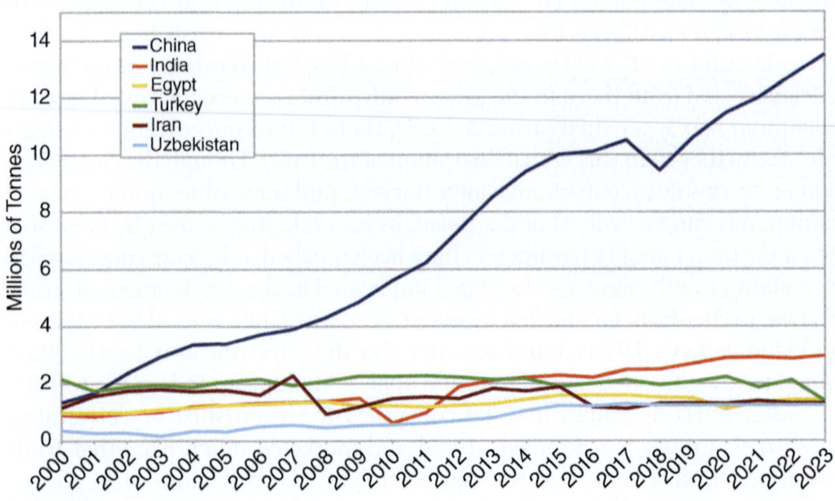

Fig. 11.1. Changes in table grape production from 2000 through 2023 (data from OIV, 2024).

concentration is a common quality measure for grapes, there is not always a good relationship between it and the palatability of grapes for fresh consumption (Winkler, 1932). The ratio of berry juice Brix to acidity can give a better measurement of the palatability of grapes to consumers than either the sugar content or acidity used alone (Winkler *et al.*, 1974), although the best measure of quality varies between cultivars (Guelfat-Reich and Safran, 1971). Climate, cultivar and harvest time affect the Brix : acid ratio. Ripening occurring in hot weather results in lower fruit acid and, generally, a lower sugar content will be sufficient for consumer acceptance (Winkler *et al.*, 1974). In cool climates, the acid will be higher relative to the sugar, so more of the latter will be required for equal taste.

Fruit colour and its uniformity within the cluster are also important (Clydesdale, 1993), and schemas for objective measurement of this have been proposed (Carreño *et al.*, 1995). Berry size is significant, as is the lack of berry shatter (separation of the berries from the rachis), and a green and firm rachis (Winkler *et al.*, 1974). No residues should be visible on the surface of the berry or the rachis. Cluster size and shape must also conform to certain norms (Cliff *et al.*, 1996), and though this is usually modified early in the season, it can be adjusted at harvest, too. Fruit texture is also important, with a crisp flesh generally being desired over those that are softer (Cliff *et al.*, 1996; Sato and Yamada, 2003).

Picking

Table grapes are all picked by hand at present. Pickers must be trained to select well-shaped clusters of mature grapes. Because not all clusters will be suitable for harvest at the same time, frequently there must be multiple harvests from each vineyard (Winkler *et al.*, 1974). Though berries can be tested to determine suitability for picking, this cannot be done for every cluster so visual cues, such as berry size and colour and rachis colour, must be used to determine whether the cluster meets the picking criteria. Any poor-quality berries (e.g. those damaged by sunburn, disease, insects or wind) must be trimmed off because they will be seen by the consumer, unlike grapes destined for processing. If too many berries have to be removed, the fruit may not be marketable due to changes in cluster shape. During picking and packing, grape clusters should be handled only by the stem, to preserve the natural bloom (waxy covering, see Fig. 11.2) as much as possible. Presence of any foreign objects, such as sand, dust, insects or webbing, is enough to cause rejection of the fruit.

Packing

Most table grapes are packed as they are picked to avoid extra handling. Storage and shipping is most common in 10 kg boxes, as larger introduces potential crushing damage and berry shatter. Boxes, which are often wax or plastic coated to improve the container's resistance to wetting, should be kept out of the sun after (and during, if possible) filling and transported as quickly as possible to a cooling facility (see Fig. 11.3). Though grapes do not ripen after they have been removed from the vine they do continue to respire, which uses stored energy in the grapes, and warm fruit loses water more quickly than cool fruit (Szulmayer, 1971). Hence, rapid cooling of the grapes and concomitant slowing of respiration and transpiration preserves their quality, particularly that of the stems.

Grapes destined for immediate delivery are sometimes shipped without pre-cooling, though they should be shipped in a refrigerated container. This is not ideal, as there is usually a considerable amount of residual heat in the grapes that is not efficiently removed in conventionally refrigerated rooms or containers (Mitchell, 1992), and high temperatures encourage disease organisms. In hot climates, delivery can be scheduled after sunset, when temperatures have dropped. Rapid cooling is extremely important for extended storage of table grapes: Nelson (1985) reported that, under Californian conditions, for every hour that grapes are held at 32°C after picking they lose 1 week of storage at 0°C. Crisosto *et al.* (2001) noted that stem browning was associated with delays in fruit cooling, possibly due to excessive water loss during that time. Therefore, grapes should be moved to a cooling facility as soon after picking as possible and be cooled to 1°C within 4–8 hours, with higher temperatures meaning less time is permissible (Thompson *et al.*, 2001).

Fig. 11.2. Close up of a cluster where two berries have been removed. Note the darker areas where the waxy cuticle has been rubbed due to berries touching, destroying the dull-looking bloom. Content author's own.

Fig. 11.3. Grapes in a packing tent in Chile. From here the boxes will be transferred to a cooling and holding facility before shipment to the UK. Content author's own.

If the fruit remains at temperatures above 0°C for too long, there are a number of disease organisms that can become established and decrease fruit quality. Among these are different types of moulds caused by botrytis or *Penicillium* and *Rhizopus* rot (Beattie and Dahlenberg, 1989). The best way to combat these problems is to bring the fruit to as close to 0°C as quickly as possible. At temperatures below this there is a risk of the berries or rachis freezing, which causes catastrophic damage. Grapes of different species have varying temperature thresholds for freezing, with *V. vinifera* grapes generally tolerating greater cold (down to −1°C) than fruit from *V. labrusca* or muscadine cultivars (Himelrick, 2003). Rather than this being a physiological resistance to freezing, it probably has more to do with the amount of solutes in the berries (Jie *et al.*, 2003), as *V. vinifera* tend to have a higher Brix than the other grape types.

Cooling at the packing house

Grape boxes should be designed to facilitate air movement through them so that heat can be removed efficiently. Cooling using water, which is used in other crops, is not feasible with grapes due to the potential for disease spread. Moderate cooling rates can be accomplished in cold rooms (0°C) with high-velocity air movement if boxes are placed with enough room between them so that air can move through easily (Mitchell and Crisosto, 1995). It is much more effective to stack the boxes in rows and pull cold air through the box slots and over the berries (Gentry and Nelson, 1964; Nelson, 1985), as this will result in cooling to a given low temperature in one-sixth to one-eighth the amount of time it takes for room cooling (Mitchell and Crisosto, 1995). After rapid cooling the grapes should be stored at 0°C and 95–98 % humidity, which minimizes respiration, water loss and development of disease.

Storage

Holding cooled grapes at 0°C and high humidity will prolong green stem colour, reduce dehydration losses, berry shatter, and maintain fruit firmness (Crisosto *et al.*, 2001). Treatment with sulphur dioxide (SO_2) is effective in preventing grape decay and also helps prevent browning of the cluster stems. The discovery that sulphur dioxide reduces the activity of many decay organisms – even botrytis, which is active at low temperatures – was made in the 1920s, and its use was rapidly taken up by the California table grape industry (Winkler *et al.*, 1974).

Continuous room fumigation at low (30 ppm) SO_2 concentrations is the most effective treatment for preserving table grapes and has been the standard procedure for many years (Winkler *et al.*, 1974). Sulphur dioxide at high concentrations, however, can cause bleaching of the berries, with a white colour being first noticed at any breaks in the berry skin (Nelson and Ahmedullah, 1970). To avoid high concentrations of sulphur dioxide during shipping, grape boxes can be fitted with in-package, two-stage sulphur dioxide generators that release SO_2 in response to water vapour released by the grapes. The first stage is quick release, high concentration, effective in killing surface disease organisms on the fruit at packing. A slow-release second stage allows for several months of storage (Nelson, 1983), increasing opportunities for lengthening the marketing season and transport to more distant markets.

Some disadvantages of using SO_2 are that it is a colourless gas with a very irritating, pungent odour and it tends to corrode steel pipes and other equipment. Because some consumers (and workers) have sensitivities to sulphur, there is interest in finding alternative decay-controlling treatments. Those being considered include the use of acetaldehyde vapours (Avissar and Pesis, 1991), carbon dioxide-enriched or otherwise modified atmospheres either in storage facilities or through packaging (Crisosto *et al.*, 2002; Candir *et al.*, 2012), ozone gas (Artés-Hernández *et al.*, 2003), hot water dipping (Sabir and Sabir, 2013), application of biocontrol organisms (Liu *et al.*, 2010) and natural volatile plant components. Examples of the latter include (E)-2-hexenal, which has been shown to reduce mould on stored table grapes (Archbold *et al.*, 1999) and volatiles from 'Isabella' grapes (*V. labrusca*), which were shown to reduce botrytis development on *V. vinifera* grapes (Kulakiotu *et al.*, 2004).

Natural essential oils, such as from thyme or fennel, were found to be effective at inhibiting fungal growth, reducing berry weight loss and maintaining rachis freshness without "considerable" effects on fruit flavour (Abdollahi *et al.*, 2012). It may be possible to isolate the most active compounds from these oils and get the beneficial effects without changing the character of the fruit.

Pre-harvest treatments of grapes with UV-C radiation have been trialled for their efficacy in delaying the development of rots in storage, because UV-C stimulates the production of antifungal resveratrol and viniferin compounds (Guerrero *et al.*, 2015). Other pre-harvest treatments have been tested, such as spraying with a chitosan solution during berry development, which was successful in reducing the development of botrytis in storage (Feliziani *et al.*, 2014), and it is possible that resveratrol was involved in this response. Postharvest application of edible chitosan coatings have been used successfully as well, though their cost may limit application to higher value markets (Sabir *et al.*, 2019). Slight vine water stress leading up to harvest has also been shown to reduce the incidence of spoilage during storage (Garganese *et al.*, 2016), but this may have been due to a slight increase in phenolics and/or berry firmness.

Some of these innovative approaches could be acceptable fungicide alternatives for use with organically grown grapes.

Dried Grapes

Preservation through drying is another method of storing grapes. Many fruits are dried, concentrating the sugars inside and making spoilage much less likely. Grapes can do this while on or off the vine. Raisins are any type of dried grapes, and many cultivars are used to make these.

Dried grapes are a heavily traded product, with almost 65% of the yearly global production total being exported to various countries in 2018 (USDA 2025c).

Turkey and the USA are the largest producers of raisins, with 280,000 and 263,000 tonnes, respectively the (45% of the total world production) (OIV, 2025b). The general trend over 2000 to 2023 is a gentle increase with a decline from 2020, where China and Argentina have the greatest percentage increases due to low initial production, and China adding over 100,000 tonnes to the market. Turkey, Iran and the USA are the largest exporters of raisins in the world. Turkey moved 250,000 tonnes to other markets in 2018 (32% of the world total), Iran 16% and the USA 14% (USDA, 2025c).

Note that to make 1 kg of raisins, you need to start with over 4 kg of fresh grapes.

Cultivars

In the marketplace raisins are referred to by generic descriptors such as golden or black, or by names such as sultana or currant. Multiple cultivars could be used to make these products, with the exception of currants, which are dried 'Black Corinth' (also known as 'Zante currant') grapes. Most seedless raisins are made from 'Sultana' (also known as 'Thompson Seedless' and other names) or similar green grapes such as 'Fiesta' or 'Selma Pete'. Seeded raisins have lost market share in recent decades (Alston *et al.*, 1997), but are often made from muscat grapes, such as 'Muscat of Alexandria', that have a distinctive fruity flavour. They can be mechanically seeded, but afterward they become sticky and difficult to eat 'out of hand', so are more frequently used in baking.

The 'Black Corinth' berries used to make currents are parthenocarpic, lacking seeds or even a seed remnant, which results in very small berries that dry more rapidly than larger ones: an advantage that has ensured it being grown in areas that have high humidity or more frequent rain late in the harvest season (Whiting, 1992).

By far, most of the raisins found in consumer markets and in world trade are made from 'Sultana' or similar cultivars, which are desirable for raisin production for a number of reasons: a noticeable and natural bloom, meaty texture and characteristic flavour, not sticky and a low tendency to form sugar crystals or cake in storage (Winkler *et al.*, 1974). With increasing adoption of dried-on-vine production of raisins, there will be a shift towards cultivars best adapted to the practice (see below).

Harvest and drying

Harvest criteria for raisins revolve around improving the final yield of dried fruit, of which sugar concentration in the fresh fruit is a major component. As the concentration of sugar in the grapes rises, so does the yield of finished raisins, according to the drying ratio, which is the mass of fresh grapes it takes to produce 1 kg of raisins at 15% moisture (Christensen and Peacock, 2000b). Because gross returns are based on the amount of raisins produced, starting with higher-Brix fruit provides more money per tonne of fresh grapes. Raisin quality also improves with increasing Brix (Winkler *et al.*, 1974), though achieving higher sugars often carries with it a greater risk of rainfall late in the season.

Traditionally, clusters of grapes were harvested by hand and laid onto trays between the grapevine rows. Exposure of the fruit to the sun hastens water loss, and berries shrivel and brown to the point where they no longer express juice easily and can be handled neatly. More recent improvements to this method have been the adoption of techniques to minimize handling, such as laying the grapes on continuous sheets of paper that could be laid down mechanically as well as picked up for harvesting of the dried fruit (Christensen, 2000). Soil between the rows can also be moved to create a slope that angles towards the sun, which enhances drying rates (Winkler *et al.*, 1974; Christensen and Peacock, 2000b). Part of this phenomenon may be that warmer grapes lose water at a faster rate than cooler grapes (Szulmayer, 1971). Turning of the grapes, once practised to improve the drying rate, appears not to be necessary in most situations (Christensen and Peacock, 2000b), saving much in labour costs. Only when fruit is damaged or clusters are unusually large is this practice worth the extra cost.

The very first mechanical harvester developed, mentioned in Chapter 9, was to help reduce labour inputs for raisin production. It was not a commercially successful design, but the promise it held led to the development of other types, culminating in the rod-type shaking harvesters that are also used in the wine grape industry. However, for this to be truly successful, a new method of drying the fruit was needed.

The beginnings of dried-on-vine (DoV) production came about with observations in California around 1965 that the fruit on canes cut from the vines was easily dislodged with mild shaking (Studer, 2000). At about the same time, Australian researchers May and Kerridge (1967) published a method for cane cutting prior to harvest followed by shaking the fruit off the vines for completion of the drying process. Both groups realized the ability to shake the partially dried fruit off the vines opened up the possibility for mechanical harvest. Notably, both did their experiments with the 'Sultana' grape, which appears to be most suitable because 1) the pedicel dries and becomes brittle rapidly after the canes are cut, meaning that it remains attached to the partially dehydrated berries, and 2) that there is little skin damage when it separates, preserving raisin quality (Studer, 2000).

Once machines could be used to harvest fruit, development of a device that could lay the grapes out on to continuous drying sheets was not far off (see Fig. 1.5; Studer and Olmo, 1971). More modern DoV techniques perform the drying almost to completion on the vine, with no need to place grapes on trays between rows (Peacock and Swanson, 2005). The procedure can be almost completely mechanized and with the development of cultivars that better allow machine harvesting of the fruit (i.e. the berry separated cleanly; see Fig. 9.5), there are several options for growers (Clingeleffer 2013). Aside from the significant savings in production costs, this method results in higher grades of raisins due to the different drying environment of the grapes, though consumer preference appears to be largely unaffected (Angulo *et al.*, 2007).

Future refinements to the DoV system have come with the development of new cultivars that do not require cane cutting, thus increasing savings. 'Sunpreme' is one such grape, patented in 2013 (Pollock, 2017), but it is not perfect: partially dried grapes fall from the vine fairly easily, meaning conventional harvesters will shake the fruit off before the harvester reaches the vine. A critical advantage is early ripening, because DoV drying occurs in the shade of the canopy and takes around four times as long as tray drying (Espíndola *et al.*, 2019).

Rain is a serious threat to raisin production, and was found to be responsible for a significant amount of the variation in production in Australia (Considine, 1973). Those sites suitable for raisin production must be good grape-growing areas and have a very low probability of rain during and after the harvest season. Raisins can be harvested or picked up from the aisles if sufficient warning of impending rain is available. They can finish drying in a dehydrator if necessary, though this is an expensive process. Because of the impact weather events have on production, forecasting is extremely important to raisin growers (Lave, 1963). In the raisin-producing regions of California, Turkey and Iran, rain is very rare during the drying period. In countries where there is even a slightly greater chance of rain, frequently raisins are dried by placing the grapes on stacked wooden racks (see Fig. 11.4). The intensity of the sun,

Fig. 11.4. Grape drying racks in Australia. If necessary, covers can be set over these racks to protect the grapes from rainfall later in the season. Content author's own.

temperature and air movement are satisfactory for production of raisins but, in case of occasional rain, the stacks can be covered. In most areas the probability of rain increases as autumn approaches, and the harvest decision for raisin grapes is a compromise between obtaining the most mature grapes and getting them dried economically before rain. Climate change will increase uncertainty around rain events, which will complicate in-vineyard drying.

Raisins dried in the sun are invariably dark coloured due to the oxidation of phenolic compounds. However, lighter-coloured raisins (often called sultanas or sultana raisins) can also be produced. For 'Sultana' (the cultivar) this used to be performed by dipping the fruit in hot caustic soda, which caused tiny cracks in the cuticle of the berries, rinsing them in cool water and then exposing them to sulphur fumes for several hours to prevent oxidation. These golden raisins are finally dried in a dehydrator at 60–70°C to reduce the water content to approximately 15%. The combination of sulphur and increased rate of drying reduces browning reactions (Grncarevic and Radler, 1971). A modification of the method, but using cold emulsions, has been used to hasten water loss commercially. The technique was originally used by the Greeks and Romans, where grapes are dipped in a mixture of olive oil and wood ash (Whiting, 1992). The drying rates of grapes so treated approximate those of peeled berries (Grncarevic, 1963), which in some ways is astonishing! Therefore, the use of emulsions brings the drying time for grapes down from 20 to 8–10 days in the sun, or in a dehydrator from 3–5 days to 24–30 hours (Grncarevic and Radler, 1971).

Golden raisins, which are very light in colour, are produced by dipping the fresh fruit (usually 'Sultana') in a very hot solution of caustic soda and sodium sulphite for just a few seconds (Winkler *et al.*, 1974). This process creates fine fissures in the berry skin, enhancing the drying rate. Following the dip, fruit are subjected to additional gaseous sulphiting and then forced-air dried. The action of the sulphites produces the yellow golden colour, which is preferable for some raisin uses. Modification to this procedure, using sulphite solutions, has been investigated for the purpose of reducing sulphite emissions and increasing the efficiency of sulphite transfer to the fruit (Lydakis *et al.*, 2003).

Other methods have been tested for drying grapes. Microwaves as a pretreatment with or without alkali dipping was found to increase the rate of drying significantly over dipping alone (Tulasidas *et al.*, 1993; Kostaropoulos and Saravacos, 1995), and also resulted in lighter colour due to the higher drying rate or possibly partial de-activation of polyphenol-oxidase, the principal cause of enzymatic browning (Kostaropoulos and Saravacos, 1995). Pre drying treatment of grapes with olive oil was also able to improve drying times while preserving raisin quality as it disrupted the cuticle layer (Shimizu-Marin *et al.*, 2024).

Processing

Once the average moisture of raisins has been reduced to 15%, they are conditioned to allow the moisture content to equilibrate, with low-moisture berries gaining some water from the higher-moisture berries. This reduces the variation in the raisins, making sorting and quality grading more efficient. Put into large boxes, they are then taken to a processing plant (see Fig. 11.5).

Raisins are usually graded in an airstream sorter (Baranek *et al.*, 1970): they are fed into a stream of air flowing upward and, due to their volume : density ratio, either fall downstream to be collected in a bin or are blown upwards to a different bin. Since raisin production often results in close proximity of the soil to the fruit, contamination with soil and sand is a possibility and the amount of grit mixed with the raisins is sometimes used as another quality parameter (Winkler *et al.*, 1974).

With optical sorting technology improving, this is a possibility for raisins as well. A proof of concept was demonstrated in 2017 (Shinde and Patil, 2017) using colour detection, and now there are optical (Raihen and Akter, 2023) and hyperspectral (Khiari *et al.*, 2025) systems that could detect size, shape, texture, composition and defects like sand contamination or disease.

Raisins (unlike grapes) do not respire and can be stored and shipped under less controlled conditions. However, they are still subject to insect infestation and disease, which must be managed mainly through exclusion, as the product is to be consumed directly.

Juice/Preserves

Grapes can be processed into juices, jellies and related products such as fruit leather. *V. labrusca* fruit tends to be high in pectin and therefore is quite amenable

Fig. 11.5. Raisins as they enter a processing plant. Note the material other than grapes, which needs to be removed before progressing. Content author's own.

to making jellies and fruit butter (a product that is one step away from fruit leather). Grapes can also be added to low-pectin fruit to help jellies or jams set. The Welch family, based in New York State, was the first to sell grape juice in the 1800s, and branched out into other products such as syrups and jellies. Today, a variety of grapes are used to make numerous grape end products, but they are also components in the manufacture of other food items to add colour, flavour or sweetness.

Fresh grape juice includes purple grape juice, traditionally made from the 'Concord' grape and common in the USA. Purple 'Concord' juice was introduced by Welch in 1869 and made possible by the commercialization of the experiments of Louis Pasteur (pasteurization), which killed spoilage organisms and allowed the juice to keep much longer. Before this, it was not possible to store juice as it would ferment naturally. The impetus behind pasteurized grape juice at that time came with the Temperance movement, as some people wished to find a substitute for wine used in religious ceremonies (Bailey, 1912). The imposition of national prohibition in the USA in 1920 created a great demand for 'Concord' grapes, which were planted heavily in the eastern states of the country. Today, they are grown mainly in the east, but Washington state produces a good portion of the total.

Compared to data collected for wine, fresh and dried grape production, there is relatively little information specifically about juice production. According to the OIV (2024), 2.7 million tonnes of grapes were used to produce about 22 million litres of juice and must, which is just under 4% of world production. In the USA in 2014, 500,000 tonnes of grapes were used for juicing, with Washington State producing over 50% of that, New York State 24%, Pennsylvania 13% and Michigan 11% (NASS, 2017). In 2023, only data from Washington state was available, which indicated 163,000 tonnes of grapes were used for juice production (USDA, 2024).

The vast majority of juice in the USA is produced from 'Concord', with around 10% being pressed from 'Niagara' (a green labrusca-type grape). Although California used to produce a large quantity of juice from *V. vinifera* cultivars (Heien and Venner, 1993), this seems to have been dropping precipitously as growers move to higher profit margin crops (Ciatti, 2016). In contrast to the *V. labrusca* grapes, those from *V. vinifera* tend to be neutrally flavoured. Because of its versatility 'Sultana' was often made into juice if the bulk wine, table grape or raisin markets were poor. In the 1990s and early 2000s as much as 65% of juice produced in California was from 'Sultana' (Paggi and Yamazaki, 2005). Much smaller quantities of juice are made from muscadine grapes (Olien, 1990) or from strongly flavoured *V. vinifera* cultivars like those in the muscat family. While the juice of these and other cultivars of grapes is sometimes sold on its own, it is more often used as an ingredient in other products, often as a natural sweetener. Juice is often concentrated to save on shipping costs and then diluted to suit at the point of use (Bates *et al.*, 2001).

Uses

Grape juice concentrate is used frequently as a food ingredient and has the advantage of being a means of adding sugar to other products in a natural form. Generally, white grape concentrates are used as sweeteners and juice stock while red grape concentrates are used to provide colour and as stock for red grape juice, although pigments from 'Concord' grape juice have been tested as colorants for other beverages and also for gelatin products (Clydesdale *et al.*, 1978). There are several commercial grape-derived concentrates on the market specifically sold to be used as colour enhancers (e.g. Mega Purple, Purple 8000).

Grape juices sweetened and mixed with grape juice concentrate can be labelled as 100% grape juice, while other fruit juices can be labelled as 100% juice (Tipton *et al.*, 1999). Concentrates and juice have also been used as wine additives as a way to sweeten and flavour the product naturally.

Since grape skins and seeds are high in phenolic antioxidants (Kanner *et al.*, 1994), there has been interest in using concentrates or extracts as a way of increasing the potential health benefits of food products or dietary supplements (Sridhar and Charles, 2021). In recent years there has been a huge increase in the use of grape-derived compounds as components of cosmetics, a trend that is predicted to continue (Castro *et al.*, 2023).

Grape juice and must are normal products of international trade; however, accurate data for the production of these are difficult to find as there is uncertainty in terms of the data collected over its eventual end use (OIV, 2003). Much of it is used as a food additive or as fresh juice. A total of 525,000 tonnes of juice and must were exported in 2023, with Italy, Spain and Argentina topping the list at 29%, 28.6% and 14.1%, respectively (ReportLinker Research, 2025). The trend from 2007 to 2023 has been gradual decline, with much less volatility in the last five years. This could be because domestic markets are absorbing more of the production as there is a shift towards natural and more healthful beverages.

Processing

Harvest of grapes for juice production is mostly by machine due to low profit margins. Red and purple grape juices are made with a hot press process that extracts many of the skin components, most importantly phenolics, which includes the pigments (anthocyanins), as well as sugar and flavour compounds (Pederson, 1971). Juice phenolics/colour and flavour can be affected by the maximum temperature to which the juice is heated, so provided that the process regulates this temperature closely, unwanted organisms can be killed and juice quality largely unaffected (Flora, 1976). Recall also that the perceived colour of a solution of anthocyanins is dependent not just on the concentration of anthocyanin pigments, but also the pH of the solution: more acidic juices with lower pH will express more intense colour than higher pH juices.

White grape juices are made with a cold press process, leaving many skin components behind. This can be advantageous if a neutrally flavoured juice is desired, as flavour components are sequestered in just the skin cells of some cultivars.

Extraction can also be enhanced with treatment of the macerated grapes, for example, with pectolytic enzymes (Bates *et al.*, 2001). There is also an increasing focus on more sustainable methods of extracting bioactives (Ilyas *et al.*, 2021), which includes the use of microwaves and ultrasound (Ghassempour *et al.*, 2008). Sulphur dioxide is commonly used to prolong the quality of wine or juice grapes waiting to be processed. This is particularly useful with mechanically harvested grapes that are susceptible to decay and oxidation (Morris *et al.*, 1972). Research suggests that UV-C radiation may be an alternative to addition of SO_2 or heat treatment to prevent spoilage. Compared to heat treatment, UV-C treatment of red juices resulted in a lower amount of colour degradation, which could be a distinct advantage in some situations (Pala and Toklucu, 2013).

It is notable that the storage temperature of 'Concord' juice was found to have an influence on its quality – at temperatures of 35°C the loss of colour and desirable flavours was greater than if the juice had been stored at 24°C (Morris *et al.*, 1986). This demonstrates the importance of postharvest factors that are sometimes out of the control of the producer or marketer.

Wine

Total world wine production was 237 million hl in 2023, with EU countries producing about around 60% of this (OIV, 2024). The major producing countries are France (48 million hl), Italy (38 million hl),, Spain (28 million hl), the USA (24 million hl). Italy and France vie for top spot, with seasonal weather determining the size of harvest, though the overall trend has been for reduced production due to decreasing vineyard area. Production varies significantly from year to year due to weather events such as frosts, drought or winter freezes. For example, drought affects those countries with greater proportions of un- irrigated vines, like Spain, and frosts those that have most of their grapes grown in marginal areas. With the increased likelihood of extreme and unexpected weather events due to climate change, we may see global, and certainly regional production vary to an even greater extent in the future.

There is little doubt that wine is an important part of many cultures, leading to its production in virtually every area that can support vine growth. Sometimes this is only because of the very significant efforts of man to make it possible. World trade in wine is significant (perhaps because other cultures appreciate that wine has a sense of place) and increased from 2001 to 2011, but has been relatively static since then (OIV, 2024). Italy, Spain and France are the top wine exporters in the world, while Germany, the UK and USA are the top importers of wine (OIV, 2024).

Harvest

The target grape maturity for wines depends on the wine being produced. Grapes for sparkling wine production, for example, are harvested at a lower Brix than those for still table wines (Zoecklein, 2002), and grapes for late-harvest wines are picked when the sugar levels are higher than for table wines. Grapes for most wines, however, require sufficient acid and sugar to result in the desired pH, alcohol content and flavour, and also acceptable varietal aroma, to produce a quality wine.

Most grapes intended for wine are harvested mechanically (Fig 6.66), although grapes for some wines may be picked by hand to select for even maturity, to avoid poor-quality grapes, and preserve the berries and clusters from damage. For example, mechanical harvesting cannot distinguish between the primary crop and second set crop that may develop on shoot laterals, resulting in a lower-quality must (Clary *et al.*, 1990). Thus, in some cases, pickers may go through a vineyard to remove immature or diseased clusters before mechanical picking. As there is mechanical damage to the fruit during the process, SO_2 may be added to the harvesting bin to slow oxidation while the fruit is transported to the winery (Ough and Berg, 1971). In some areas where the wineries are distant from the vineyards, grapes are harvested, crushed and the juice pressed out near the vineyard and then transported by tanker truck to the winery, which can take upwards of 24 hours or more (Pocock *et al.*, 1998). Adding SO_2 to the juice at crushing prevents oxidation and microbial spoilage during the long transport times.

Hand-harvested grapes are picked into small baskets or bins that are carried to larger bins of up to 500 kg capacity – the size of these being limited by how much weight the grape clusters at the bottom can carry before crushing of the berries (Hamilton and Coombe, 1992). Bins are usually lined with plastic to retain any juice that may escape from the fruit, e.g. during transport.

Machine harvesting has the advantage that it is much faster and cheaper than hand picking and that harvest can continue through the night without the need for large flood-lamps for people to be able to see which clusters they are choosing (and where their fingers are in relation to the picking shears!). Night picking in warm climates is desirable, as it means that the fruit arrives at the winery much cooler than if picked during the day, which results in less loss of quality and cost to cool the fruit (Fairbank *et al.*, 1987).

Processing

Hand-harvested fruit should arrive as whole clusters at the winery, which can be transferred directly to a press (see Fig. 11.6) and juice extracted from the

Fig. 11.6. Left: a hydraulic, basket-type press, which is often used for whole- bunch fruit. Right: a large bag-press, which in this case is mounted on rails so it can be moved along a row of tanks. Content author's own.

undamaged fruit. This yields a lighter juice, usually of more delicate quality and with fewer phenolics, though there is less volume per tonne of grapes obtained. This process is widely used for sparkling wine grapes and some white wine styles. The juice is then fermented with yeasts, either added or naturally occurring, which convert the sugar to alcohol and carbon dioxide.

In some cases, hand-harvested grapes are sorted before they are put into the press or crusher/de-stemmer. This is a last point of quality control, where, on sorting tables a crew removes diseased fruit, material other than grapes (MOG), or fruit that is not ripe enough. However, if the fruit is hand-harvested, it is much more efficient to instruct and train the pickers to send only desirable fruit rather than sorting at the winery. Post-harvest optical sorting successfully removed less-colored berries and MOG and has the potential to improve wine quality (Bruce *et al.*, 2021).

Fruit entering the winery is usually sampled to acquire information about Brix, acidity, pH, MOG and disease. Payment for grapes is mainly dependent on these factors, so obtaining an accurate and representative sample is important for both the grower and winery. A remote-controlled sampling arm can be used in large bins of fruit, which can take a sample spanning the top to the bottom of the load (see Fig. 11.7). Some of the sample is used for testing, another portion is frozen for later reference if needed, for example, if pesticide residue is found in a wine the grapes that went into that wine can be examined to see which batch was responsible.

Fig. 11.7. An example of a remote sampling arm, which is used to take a representative sample of grapes from a large bin on arrival at the winery. Content author's own.

Fruit destined for white or rosé wines are generally destemmed and crushed (in a machine with rotating tines and a grating that removes the berries from the rachis, followed by closely placed rollers that break open the berries, Fig. 11.8), and then pressed either right away or after a short time to allow for extraction of aroma and flavour compounds out of the skin. For rosé wines the longer the time on skins, the greater the colour extraction. The juice is pumped

Fig. 11.8. A crusher/de-stemmer. Grapes are fed into the chute at top right and the rachides are removed in the cylindrical section at middle left. Below the de- stemmer are rollers that crush the fruit, which is then emptied out into the vessel below, where the must is pumped into a press or temporary holding tank. Content author's own.

to a settling tank, where solids suspended in it have a chance to collect at the bottom through gravity, or the juice can be filtered or treated with proteins in a process called flotation. Mostly clear juice is racked off the top prior to fermentation, as if too many solids are retained off odours can develop in the wine and quality is reduced (Singleton *et al.*, 1975), but the presence of some solids is beneficial for yeast-mediated fermentation and can contribute positive characteristics, like improved mouthfeel or 'struck match' aromas, which are preferred by a segment of the market (Espinase Nandorfy *et al.*, 2023).

As for white wine grapes, clusters of grapes for red wines are usually crushed and de-stemmed, resulting in a must, which is the soupy mix of the skins, seeds and pulp of the fruit. However, rather than pressing the juice out at this state, yeast is added and fermentation is carried out with the skins and seeds in contact with the juice . When fermentation has finished the alcoholic wine is drained away, leaving the skins and seeds to be transferred to a press where the remaining wine is extracted.

The cake of pressed grape skins, either those from fresh grapes destined for white wine or those for red wine that have been through fermentation, are a source of many antioxidant compounds (Larrauri *et al.*, 1996). As winemaking does not extract everything from the grapes, sugar and water can be added back to the pressed grapes, which are then re-fermented and the resulting wine distilled to make grappa or marc. These high-alcohol spirits often contain flavours quite reminiscent of the grape itself. Or, the solids may be extracted again to produce antioxidant materials useful in other markets.

Other Wine Styles

Passito wines, or wine made from dried grapes as the term is used in Italy, are another category of wine types. It is not unusual to make wine from raisined grapes, as is the case for wines made from botrytis-affected 'noble rot' fruit, but it is rarely made using red grapes. The exceptions to the rule are recioto (sweet)- and amarone (dry)-style wines made in Italy, where red-skinned grapes are harvested at ripeness and laid out on trays (much like those in Fig. 11.4, but under shelter) to lose water and concentrate sugars and flavour compounds at moderate temperatures (Versari *et al.*, 2001). Because the grapes are under shelter and do not dry out as quickly as in raisin production, the flavour profile is said to be unique (Belfrage and Loftus, 1993). Under conditions like these, botrytis is likely to become established in the fruit, which is fine for white grapes as long as it is of the 'noble rot' type, but it is not desirable in red grapes (Belfrage and Loftus, 1993) because of the negative effects on flavour and color.

Icewine is another style of sweet wine that is out of the ordinary. The special characteristics of this product come from the fact that, as water freezes in a berry, it concentrates the remaining solutes. So, if grapes are picked and pressed at very cold temperatures, such as –8°C or below (Pickering, 2006), a very high-Brix juice can be extracted and fermented into a very sweet wine. Understandably, there are few areas in the world that have the right combination of climate and cultivars, as well as hardy people available to work under such conditions. Traditionally, these wines were made in Germany and Austria, but more recently Canada has made it a focus and is producing more and more (Pickering, 2006), and China has areas with appropriate climate, too, and is ramping up production (Patton, 2006). Ice wines (note the space between the words) may also be produced, but these are made from grapes picked before winter and then put in a freezer to concentrate the sugars prior to pressing and fermentation.

Interest in lower-alcohol wines is increasing, and while some of this market is made up of conventionally produced wines that have the alcohol removed using various types of equipment, there is a growing segment where the grapevines are specially managed during the ripening period to result in grapes that have lower sugar (and therefore lower-alcohol potential), but flavours and aromas that are similar to traditionally produced wines. This is most commonly achieved through shoot trimming around the time of veraison, which changes the leaf area to fruit weight ratio. This is thought to de-synchronize the ripening of the various components in the berry: slow the rate of sugar accumulation in the fruit, but allow time for organic acids concentrations to drop and for aroma compounds to develop (Parker *et al.*, 2015; Caccavello *et al.*, 2017). The market for this type of wine is expected to grow (Sekhon *et al.*, 2024), so we can expect to see further innovation in this sector.

References

Aballay, E., Prodan, S., Correa, P. and Allende, J. (2020) Assessment of rhizobacterial consortia to manage plant parasitic nematodes of grapevine. *Crop Protection* 131, 105103. https://doi.org/10.1016/j.cropro.2020.105103.

Abdollahi, A., Hassani, A., Ghosta, Y., Bernousi, I., Meshkatalsadat, M.H., Shabani, R. and Ziaee, S.M. (2012) Evaluation of essential oils for maintaining postharvest quality of Thompson seedless table grape. *Natural Product Research* 26, 77–83. https://doi.org/10.1080/14786419.2010.541887.

Abel, N., Baxter, J., Campbell, A., Cleugh, H., Fargher, J., Lambeck, R., Prinsley, R., Prosser, M., Reid, R., Revell, G., Schmidt, C., Stirzaker, R. and Thorburn, P. (1997) *Design Principles for Farm Forestry*. Rural Industries Research & Development Corporation/Land and Water Resources R&D Corporation/Forest and Wood Products R&D Corporation Joint Venture Agroforestry Program, Canberra, Australia. ISBN: 0642246556.

Adams, L.D. (1985) *The Wines of America*. McGraw-Hill Book Company, New York. ISBN: 9780070003194.

Adrian, M., Rajaei, H., Jeandet, P., Veneau, J. and Bessis, R. (1998) Resveratrol oxidation in *Botrytis cinerea* conidia. *Phytopathology* 88, 472–476. DOI 10.1094/PHYTO.1998.88.5.472.

Agüero, C.B., Uratsu, S.L., Greve, C.L., Powell, A.L.T., Labavitch, J.M., Meredith, C.P. and Dandekar, A.M. (2005) Evaluation of tolerance to Pierce's disease and Botrytis in transgenic plants of *Vitis vinifera* L. expressing the pear PGIP gene. *Molecular Plant Pathology* 6(1), 43–51. DOI 10.1111/j.1364-3703.2004.00262.x.

Aguín, O., Mansilla, J.P., Vilarino, A. and Sainz, M.J. (2004) Effects of mycorrhizal inoculation on root morphology and nursery production of three grapevine rootstocks. *American Journal of Enology and Viticulture* 55(1), 108–111. DOI: 10.5344/ajev.2004.55.1.108.

Ahmad, M.F. and Zargar, G.H. (2005) Effect of trunk girdling, flower thinning, GA3 and ethephon application on quality characteristics in grape cv. Perlette under temperate Kashmir valley conditions. *Indian Journal of Horticulture* 62(3), 285–287.

Al Rwahnih, M., Dave, A., Anderson, M.M., Rowhani, A., Uyemoto, J.K. and Sudarshana, M.R. (2013) Association of a DNA virus with grapevines affected by Red Blotch Disease in California. *Phytopathology* 103, 1069–1076. DOI 10.1094/PHYTO--10-12-0253-R.

Allen, M., Lacey, M.J. and Boyd, S.J. (1997) Methoxypyrazines of grapes and wines: new insights into their biosynthesis and occurrence. *Proceedings of the Fourth International Symposium for Cool Climate Viticulture and Enology*, 16–20 July 1996, Rochester, New York. Agricultural and Experimental Station, Geneva, New York, pp. V, 36–39.

Alleweldt, G. and Possingham, J.V. (1988) Progress in grapevine breeding. TAG Theoretical and Applied Genetics 75(5), 669–673. DOI 10.1007/BF00265585.

Alston, J.M., Chalfant, J.A., Christian, J.E., Meng, E. and Piggott, N. (1997) *The California Table Grape Commission's Promotion Program: an Evaluation*. Paper 43, Giannini Foundation of Agricultural Economics, University of California, 133 pp.

Amerine, M.A. and Roessler, E.B. (1976) *Wines. Their Sensory Evaluation*. W.H. Freeman and Company, San Francisco, California. ISBN 978-0716705536.

Amerine, M.A. and Root, G.A. (1960) Carbohydrate content of various parts of the grape cluster. II. *American Journal of Enology and Viticulture* 11(3), 137–139.

Amerine, M.A. and Winkler, A.J. (1944) Composition and quality of musts and wines of California grapes. *Hilgardia* 15, 493–675. DOI 10.3733/hilg.

Amerine, M.A., Roessler, E.B. and Filipello, F. (1959) Modern sensory methods of evaluating wine. *Hilgardia* 28, 477–567. DOI 10.3733/hilg.v28n18p477.

Anderson, L.J., Comas, L.H., Lakso, A.N. and Eissenstat, D.M. (2003) Multiple risk factors in root survivorship: a 4-year study in Concord grape. *New Phytologist* 158, 489–501. DOI 10.1046/j.1469-8137.2003.00757.x.

Angot, A. (1883) Étude sur les vendanges en France. *Annales du Bureau central météorologique de France*, 1, B29-B120.

Angulo, O., Fidelibus, M.W. and Heymann, H. (2007) Grape cultivar and drying method affect sensory characteristics and consumer preference of raisins. *Journal of the Science of Food and Agriculture* 87(5), 865–870. DOI 10.1002/jsfa.2790.

Annabell, R. (2001) *Novel Way to Fight Frost Threat*. Wairarapa Times-Age, Masterton, New Zealand.

Anon. (2007) *Phylloxera resistant Rootstocks for Grapevines*. Hellman, E., Northwest Berry and Grape Information Network, 2008.

Anon. (2008) *Yalumba Nursery Rootstock Selector*, Yalumba Nursery, Australia, 2008.

Anon. (2020) Code de l'Environnement Article L213-10-8. *Journal Officiel "Lois et Décrets"* 0315. 30 December 2020 https://www.legifrance.gouv.fr/codes/article_lc/LEGIARTI000042912003/2021-02-14 (accessed 15 April 2025).

Anstey, T.H. (1966) Prediction of full bloom date for apple, pear, cherry, peach, and apricot from temperature data. *Proceedings of the American Horticultural Society* 88, 57–66.

Antcliff, A.J. and Webster, W.J. (1955) Studies on the Sultana vine I. Fruit bud distribution and bud burst with reference to forecasting potential crop. *Australian Journal of Agricultural Research* 6(4), 565–588. DOI 10.1071/AR9550565.

Antcliff, A., Webster, W. and May, P. (1957) Studies on the Sultana vine. V. Further studies on the course of bud burst with reference to time of pruning. *Australian Journal of Agricultural Research* 8(1), 15–23. DOI 10.1071/AR9570015.

Aravind, K.R., Raja, P. and Pérez-Ruiz, M. (2017) Task-based agricultural mobile robots in arable farming: A review. *Spanish Journal of Agricultural Research* 15, 02–01. DOI 10.5424/sjar/2017151-9573.

Arbabzadeh, F. and Dutt, G. (1987) Salt tolerance of grape rootstocks under greenhouse conditions. *American Journal of Enology and Viticulture* 38(2), 95–99.

Archbold, D.D., Hamilton-Kemp, T.R., Clements, A.M. and Collins, R.W. (1999) Fumigating 'Crimson Seedless' table grapes with (E)-2-hexenal reduces mold during long-term postharvest storage. *HortScience* 34(4), 705–707.

Aroca, A., García-Figueres, F., Bracamonte, L., Luque, J. and Raposo, R. (2006) A survey of trunk disease pathogens within rootstocks of grapevines in Spain. *European Journal of Plant Pathology* 115, 195. DOI 10.1007/s10658-006-9008-5.

Artés-Hernández, F., Artés, F. and Tomás-Barberán, F.A. (2003) Quality and enhancement of bioactive phenolics in Cv. Napoleon table grapes exposed to different postharvest gaseous treatments. *Journal of Agricultural and Food Chemistry* 51, 5290–5295. DOI 10.1002/jsfa.2780.

Ascard, J. (1998) Comparison of flaming and infrared radiation techniques for thermal weed control. *Weed Research* 38(1), 69–76. DOI 10.1046/j.1365-3180.1998.00073.x.

Aubert, C., Baumes, R., Gunata, Z., Lepoutre, J.P., Cooper, J.F. and Bayonove, C. (1998) Effects of sterol biosynthesis inhibiting fungicides on the aroma of grape. *Vitis* 18(1), 41–58.

Aubin, T. (1990) Synthetic bird calls and their application to scaring methods. *Ibis* 132(2), 290–299. DOI 10.1111/j.1474-919X.1990.tb01046.x.

Augsburger, H.K.M. (2000) Frost control in temperate climates through dissipation of cold air. *Aspects of Applied Biology* 61, 201–204.

Avissar, I. and Pesis, E. (1991) The control of postharvest decay in table grapes using acetaldehyde vapours. *Annals of Applied Biology* 118(1), 229–237. DOI 10.1111/j.1744-7348.1991.tb06101.x.

Ayres, M.R., Billones-Baaijens, R., Savocchia, S., Scott, E.S. and Sosnowski, M.R. (2022) Critical timing of fungicide application for pruning wound protection to control grapevine trunk diseases. *Australian Journal of Grape and Wine Research* 28, 70–74. https://doi.org/10.1111/ajgw.12517.

Baccari, C. and Lindow, S.E. (2010) Assessment of the process of movement of *Xylella fastidiosa* within susceptible and resistant grape cultivars. *Phytopathology* 101, 77–84. DOI 10.1094/PHYTO-04-10-0104.

Bailey, L.H. (1912) *Cyclopedia of American Agriculture: Vol. II – crops*. Ayer Company Publishers, Manchester, New Hampshire.

Bailey, L.H. (1933) The species of grapes peculiar to North America. *Gentes Herbarium* 3, 150–244. http://hdl.handle.net/1813/17799.

Bailey, P.T. (1989) The millipede parasitoid *Pelidnoptera nigripennis* (F.) (Diptera: Sciomyzidae) for the biological control of the millipede *Ommatoiulus moreleti* (Lucas) (Diplopoda: Julida: Julidae) in Australia. *Bulletin of Entomological Research*, 79(3), 381–391. https://doi.org/10.1017/S0007485300018381.

Bailey, P. (2016) *Fact sheet: Portuguese millipedes* (p. 7). South Australian Research and Development Institute. https://pir.sa.gov.au/__data/assets/pdf_file/0009/285516/Portuguese_Millipedes_Fact_Sheet.pdf (accessed12 April 2025).

Bajwa, A.A., Mahajan, G. and Chauhan, B.S. (2015) Nonconventional weed management strategies for modern agriculture. *Weed Science* 63, 723–747. DOI 10.1614/WS-D-15-00064.1.

Baker, N.R. (2008) Chlorophyll Fluorescence: A probe of photosynthesis *in vivo*. *Annual Review of Plant Biology* 59, 89–113. DOI 10.1146/annurev.arplant.59.032607.092759.

Baranek, P., Miller, M.W., Kasimatis, A.N. and Lynn, C.D. (1970) Influence of soluble solids in 'Thompson Seedless' grapes on airstream grading for raisin quality. *American Journal of Enology and Viticulture* 21(1), 19–25.

Barbaro, L., Rusch, A., Muiruri, E.W., Gravellier, B., Thiery, D. and Castagneyrol, B. (2017) Avian pest control in vineyards is driven by interactions between bird functional diversity and landscape heterogeneity. *Journal of Applied Ecology* 54, 500–508. https://doi.org/10.1111/1365-2664.12740.

Barjolle, D., Lehmann, B., Chappuis, J.-M., and Dufour, M. (1997) Protected designation of origin and institutions (France, Spain and Italy). *Proceedings of the 52nd EAAE Seminar*, 483–502. https://papers.ssrn.com/abstract=11079 (accessed 8 April 2025).

Barnes, P.W., Flint, S.D. and Caldwell, M.M. (1990) Morphological responses of crop and weed species of different growth forms to ultraviolet-B radiation. *American Journal of Botany* 77(10), 1354–1360. DOI 10.2307/2444596.

Barnes, P.W., Robson, T.M., Zepp, R.G., Bornman, J.F., Jansen, M.A.K., Ossola, R., Wang, Q.-W., Robinson, S.A., Foereid, B., Klekociuk, A.R., Martinez-Abaigar, J., Hou, W.-C., Mackenzie, R. and Paul, N.D. (2023) Interactive effects of changes in UV radiation and climate on terrestrial ecosystems, biogeochemical cycles, and feedbacks to the climate system. *Photochemical & Photobiological Sciences* 22, 1049–1091. https://doi.org/10.1007/s43630-023-00376-7.

Barrios, E. (2007) Soil biota, ecosystem services and land productivity. *Ecological Economics, Special Section – Ecosystem Services and Agriculture* 64, 269–285. DOI 10.1016/j.ecolecon.2007.03.004.

Bates, R.P., Morris, J.R. and Crandall, P.G. (2001) *Principles and practices of small – and medium – scale fruit juice processing.* Food and Agriculture Organization of the United Nations, Rome, Italy, 226 pp.

Bates, T.R., Dunst, R.M., Taft, T. and Vercant, M. (2002) The vegetative response of 'Concord' grapevines to soil pH. *HortScience* 37(6), 890–893.

Baudoin, A.B., Finger, S.A. and Wolf, T.K. (2006) Factors affecting reductions in photosynthesis caused by applying horticultural oil to grapevine leaves. *HortScience* 41(2), 346–351.

Bazzoffi, P., Francesco, A., Vanino, S. and Pellegrini, S. (2006) Impact of land levelling for vineyard plantation on soil degradation in Italy. *Bollettino-Societa Geologica Italiana* Vol. 6, 191–199.

Beattie, B.B. and Dahlenberg, A.P. (1989) Grapes. In: Beattie, B.B., McGlasson, W.B. and Wade, N.L. (eds) *Postharvest Diseases of Horticultural Produce Vol. 1. Temperate Fruit.* CSIRO Publications, East Melbourne, Victoria, Australia, pp. 67–73. ISBN 9780643050518.

Becker, N.J. (1977) Experimental research on the influence of microclimate on grape constituents and on the quality of the crop. *Proceedings of the OIV Symposium on Quality and Vintage*, Cape Town, South Africa, Oenological and Viticulture Research Institute, pp. 181–188.

Belancic, A., Agosin, E., Ibacache, A., Bordeu, E., Baumes, R., Razungles, A. and Bayonove, C. (1997) Influence of sun exposure on the aromatic composition of Chilean Muscat grape cultivars Moscatel de Alejandria and Moscatel rosada. *American Journal of Enology and Viticulture* 48(2), 181–186.

Bélanger, R.R., Dufour, N., Caron, J. and Benhamou, N. (1995) Chronological events associated with the antagonistic properties of *Trichoderma harzianum* against *Botrytis cinerea*: indirect evidence for sequential role of antibiosis and parasitism. *Biocontrol Science and Technology* 5, 41–54. DOI 10.1080/09583159550040006.

Belfrage, N. and Loftus, S. (1993) Dried grapes: the classic wines of antiquity. *Journal of Wine Research* 4(3), 205–225. DOI 10.1080/09571269308717970.

Bell, V.A. (2015) *An integrated strategy for managing Grapevine leafroll-associated virus 3 in red berry cultivars in New Zealand vineyards.* PhD thesis, University of Victoria 118 pp. http://hdl.handle.net/10063/4147.

Bell, V.A., Bonfiglioli, R.G.E., Walker, J.T.S., Lo, P.L., Mackay, J.F. and McGregor, S.E. (2009) Grapevine leafroll-associated virus 3 persistence in *Vitis vinifera* remnant roots. *Journal of Plant Pathology* 91, 527–533. DOI 10.4454/jpp.v91i3.543.

Benito, A., Romero, I., Domínguez, N., García-Escudero, E. and Martín, I. (2013) Leaf blade and petiole analysis for nutrient diagnosis in *Vitis vinifera* L. cv. Garnacha tinta. *Australian Journal of Grape and Wine Research* 19, 285–298. DOI 10.1111/ajgw.12022.

Bennett, J. (2002) *Relationships between Carbohydrate Supply and Reserves and the Reproductive Growth of Grapevines* (Vitis vinifera *L.*). Soil, Plant and Ecological Sciences Division, Lincoln University, Christchurch, New Zealand, 204 pp. https://hdl.handle.net/10182/1597.

Bennett, J., Jarvis, P., Creasy, G.L. and Trought, M.C.T. (2005) Influence of defoliation on overwintering carbohydrate reserves, return bloom, and yield of mature Chardonnay grapevines. *American Journal of Enology and Viticulture* 56(4), 386–393.

Beresford, R.M., Wright, P.J., Wood, P.N. and Agnew, R.H. (2016) Sensitivity of grapevine powdery mildew (*Erysiphe necator*) to demethylation inhibitor and quinone outside inhibitor fungicides in New Zealand. *New Zealand Plant Protection* 69, 1–10. http://www.nzpps.org/nzpp_abstract.php?paper=690010 (accessed 20 Apr 2025).

Bergström, L. and Brink, N. (1986) Effects of differentiated applications of fertilizer N on leaching losses and distribution of inorganic N in the soil. *Plant and Soil* 93(3), 333–345. DOI 10.1007/BF02374284.

Berisha, B., Chen, Y.D., Zhang, G.Y., Xu, B.Y. and Chen, T.A. (1998) Isolation of Pierce's disease bacteria from grapevines in Europe. *European Journal of Plant Pathology* 104(5), 427–433. DOI 10.1023/A:1008655621235.

Berlinger, M.J. (1977) The Mediterranean vine mealybug and its natural enemies in southern Israel. *Phytoparasitica* 5, 3–14. DOI 10.1007/BF03179429.

Bertazzon, N., Forte, V., Filippin, L., Causin, R., Maixner, M. and Angelini, E. (2017) Association between genetic variability and titre of Grapevine Pinot gris virus with disease symptoms. *Plant Pathology* 66, 949–959. DOI 10.1111/ ppa.12639.

Bettiga, L.J., Dokoozlian, N.K. and Williams, L.E. (1996) Windbreaks improve the growth and yield of Chardonnay grapevines grown in a cool climate. *Proceedings of the Fourth International Symposium for Cool Climate Viticulture and Enology*, New York State Agricultural Experimental Station, Rochester, New York, pp. II, 43–46.

Bioletti, F.T. (1929) *Elements of grape growing in California.* Berkeley, Calif.: University of California Print. Off. https://babel.hathitrust.org/cgi/pt?id=uc2.ark:/13960/t3b00544h (accessed 13 March 2025).

Bioletti, F. (1938) Outline of ampelography for the *vinifera* grapes in California. *Hilgardia* 11, 227–293. DOI 10.3733/hilg.v11n06p227.

Biricolti, S., Ferrini, F., Rinaldelli, E., Tamantini, I. and Vignozzi, N. (1997) VAM fungi and soil lime content influence rootstock growth and nutrient content. *American Journal of Enology and Viticulture* 48, 93–99.

Bisson, L.F. and Butzke, C.E. (2000) Diagnosis and rectification of stuck and sluggish fermentations. *American Journal of Enology and Viticulture* 51(2), 168–177.

Blaauw, B. and Shapiro-Ilan, D., 2023. Getting to the root of the problem: Entomopathogenic nematodes as a management tool for grape root borer. *Southern Region Small Fruit*

Consortium. Small Fruit News 23(1). https://smallfruits.org/2023/01/getting-to-the-root-of-the-problem-entomopathogenic-nematodes-as-a-management-tool-for-grape-root-borer/ (accessed 2 April 2025).

Blanco-Ulate, B., Hopfer, H., Figueroa-Balderas, R., Ye, Z., Rivero, R.M., Albacete, A., Pérez-Alfocea, F., Koyama, R., Anderson, M.M., Smith, R.J. *et al.* (2017) Red blotch disease alters grape berry development and metabolism by interfering with the transcriptional and hormonal regulation of ripening. *Journal of Experimental Botany* 68, 1225–1238. DOI 10.1093/jxb/erw506.

Blanke, M.M. and Leyhe, A. (1987) Stomatal activity of the grape berry cv. Riesling, Müller–Thurgau and Ehrenfelser. *Plant Physiology and Biochemistry* 127(5), 451–460. DOI 10.1016/S0176-1617(87)80253-5.

Blanke, M.M. and Leyhe, A. (1988) Stomatal and cuticular transpiration of the cap and berry of grape. *Journal of Plant Physiology* 132, 250–253. DOI 10.1016/S0176-1617(88)80170-6.

Bodin, F. and Morlat, R. (2006) Characterization of viticultural terroirs using a simple field model based on soil depth I. Validation of the water supply regime, phenology and vine vigour, in the Anjou Vineyard (France). *Plant and Soil* 281(1–2), 37–54. DOI 10.1007/s11104-005-3768-0.

Boiling, C. (2025) Latest vineyard robots. *Canopy. Insights for winemakers around the world.* URL https://www.internationalwinechallenge.com/Canopy-Articles/latest-vineyard-robots.html (accessed 26 March 2025).

Bondada, B.R., and Keller, M. (2012) Not all shrivels are created equal - morpho-anatomical and compositional characteristics differ among different shrivel types that develop during ripening of grape (*Vitis vinifera* L.) berries. *American Journal of Plant Sciences*, 3(7), 879–898.

Bonilla, F., Mayen, M., Merida, J. and Medina, M. (1999) Extraction of phenolic compounds from red grape marc for use as food lipid antioxidants. *Food Chemistry* 66(2), 209–215. DOI 10.1016/S0308-8146(99)00046-1.

Borgo, M. and Angelini, E. (2002) Influence de l'enroulement foliaire GLRaV3 sur les paramètres de la production du Merlot. *Bulletin de l'OIV* 72, 611–622.

Borkar, S.G. and Yumlembam, R.A. (2016) Bacterial Diseases of Fruit Crops. In: *Bacterial Diseases of Crop Plants*. CRC Press, Milton, USA, pp. 117–284. ISBN 9781498755993.

Boss, P.K., Davies, C. and Robinson, S.P. (1996) Anthocyanin composition and anthocyanin pathway gene expression in grapevine sports differing in berry skin colour. *Australian Journal of Grape and Wine Research* 2(3), 163–170. DOI 10.1111/j.1755-0238.1996.tb00104.x.

Botelho, R.V., Pavanello, A.P., Pires, E.J.P., Terra, M.M. and Muller, M.M.L. (2007) Effects of chilling and garlic extract on bud dormancy release in Cabernet Sauvignon grapevine cuttings. *American Journal of Enology and Viticulture* 58(3), 402–404.

Botero-Garcés, N. and Isaacs, R. (2004) Movement of the grape berry moth, *Endopiza viteana*: displacement distance and direction. *Physiological Entomology* 29(5), 443–452. DOI 10.1111/j.0307-6962.2004.00415.x.

Botterill, T., Paulin, S., Green, R., Williams, S., Lin, J., Saxton, V., Mills, S., Chen, X., and Corbett-Davies, S. (2017). A robot system for pruning grape vines. *Journal of Field Robotics* 34, 1100–1122. DOI 10.1002/rob.21680.

Boubals, D. (1989) La maladie de Pierce arrive dans les vignobles d'Europe. *Bulletin de l'OIV* 62, 309–314.

Boulton, R. (2001) The copigmentation of anthocyanins and its role in the color of red wine: a critical review. *American Journal of Enology and Viticulture* 52, 67–87.

Bouquet, A. (1986) Introduction dans l'espèce *Vitis vinifera* L. d'un caractère de résistance à l'oidium (*Uncinula necator* Schw. Burr.) issu de l'espèce *Muscadinia rotundifolia* (Michx.) Small. *Vignevini*, 12(suppl), 141-146.

Bouquet, A., Pauquet, J., Adam-Blondon, A.-F., Torregrosa, L., Merdinoglu, D. and Wiedemann-Merdinoglu, S. (2000) Vers l'obtention de variétés de vigne résistantes à l'oïdium et au mildiou par les méthodes conventionnelles et biotechnologiques. *Bulletin de l'OIV* No. 2000_07-08_n833-n834_art01.

Bridle, P. and Garcia-Viguera, C. (1996) A simple technique for the detection of red wine adulteration with elderberry pigments. *Food Chemistry* 55(2), 111–113. DOI 10.1016/0308-8146(95)00179-4.

Broome, J.C., English, J.T., Marois, J.J., Latorre, B.A. and Aviles, J.C. (1995) Development of an infection model for botrytis bunch rot of grapes based on wetness duration and temperature. *Phytopathology* 85, 97–102. DOI 10.1094/Phyto-85-97.

Brossaud, F., Cheynier, V., Asselin, C. and Moutounet, M. (1999) Flavonoid compositional differences of grapes among site test plantings of Cabernet franc. *American Journal of Enology and Viticulture* 50(3), 277–284.

Bruce, R.C., Lestringant, P., Brenneman, C.A., Heymann, H. and Oberholster, A. (2021) The impact of optical berry sorting on red wine composition and sensory properties. *Foods* 10, 402. https://doi.org/10.3390/foods10020402.

Bruez, E., Baumgartner, K., Bastien, S., Travadon, R., Guérin-Dubrana, L. and Rey, P. (2016) Various fungal communities colonise the functional wood tissues of old grapevines externally free from grapevine trunk disease symptoms. *Australian Journal of Grape and Wine Research* 22, 288–295. DOI 10.1111/ajgw.12209.

Bruez, E., Cholet, C., Giudici, M., Simonit, M., Martignon, T., Boisseau, M., Weingartner, S., Poitou, X., Rey, P., and Geny-Denis, L. (2022) Pruning quality effects on desiccation cone installation and wood necrotization in three grapevine cultivars in France. *Horticulturae*, 8(8), Article 8. https://doi.org/10.3390/horticulturae8080681.

Brulé, D., Héloir, M.-C., Roudaire, T., Villette, J., Bonnet, S., Pascal, Y., Darblade, B., Crozier, P., Hugueney, P., Coma, V. and Poinssot, B. (2024) Increasing vineyard sustainability: innovating a targeted chitosan-derived biocontrol solution to induce grapevine resistance against downy and powdery mildews. *Frontiers in Plant Science* 15. https://doi.org/10.3389/fpls.2024.1360254.

Brun, L.A., Maillet, J., Hinsinger, P. and Pepin, M. (2001) Evaluation of copper availability to plants in copper-contaminated vineyard soils. *Environment and Pollution* 111(2), 293–302. DOI 10.1016/S0269-7491(00)00067-1.

Bush, D. (2011). Selecting and breeding eucalypts for natural durability. *Proceedings of a Workshop: Developing a Eucalypt Resource – Learning from Australia and Elsewhere* Blenheim, New Zealand, pp. 125–136.

Buttery, R.G., Seifert, R.M., Guadagni, D.G. and Ling, L.C. (1969) Characterization of some volatile constituents of bell peppers. *Journal of Agricultural and Food Chemistry* 17, 1322–1327. DOI 10.1021/jf60166a061.

Buttrose, M.S. (1969) Fruitfulness in grapevines: effects of light intensity and temperature. *Botanical Gazette* 130(6), 166–173. DOI 10.1086/336486.

Buxó, R. 2008. The agricultural consequences of colonial contacts on the Iberian Peninsula in the first millennium B.C. *Vegetation History and Archaeobotany* 17(1), 145–154. DOI 10.1007/s00334-007-0133-7.

Cabaleiro, C. and Segura, A. (1997) Field transmission of grapevine leafroll associated virus 3 (GLRaV-3) by the mealybug *Planococcus citri*. *Plant Disease* 81(3), 283–287. DOI 10.1094/PDIS.1997.81.3.283.

Cabaleiro, C., Segura, A. and Garcia-Berrios, J.J. (1999) Effects of grapevine leafroll-associated Virus 3 on the physiology and must of *Vitis vinifera* L. cv. Albarino following contamination in the field. *American Journal of Enology and Viticulture* 50(1), 40–44.

Cabaroglu, T., Canbas, A., Baumes, R., Bayonove, C., Lepoutre, J. P. and Günata, Z. (1997) Aroma composition of a white wine of *Vitis vinifera* L. cv. Emir as affected by skin contact. *Journal of Food Science* 62, 680–683. DOI 10.1111/j.1365-2621.1997.tb15434.x.

Caccavello, G., Giaccone, M., Scognamiglio, P., Forlani, M. and Basile, B. (2017) Influence of intensity of post-veraison defoliation or shoot trimming on vine physiology, yield components, berry and wine composition in Aglianico grapevines. *Australian Journal of Grape and Wine Research* 23, 226–239. DOI 10.1111/ajgw.12263.

Cakmak, I., Hengeler, C. and Marschner, H. (1994) Partitioning of shoot and root dry matter and carbohydrates in bean plants suffering from phosphorus, potassium and magnesium deficiency. *Journal of Experimental Botany* 45(9), 1245–1250. DOI 10.1093/jxb/45.9.1245.

Camilleri, M. A. (2022) The rationale for ISO 14001 certification: A systematic review and a cost–benefit analysis. *Corporate Social Responsibility and Environmental Management*, 29(4), 1067–1083. https://doi.org/10.1002/csr.2254.

Camp, C.R. (1998) Subsurface drip irrigation: a review. *Transactions of the ASAE* 41(5), 1353–1367. DOI 10.13031/2013.17309.

Campbell, C. (2004) *Phylloxera: How Wine was Saved for the World.* Harper Perennial, London. ISBN 9780007325863.

Campbellclause, J.M. (1998) Stomatal response of grapevines to wind. *Australian Journal of Experimental Agriculture* 38, 77–82. DOI 10.1071/EA91220.

Candir, E., Ozdemir, A.E., Kamiloglu, O., Soylu, E.M., Dilbaz, R. and Ustun, D. (2012) Modified atmosphere packaging and ethanol vapor to control decay of 'Red Globe' table grapes during storage. *Postharvest Biology and Technology* 63, 98–106. DOI 10.1016/j.postharvbio.2011.09.008.

Candolfi, M., Jermini, M., Carrera, E. and Candolfi-Vasconcelos, M. (1993) Grapevine leaf gas exchange, plant growth, yield, fruit quality and carbohydrate reserves influenced by the grape leafhopper, *Empoasca vitis*. *Entomologia Experimentalis et Applicata* 69(3), 289–296. DOI 10.1111/j.1570-7458.1993.tb01751.x.

Caputi, L., Carlin, S., Ghiglieno, I., Stefanini, M., Valenti, L., Vrhovsek, U. and Mattivi, F. (2011) Relationship of changes in rotundone content during grape ripening and winemaking to manipulation of the 'peppery' character of wine. *Journal of Agricultural and Food Chemistry* 59, 5565–5571. DOI 10.1021/jf200786u.

Carbonneau, A. (1985) The early selection of grapevine rootstocks for resistance to drought conditions. *American Journal of Enology and Viticulture* 36, 195–198.

Carbonneau, A. and Casteran, P. (1987) Optimization of vine performance by the lyre training systems. *Proceedings of the 6th Australian Wine Industry Technical Conference*, Adelaide, Australia. Winetitles, Underdale, South Australia pp. 194–204.

Carreño, J., Martínez, A., Almela, L. and Fernández-López, J.A. (1995) Proposal of an index for the objective evaluation of the colour of red table grapes. *Food Research International* 28(4), 373–377. DOI 10.1016/0963-9969(95)00008-A.

Carey, R. (2024, August 1). Evolution of freeze and frost protection continues with infrared, electromagnetic technologies. *WineBusiness Monthly*, 31(8), 60–67.

Caspari, H.W., Neal, S., Naylor, A., Trought, M.C.T. and Tannock, S. (1997) Use of cover crops and deficit irrigation to reduce vegetative vigour of 'Sauvignon blanc' grapevines in a humid climate. *Proceedings of the Fourth International Symposium for Cool Climate Viticulture and Enology*, Rochester, New York, pp. II, 63–66.

Castro, M.L., Ferreira, J.P., Pintado, M., Ramos, O.L., Borges, S. and Baptista-Silva, S. (2023) Grape by-products in sustainable cosmetics: nanoencapsulation and market trends. *Applied Sciences* 13, 9168. https://doi.org/10.3390/app13169168.

Cavalieri, D., McGovern, P., Hartl, D., Mortimer, R. and Polsinelli, M. (2003) Evidence for *S. cerevisiae* fermentation in ancient wine. *Journal of Molecular Evolution* 57, S226–S232. DOI 10.1007/s00239-003-0031-2.

Cervera, M.-T., Cabezas, J.A., Sancha, J.C., Toda, F.M.D. and Martínez-Zapater, J.M. (1998) Application of AFLPs to the characterization of grapevine *Vitis vinifera* L. genetic resources. A case study with accessions from Rioja (Spain). *Theoretical Applied Genetics* 97(1–2), 51–59. DOI 10.1007/s001220050866.

Cha, D.H., Linn Jr, C.E., Teal, P.E.A., Zhang, A., Roelofs, W.L. and Loeb, G.M. (2011) Eavesdropping on plant volatiles by a specialist moth: significance of ratio and concentration. *PLOS ONE* 6, e17033. DOI 10.1371/journal.pone.0017033.

Chalker-Scott, L. (2013) The science behind biodynamic preparations: A literature review. *HortTechnology*, 23(6), 814–819.

Chalmers, D.J., Mitchell, P.D. and van Heek, L. (1981) Control of peach tree growth and productivity by regulated water supply, tree density and summer pruning. *Journal of the American Society for Horticultural Science* 106, 307–312.

Chardonnet, C. and Donéche, B. (1995) Relation between calcium content and resistance to enzymatic digestion of the skin during grape ripening. *Vitis* 34, 95–98.

Charles, J.G., Cohen, D., Walker, J.T.S., Forgie, S.A., Bell, V.A. and Breen, K.C. (2006) A review of the ecology of grapevine leafroll associated virus type 3 (GLRaV-3). *New Zealand Plant Protection* 59, 330–337.

Chen, L.S. and Cheng, L. (2003) Carbon assimilation and carbohydrate metabolism of 'Concord' grape (*Vitis labrusca* L.) leaves in response to nitrogen supply. *Journal of the American Society for Horticultural Science* 128(5), 754–760.

Cheynier, V. and Rigaud, J. (1986) HPLC separation and characterization of flavonols in the skins of *Vitis vinifera* var. Cinsault. *American Journal of Enology and Viticulture* 37, 248–252.

Cheynier, V., Fulcrand, H., Sarni, P. and Moutounet, M. (1998) Progress in phenolic chemistry in the last ten years. *Proceedings of the ASVO Seminar Phenolics and Extraction*, pp. 12–17.

Chitkowski, R.L. and Fisher, J.R. (2005) Effect of soil type on the establishment of grape phylloxera colonies in the Pacific Northwest. *American Journal of Enology and Viticulture* 56(3), 207–211.

Choné, X., Van Leeuwen, C., Dubourdieu, D. and Gaudillére, J.P. (2001) Stem water potential is a sensitive indicator of grapevine water status. *Australian Journal of Experimental Agriculture* 87(4), 477–483. DOI 10.1006/anbo.2000.1361.

Christen, E.W., Moll, J.L. and Muirhead, W.A. (1995) Furrows that trickle? *Proceedings of the 9th Australian Wine Industry Technical Conference*, July 16–19, Adelaide, South Australia. Winetitles, Underdale, South Australia, pp. 118.

Christensen, L.P. (2000) Current developments in harvest mechanization and DOV. *Raisin Production Manual*, University of California, Agricultural and Natural Resources, Oakland, California, pp. 252–263. ISBN 9781879906440.

Christensen, L.P. and Peacock, W.L. (2000a) Raisin grape varieties. *Raisin Production Manual*. University of California, Agricultural and Natural Resources, Oakland, California, pp. 38–47. ISBN 9781879906440.

Christensen, L.P. and Peacock, W.L. (2000b) Harvesting and handling. *Raisin Production Manual*. University of California, Agricultural and Natural Resources, Oakland, California, pp. 193–206. ISBN 9781879906440.

Christensen, L.P., Boggero, J. and Bianchi, M. (1990) Comparative leaf tissue analysis of potassium deficiency and a disorder resembling potassium deficiency in Thompson Seedless grapevines. *American Journal of Enology and Viticulture* 41(1), 77–83.

Christensen, L.P., Leavitt, G.M., Hirschfelt, D.J. and Bianchi, M.L. (1994) The effects of pruning level and post-budbreak cane adjustment on Thompson Seedless raisin production and quality. *American Journal of Enology and Viticulture* 45(2), 141–149.

Christensen, P. (1984) Nutrient level comparisons of leaf petioles and blades in twenty-six grape cultivars over three years (1979 through 1981). *American Journal of Enology and Viticulture* 35(3), 124–133.

Chuche, J., and Thiéry, D. (2014) Biology and ecology of the Flavescence dorée vector *Scaphoideus titanus*: A review. *Agronomy for Sustainable Development*, 34(2), 381–403. https://doi.org/10.1007/s13593-014-0208-7.

Ciatti, J. (2016) *Global Market Update – 2016*. Ciatti Global Wine and Grape Brokers, San Rafael, California, USA, 7(5), 1–14.

Cieniewicz, E.J., Pethybridge, S.J., Loeb, G., Perry, K. and Fuchs, M. (2018) Insights into the ecology of grapevine red blotch virus in a diseased vineyard. *Phytopathology* 108, 94–102. https://doi.org/10.1094/PHYTO-07-17-0239-R.

Clarke, C.W., Henneken, J., Carmody, B.M. and Cunningham, J.P. (2023) Performance of six genetically diverse phylloxera strains on 5C Teleki (*V. berlandieri* × *V. riparia*) rootstock. *Australian Journal of Grape and Wine Research* 2259967. https://doi.org/10.1155/2023/2259967.

Clarke, J.H., Clark, W.S. and Hancock, M. (1997) Strategies for the prevention of development of pesticide resistance in the UK – lessons for and from the use of herbicides, fungicides and insecticides. *Pesticide Science* 51(3), 391–397. DOI 10.1002/(SICI)1096-9063(199711)51:3<391::AID-PS654>3.0.CO;2-M.

Clary, C.D., Steinhauer, R.E., Frisinger, J.E. and Peffer, T.E. (1990) Evaluation of machine- vs. hand-harvested Chardonnay. *American Journal of Enology and Viticulture* 41(2), 176–181.

Claverie, M., Lecomte, P., Delorme, G., Dumot, V., Jacquet, O., and Cochard, H. (2023) Xylem water transport is influenced by age and winter pruning characteristics in grapevine (*Vitis vinifera*). *OENO One*, 57(3), Article 3. https://doi.org/10.20870/oeno-one.2023.57.3.7452.

Cleugh, H. (1998) Effects of windbreaks on airflow, microclimates and crop yields. *Agroforestry Systems* 41(1), 55–84. DOI 10.1023/A:1006019805109.

Cliff, M.A., Dever, M.C. and Reynolds, A.G. (1996) Descriptive profiling of new and commercial British Columbia table grape cultivars. *American Journal of Enology and Viticulture* 47(3), 301–308.

Clingeleffer, P.R. (1993a) Vine response to modified pruning practices. *Proceedings of the 2nd New Jersey Shaulis Grape Symposium on Pruning Mechanization and Crop Control*, Fredonia, New York. New York State Agriculture Experiment Station, Geneva, New York, pp. 20–30.

Clingeleffer, P.R. (1993b) Development of management systems for low cost, high quality wine production and vigour control in cool climate Australian vineyards. *Die WeinWissenschaft* 48, 130–134.

Clingeleffer, P.R. (1999) Developments in Australian winegrape production. *Proceedings du Groupe d' Étude des Systèmes de Conduite de la Vigne, 11th Meeting*, University of Palermo, Marsala, Italy, Vol. I, pp. 56–69.

Clingeleffer, P.R. (2013) Mechanization in Australian vineyards. *Acta Horticulturae* 978, 169–177. DOI 10.17660/ActaHortic.2013.978.17.

Clingeleffer, P.R. and May, P. (1981) The swing-arm trellis for Sultana grapevine management. *South African Journal of Enology and Viticulture* 2, 37–44. DOI 10.21548/2-2-2394.

Clore, W.J., Wallace, M.A. and Fay, R.D. (1974) Bud survival of grape varieties at sub-zero temperatures in Washington. *American Journal of Enology and Viticulture* 25(1), 24–29.

Clydesdale, F.M. (1993) Color as a factor in food choice. *Critical Reviews in Food Science and Nutrition* 33(1), 82–101. DOI 10.1080/10408399309527614.

Clydesdale, F.M., Main, J.H., Francis, F.J. and Damon, R.A. (1978) Concord grape pigments as colorants for beverages and gelatin desserts. *Journal of Food Science* 43(6), 1687–1692. DOI 10.1111/j.1365-2621.1978.tb07389.x.

Coertze, S., Holz, G. and Sadie, A. (2001) Germination and establishment of infection on grape berries by single airborne conidia of *Botrytis cinerea*. *Plant Disease* 85(6), 668–677. DOI 10.1094/PDIS.2001.85.6.668.

Cohen, S.D., Tarara, J.M. and Kennedy, J.A. (2008) Assessing the impact of temperature on grape phenolic metabolism. *Analytica Chimica Acta* 621, 57–67. DOI 10.1016/j.aca.2007.11.029.

Coipel, J., Lovelle, B.R., Sipp, C. and Van Leeuwen, C. (2006) 'Terroir' effect, as a result of environmental stress, depends more on soil depth than on soil type (*Vitis vinifera* L. cv. Grenache noir, Côtes du Rhône, France, 2000). *Journal International des Sciences de la Vigne et du Vin* 40(4), 177–185.

Collins, R.M., Bertram, A., Roche, J.-A. and Scott, M.E. (2002) Preliminary studies in the comparison of hot water and hot foam for weed control. *Proceedings of the 5th EWRS Workshop on Physical Weed Control*, Pisa, Italy, 11–13 March, pp. 207–215.

Comas, L.H., Eiseeenstat, D.M. and Lakso, A.N. (2000) Assessing root death and root system dynamics in a study of grape canopy pruning. *New Phytologist* 147(1), 171–178.

Conradie, W.J. (1980) Seasonal uptake of nutrients by Chenin blanc in sand culture: I. Nitrogen. *South African Journal of Enology and Viticulture* 1, 59–65. DOI 10.21548/2-1-2403.

Conradie, W.J. (1990) Distribution and translocation of nitrogen absorbed during late spring by two-year-old grapevines grown in sand culture. *American Journal of Enology and Viticulture* 41(3), 241–250. DOI: 10.5344/ajev.1990.41.3.241.

Conradie, W.J. (2005) Partitioning of mineral nutrients and timing of fertilizer application for optimum efficiency. *Proceedings of the Soil, Environment, Vine and Mineral Nutrition Symposium*, American Society for Enology and Viticulture, Davis, California, pp. 69–81.

Conradie, W.J. and Saayman, D. (1989) Effects of long-term nitrogen, phosphorus, and potassium fertilization on Chenin blanc vines. II. Leaf analyses and grape composition. *American Journal of Enology and Viticulture* 40(2), 91–98.

Considine, J.A. (1973) A statistical study of rain damage of grapes grown for drying in Victoria. *Australian Journal of Experimental Agriculture and Animal Husbandry* 13, 604–611. DOI 10.1071/EA9730604.

Considine, J.A. and Knox, R.B. (1979) Development and histochemistry of the cells, cell walls, and cuticle of the dermal system of fruit of the grape, *Vitis vinifera* L. *Protoplasma* 99(4), 347–365. DOI 10.1007/BF01275807.

Cook, J.A., Lynn, C.D. and Kissler, J.J. (1960) Boron deficiency in California vineyards. *American Journal of Enology and Viticulture* 11(4), 185–194.

Coombe, B.G. (1959) Fruit set and development in seeded grape varieties as affected by defoliation, topping, girdling, and other treatments. *American Journal of Enology and Viticulture* 10, 85–100.

Coombe, B.G. (1960) Relationship of growth and development to changes in sugars, auxins, and gibberellins in fruit of seeded and seedless varieties of *Vitis vinifera*. *Plant Physiology* 35, 241–250. DOI 10.1104/pp.35.2.241.

Coombe, B.G. (1973) The regulation of set and development of the grape berry. *Acta Horticulturae* 34, 261–273. DOI 10.17660/ActaHortic.1973.34.36.

Coombe, B.G. (1976) The development of fleshy fruits. *Annual Review of Plant Physiology* 27, 507–528. DOI 10.1146/annurev.pp.27.060176.001231.

Coombe, B.G. (1995) Adoption of a system for identifying grapevine growth stages. *Australian Journal of Grape and Wine Research* 1(2), 104–110. DOI 10.1111/j.1755-0238.1995.tb00086.x.

Coombe, B.G. and McCarthy, M.G. (1997) Identification and naming of the inception of aroma development in ripening grape berries. *Australian Journal of Grape Wine Research* 3(1), 18–20. DOI 10.1111/j.1755-0238.1997.tb00111.x.

Coombe, B.G. and McCarthy, M.G. (2000) Dynamics of grape berry growth and physiology of ripening. *Australian Journal of Grape and Wine Research* 6(2), 131–135. DOI 10.1111/j.1755-0238.2000.tb00171.x.

Corbett-Davies, S., Botterill, T., Green, R. and Saxton, V. (2012) An expert system for automatically pruning vines. In *Proceedings of the 27th Conference on Image and Vision Computing New Zealand*, (Nov 26-28, Dunedin, New Zealand), p. 6. DOI 10.1145/2425836.2425849.

Cornford, C.E. (1938) Katabatic winds and the prevention of frost damage. *Quarterly Journal of the Royal Meteorological Society* 64, 553–587. DOI 10.1002/qj.49706427702.

Corso, M. and Bonghi, C. (2014) Grapevine rootstock effects on abiotic stress tolerance. *Plant Science Today* 1, 108–113. DOI 10.14719/pst.2014.1.3.64.

Costa, H.S., Raetz, E., Pinckard, T.R., Gispert, C., Hernandez-Martinez, R., Dumenyo, C.K. and Cooksey, D.A. (2004) Plant hosts of *Xylella fastidiosa* in and near southern California vineyards. *Plant Disease* 88(11), 1255–1261. DOI 10.1094/PDIS.2004.88.11.1255.

Creasy, G.L. (1991) *Xylem Discontinuity in* Vitis vinifera *L. Berries*. Department of Horticulture, Oregon State University, Corvallis, Oregon, 89 pp. http://hdl.handle.net/1957/25736.

Creasy, G.L. (2004) Soil and air temperatures as affected by soil cultivation in a Canterbury vineyard. *Proceedings of the 12th Australian Wine Industry Technical Conference*, Melbourne, Victoria, Australia. Australian Wine Industry Technical Conference, Inc, pp. 245.

Creasy, G.L. and Qi, G. (2005) Grapevine leaf and fruit tissue responses to fosetyl-Al and UV radiation. *American Journal of Enology and Viticulture* 56(3), 303A.

Creasy, G.L., Crawford, M., Ibbotson, L., Gladstone, P., Kavanaugh, J. and Sutherland, A. (2006) Mussel shell mulch alters Pinot noir grapevine development and wine qualities. *Australian Grapegrower and Winemaker* 509a, 12–18.

Creasy, L.L. and Coffee, M. (1988) Phytoalexin production potential of grape berries. *Journal of the American Society of Horticultural Science* 113, 230–234.

Creasy, L.L. and Creasy, M.T. (2016) Grape anatomy and physiology. In *Grapes and Health*, J.M Pezzuto Ed. Springer. DOI 10.1007/978-3-319-28995-3.

Crisosto, C.H., Smilanick, J.L. and Dokoozlian, N.K. (2001) Table grapes suffer water loss, stem browning during cooling delay. *California Agriculture* 55(1), 39–42. DOI 10.3733/ca.v055n01p39.

Crisosto, C.H., Garner, D. and Crisosto, G. (2002) Carbon dioxide-enriched atmospheres during cold storage limit losses from Botrytis but accelerate rachis browning of 'Redglobe' table grapes. *Postharvest Biology and Technology* 26(2), 181–189. DOI 10.1016/S0925-5214(02)00013-3.

Crisp, P., Scott, E.S. and Wicks, T.J. (2002) Novel control of grapevine powdery mildew. *Proceedings of the 4th International Workshop on Grapevine Powdery and Downy Mildew*, Napa, California. University of California, Davis, California, pp. 78–79.

Crisp, P., Wicks, T.J., Troup, G. and Scott, E.S. (2006) Mode of action of milk and whey in the control of grapevine powdery mildew. *Australian Plant Pathology* 35(5), 487–493. DOI 10.1071/AP06052.

Cristofolini, F. and Gottardini, E. (2000) Concentration of airborne pollen of *Vitis vinifera* L. and yield forecast: a case study at S. Michele all'Adige, Trento, Italy. *Aerobiologia* 16(1), 125–129. DOI 10.1023/A:1007690802901.

Cucuzza, J.D. and Sall, M.A. (1982) Phomopsis cane and leaf spot disease of grape vine: effect of chemical treatments on inoculum level, disease severity, and yield. *Plant Disease* 66, 794–797. DOI 10.1094/PD-66-794.

Cunha, M., Abreu, I., Pinto, P. and Castro, R.D. (2003) Airborne pollen samples for early-season estimates of wine production in a Mediterranean climate area of northern Portugal. *American Journal of Enology and Viticulture* 54(3), 189–194.

D'Addabbo, T., Ladurner, E. and Troccoli, A. (2023) Nematicidal activity of a garlic extract formulation against the grapevine nematode *Xiphinema index*. *Plants* 12, 739. https://doi.org/10.3390/plants12040739.

Dadic, M. and Belleau, G. (1973) Polyphenols and beer flavor. *Proceedings of the American Society of Brewing Chemists* 4, 107–114.

Das, M., Bhowmick, T.S., Ahern, S.J., Young, R., and Gonzalez, C.F. (2015) Control of Pierce's Disease by phage. *PLOS ONE*, 10(6), e0128902. https://doi.org/10.1371/journal.pone.0128902.

Davenport, J.R. and Stevens, R.G. (2006) High soil moisture and low soil temperature are associated with chlorosis occurrence in Concord grape. *HortScience* 41(2), 418–422.

Deal, D.R., Boothroyd, C.W. and Mai, W.F. (1972) Replanting of vineyards and its relationship to vesicular–arbuscular mycorrhiza. *Phytopathology* 62, 172–175. DOI 10.1094/Phyto-62-172.

De Bei, R., Fuentes, S., Gilliham, M., Tyerman, S., Edwards, E., Bianchini, N., Smith, J. and Collins, C. (2016) VitiCanopy: a free computer app to estimate canopy vigor and porosity for grapevine. *Sensors* 16, 585. DOI 10.3390/s16040585.

De Boubée, D.R. (2003) *Research on 2-methyl-3-isobutylpyrazine in Grapes and Wines*. Amorim Academy competition, Paris, 21 pp.

De Boubée, R. (2004) Recherches sur le caractère végétal-poivron vert dans les raisins et dans les vins. *Revue des Oenologues* 110, 6–10.

Deckers, T., Daemen, E., Lemmens, K. and Missotten, C. (1997) Influence of foliar applications of Mn during summer on the fruit quality of Jonagold. *Acta Horticulturae* 448, 467–473. DOI 10.17660/ActaHortic.1997.448.84.

De Kok, L.J. (1990) Sulfur metabolism in plants exposed to atmospheric sulfur. In: Rennenberg, H., Brunold, C., De Kok, L.J. and Stulen, I. (eds) *Sulfur Nutrition and Sulfur Assimilation in Higher Plants: Fundamental, Environmental and Agricultural*

Aspects. SPB Academic Publishing, The Hague, Netherlands, pp. 111–130. ISBN 905103038X.

Delcroix, A., Gunata, Z., Sapis, J.-C., Salmon, J.–M. and Bayonove, C. (1994) Glycosidase activities of three enological yeast strains during winemaking: effect on the terpenol content of Muscat wine. *American Journal of Enology and Viticulture* 45(3), 291–296.

Dell, K.J., Gubler, W.D., Krueger, R., Sanger, M. and Bettiga, L.J. (1998) The efficacy of JMS stylet-oil on grape powdery mildew and botrytis bunch rot and effects on fermentation. *American Journal of Enology and Viticulture* 49(1), 11–16.

Delmas, C.E.L., Fabre, F., Jolivet, J., Mazet, I.D., Richart Cervera, S., Delière, L. and Delmotte, F. (2016) Adaptation of a plant pathogen to partial host resistance: selection for greater aggressiveness in grapevine downy mildew. *Evolutionary Applications* 9, 709–725. 10.1111/eva.12368.

Deloire, A., Dumont, C., Giudici, M., Rogiers, S., and Pellegrino, A. (2022) A few words on grapevine winter buds and pruning in consideration of sap flow. *IVES Technical Reviews, Vine and Wine*. https://doi.org/10.20870/IVES-TR.2022.5512.

De Meyer, G., Bigirimana, J., Elad, Y. and Hafte, M. (1998) Induced systemic resistance in *Trichoderma harzianum* T39 biocontrol of *Botrytis cinerea*. *European Journal of Plant Pathology* 104(3), 279–286. DOI 10.1023/A:1008628806616.

Demian, E., Jaksa-Czotter, N., and Varallyay, E. (2022) Grapevine Pinot gris virus is present in different non-*Vitis* hosts. *Plants*, 11(14), Article 14. https://doi.org/10.3390/plants11141830.

Derksen, R.C., Coffman, C.W., Jiang, C. and Gulyas, S.W. (1999) Influence of hooded and air-assist vineyard applications on plant and worker protection. *Transactions of the American Society of Agricultural Engineers* 42(1), 31–36. DOI 10.13031/2013.13205.

Dermen, H. (1954) Colchiploidy in grapes. *Journal of Heredity* 45(4), 159–172. DOI 10.1093/oxfordjournals.jhered.a106467.

Dermen, H. (1964) Cytogenetics in hybridization of bunch and muscadine-type grapes. *Economic Botany* 18(2), 137–148. DOI 10.1007/BF02862710.

Dethier, B.E. and Shaulis, N. (1964) Minimizing the hazard of cold in New York vineyards. *New York Agricultural Experiment Station Bulletin*, Geneva, New York, 8 pp.

Detjen, L. R. (1919). *Some F1 hybrids of* Vitis rotundifolia *with related species and genera* (No. 18). North Carolina Agricultural Experiment Station.

de Toda, F.M. and Sancha, J.C. (1999) Long-term effects of simulated mechanical pruning on Grenache vines under drought conditions. *American Journal of Enology and Viticulture* 50(1), 87–90.

Deytieux-Belleau, C., Geny, L., Roudet, J., Mayet, V., Donèche, B. and Fermaud, M. (2009) Grape berry skin features related to ontogenic resistance to *Botrytis cinerea*. *European Journal of Plant Pathology* 125, 551–563. DOI 10.1007/s10658-009-9503-6.

Diago, M.P., Krasnow, M., Bubola, M., Millan, B. and Tardaguila, J. (2016) Assessment of vineyard canopy porosity using machine vision. *American Journal of Enology and Viticulture* 67, 229–238. DOI 10.5344/ajev.2015.15037.

Dirr, M.A. and Heuser, C.W. (1987) *The Reference Manual of Woody Plant Propagation*. Varsity Press, Athens, Georgia. ISBN 9780942375008.

Dobrowski, S.Z., Ustin, S.L. and Wolpert, J.A. (2002) Remote estimation of vine canopy density in vertically shoot positioned vineyards: determining optimal vegetation indices. *Australian Journal of Grape and Wine Research* 8(2), 117–125. DOI 10.1111/j.1755-0238.2002.tb00220.x.

Dokoozlian, N.K. and Hirschfelt, D.J. (1995) The influence of cluster thinning at various stages of fruit development on Flame Seedless table grapes. *American Journal of Enology and Viticulture* 46(4), 429–436.

Doran, J.W. (1980) Soil microbial and biochemical changes associated with reduced tillage. *Soil Science Society of America Journal* 44(4), 765–771. DOI 10.2136/sssaj1980.03615995004400040022x.

Downey, M.O., Harvey, J.S. and Robinson, S.P. (2003) Analysis of tannins in seeds and skins of Shiraz grapes throughout berry development. *Australian Journal of Grape and Wine Research* 9(1), 15–27. DOI 10.1111/j.1755-0238.2003.tb00228.x.

Downton, W.J.S. (1985) Growth and mineral composition of the Sultana grapevine as influenced by salinity and rootstock. *Australian Journal of Agricultural Research* 36(3), 425–434. DOI 10.1071/AR9850425.

Dry, P.R. (1997) *The Response of Grapevines to Partial Drying of the Rootzone*. PhD thesis, University of Adelaide, Adelaide, South Australia. http://hdl.handle.net/2440/19152.

Dry, P.R., Reed, S. and Potter, G. (1989) The effect of wind on the performance of Cabernet Franc grapevines. *Acta Horticulturae* 240, 143–146. DOI 10.17660/ActaHortic.1989.240.24.

Dubuis, P.-H., Viret, O., Bloesch, B., Fabre, A.-L., Naef, A., Kassemeyer, H.-H. and Krause, R. (2012) Lutte contre le mildiou de la vigne avec le modèle VitiMeteo-Plasmopara. *Revue Suisse de Viticulture Arboriculture Horticulture* 44, 192–198.

Duchêne, E. and Schneider, C. (2005) Grapevine and climatic changes: a glance at the situation in Alsace. *Agronomy for Sustainable Development* 25, 93–99. DOI 10.1051/agro:2004057.

Due, G. (1990) The use of polypropylene shelters in grapevine establishment – a preliminary trial. *Australian Grapegrower and Winemaker* 318, 29–33.

Dutcher, J.D. and All, J.N. (1978) Survivorship of the grape root borer in commercial grape vineyards with contrasting cultural practices. *Journal of Economic Entomology* 71, 751–754. DOI 10.1093/jee/71.5.751.

Ebadi, A., Coombe, B.G. and May, P. (1995) Fruit-set on small Chardonnay and Shiraz vines grown under varying temperature regimes between budburst and flowering. *Australian Journal of Grape and Wine Research* 1(1), 3–10. DOI 10.1111/j.1755-0238.1995.tb00071.x.

Ebadi, A., May, P. and Coombe, B.G. (1996) Effect of short-term temperature and shading on fruit-set, seed and berry development in model vines of *V. vinifera*, cvs Chardonnay and Shiraz. *Australian Journal of Grape and Wine Research* 2, 2–9. DOI 10.1111/j.1755-0238.1996.tb00087.x.

Edson, C.E., Howell, G.S. and Flore, J.A. (1993) Influence of crop load on photosynthesis and dry matter partitioning of Seyval grapevines I. Single leaf and whole vine response pre- and post-harvest. *American Journal of Enology and Viticulture* 44(2), 139–147. DOI: 10.5344/ajev.1993.44.2.139.

Einset, J. and Pratt, C. (1975) Grapes. In: Janick, J. and Moore, J.N. (eds) *Advances in Fruit Breeding*. Purdue University Press, West Lafayette, Indiana, pp. 130–153. ISBN 978-0911198362.

Elad, Y. and Kapat, A. (1999) The role of *Trichoderma harzianum* protease in the biocontrol of *Botrytis cinerea*. *European Journal of Plant Pathology* 105(2), 177–189. DOI 10.1023/A:1008753629207.

Ellis, R.J. (1979) The most abundant protein in the world. *Trends in Biochemical Science* 4, 241–244. DOI 10.1016/0968-0004(79)90212-3.

El-Sharkawy, H.H.A., Abo-El-Wafa, T.S.A. and Ibrahim, S.A.-A. (2018) Biological control agents improve the productivity and induce the resistance against downy mildew of grapevine. *Jouurnal of Plant Pathology* 100, 33–42. https://doi.org/10.1007/s42161-018-0007-0.

English-Loeb, G.M., Flaherty, D.L., Wilson, L.T., Barnett, W.W., Leavitt, G.M. and Settle, W.H. (1986) Pest management affects spider mites in vineyards. *California Agriculture* 40, 28–30.

English-Loeb, G., Norton, A.P., Gadoury, D.M., Seem, R.C. and Wilcox, W.F. (1999a) Control of powdery mildew in wild and cultivated grapes by a tydeid mite. *Biological Control* 14(2), 97–103. DOI 10.1006/bcon.1998.0681.

English-Loeb, G., Villani, M., Martinson, T., Forsline, A. and Consolie, N. (1999b) Use of entomopathogenic nematodes for control of grape *Phylloxera* (Homoptera: Phylloxeridae): a laboratory evaluation. *Environmental Entomology* 28(5), 890–894. DOI 10.1093/ee/28.5.890.

English-Loeb, G., Norton, A.P., Gadoury, D., Seem, R. and Wilcox, W. (2007) Biological control of grape powdery mildew using mycophagous mites. *Plant Disease* 91(4), 421–429. DOI 10.1094/PDIS-91-4-0421.

Epstein, E. (1961) The essential role of calcium in selective cation transport by plant cells. *Plant Physiology* 36(4), 437–444. DOI 10.1104/pp.36.4.437.

Erincik, O., Madden, L.V., Ferree, D.C. and Ellis, M.A. (2001) Effect of growth stage on susceptibility of grape berry and rachis tissues to infection by *Phomopsis viticola*. *Plant Disease* 85(5), 517–520. DOI 10.1094/PDIS.2001.85.5.517.

Escalona, J., Flexas, J. and Medrano, H. (2002) Drought effects on water flow, photosynthesis and growth of potted grapevines. *Vitis* 41(2), 57–62.

Espinase Nandorfy, D., Siebert, T., Bilogrevic, E., Likos, D., Watson, F., Barter, S., Pisaniello, L., Kulcsar, A., Shellie, R.A., Keast, R., Francis, L. and Bekker, M. (2023) The role of potent thiols in “empyreumatic” flint/struck-match/mineral odours in Chardonnay wine. *Australian Journal of Grape and Wine Research* 8847476. https://doi.org/10.1155/2023/8847476.

Espíndola, R.S., Gutierrez, A. and Suero, E. (2019) Features of raisins grape quality according to the drying process method used. *IJRDO-Journal of Agriculture and Research* 5, 01–07. https://doi.org/10.53555/ar.v5i2.2643.

Evans, R.G. (2000) The art of protecting grapevines from low temperature injury. *Proceedings of the American Society for Enology and Viticulture 50th Annual Meeting, Seattle, Washington*, pp. 60–72.

Ewart, A.J. (1987) Influence of vineyard site and grape maturity on juice and wine quality of *Vitis vinifera*, cv. Riesling. *Proceedings of the 6th Australian Wine Industry Technical Conference*, Adelaide, South Australia. Australian Industrial Publishers, Australia, pp. 71–74.

Fairbank, W., Ede, L., Johnson, H., Luvisi, D. and Neja, R. (1987) Night picking. *California Agriculture* 41, 13–16.

Falks, S.P., Gadoury, D.M., Pearson, R.C. and Seem, R.C. (1995) Partial control of grape powdery mildew by the mycoparasite *Ampelomyces quisqualis*. *Plant Disease* 75, 483–490. DOI 10.1094/PD-79-0483.

Fang, Y. and Qian, M.C. (2006) Quantification of selected aroma-active compounds in Pinot noir wines from different grape maturities. *Journal of Agricultural and Food Chemistry* 54, 8567–8573. DOI 10.1021/jf061396m.

Faralli, M., Mallucci, S., Bignardi, A., Varner, M. and Bertamini, M. (2024) Four decades in the vineyard: the impact of climate change on grapevine phenology and wine quality in northern Italy. *OENO One* 58. https://doi.org/10.20870/oeno-one.2024.58.3.8083.

Feil, H. and Purcell, A.H. (2001) Temperature-dependent growth and survival of *Xylella fastidiosa in vitro* and in potted grapevines. *Plant Disease* 85(12), 1230–1234. DOI 10.1094/PDIS.2001.85.12.1230.

Feliziani, E., Romanazzi, G., Margosan, D.M., Mansour, M.F., Smilanick, J.L., Gu, S., Gohil, H.L., Ames, Z.R. and Lichter, A. (2014) Effect of field treatments with fungicide, potassium sorbate, or chitosan on postharvest rots and quality of table grapes. *Acta Horticulturae* 1053, 257–264. DOI 10.17660/ActaHortic.2014. 1053.28.

Fennell, A. and Hoover, E. (1991) Photoperiod influences growth, bud dormancy, and cold acclimation in *Vitis labruscana* and *V. riparia*. *Journal of the American Society for Horticultural Science* 116, 270–273.

Fereres, E. and Soriano, M.A. (2007) Deficit irrigation for reducing agricultural water use. *Journal of Experimental Botany* 58(2), 147–159. DOI 10.1093/jxb/erl165.

Fernandes, A.M., Melo-Pinto, P., Millan, B., Tardaguila, J. and Diago, M.P. (2015) Automatic discrimination of grapevine (*Vitis vinifera* L.) clones using leaf hyperspectral imaging and partial least squares. *The Journal of Agricultural Science* 153: 455–465. DOI 10.1017/S0021859614000252.

Ferreira, J.H.S., Matthee, F.N. and Thomas, A.C. (1991) Biological control of *Eutypalata* on grapevine by an antagonistic strain of *Bacillus subtilis*. *Phytopathology* 81, 283–287. DOI 10.1094/Phyto-81-283.

Ferris, H. and McKenry, M.V. (1974) Seasonal fluctuation in the spatial distribution of nematode populations in a California vineyard. *Journal of Nematology* 6, 203–210.

Ficiu, L., Brinduse, E., and Ion, M. (2015). Biological control of adult populations of grape phylloxera with enthomopathogenic fungus. *Bulletin of University of Agricultural Sciences and Veterinary Medicine ClujNapoca, Horticulture* 72, 68–73.

Fidelibus, M.W. (2014) Grapevine cultivars, trellis systems, and mechanization of the California raisin industry. *HortTechnology* 24, 285–289.

Filippini Alba, J.M., Flores, C.A. and Miele, A. (2017). Geotechnologies and soil mapping for delimitation of management zones as an approach to precision viticulture. *Applied and Environmental Soil Science* 2017, Article ID 4180965, 10 pp. DOI 10.1155/2017/4180965.

Finger, S.A., Wolf, T.K. and Baudoin, A.B. (2002) Effects of horticultural oils on the photosynthesis, fruit maturity, and crop yield of winegrapes. *American Journal of Enology and Viticulture* 53(2), 116–124. DOI: 10.5344/ajev.2002.53.2.116

Fiore, M., Galati, A., Crescimanno, M., Contò, F., Giacomarra, M. and Tinervia, S. (2016) Managerial suggestions to sustainable market choices: A business profitability assessment on the adoption of voluntary certification in the wine industry of the Italian "Mezzogiorno" regions. *Calitatea-Acces La Succes*, 17(154), 71–79.

Fisher, K.H., Piott, B. and Barkovic, J. (1997) Adaptability of labrusca and French Hybrid grape varieties to mechanical pruning and mechanical thinning. *Proceedings of the Fourth International Symposium for Cool Climate Viticulture and Enology*, Rochester, New York. New York State Agricultural and Experimental Station, Geneva, New York, pp. IV, 33–39.

Flora, L.F. (1976) Time–temperature influence on muscadine grape juice quality. *Journal of Food Science* 41(6), 1312–1315. DOI 10.1111/j.1365-2621.1976. tb01160.x.

Flores-Vélez, L.M., Ducaroir, J., Jaunet, A.M. and Robert, M. (1996) Study of the distribution of copper in an acid sandy vineyard soil by three different methods.

European Journal of Soil Science 47(4), 523–532. DOI 10.1111/j.1365-2389.1996.tb01852.x.

Forsline, P.L., Musselman, R.C., Dee, R.J. and Kender, W.J. (1983) Effects of acid rain on grapevines. *American Journal of Enology and Viticulture* 34(1), 17–22.

Freeman, B.M., Kliewer, W.M. and Stern, P. (1982) Influence of windbreaks and climatic region on diurnal fluctuation of leaf water potential, stomatal conductance, and leaf temperature of grapevines. *American Journal of Enology and Viticulture* 33, 233–236.

Freeman, B.M., Tassie, E. and Rebbechi, M.D. (1992) Training and trellising. In: Coombe, B.G. and Dry, P.R. (eds) *Viticulture Volume 2. Practices*. Winetitles, Underdale, South Australia, pp. 42–65. ISBN 9781875130016.

Friend, A.P. (2005) *Berry Set and Development in* Vitis vinifera *L.* PhD thesis, Food and Wine Group, Agriculture and Life Sciences Division, Lincoln University, Christchurch, New Zealand, 193 pp. http://hdl.handle.net/10182/3441.

Friend, A.P. and Trought, M.C.T. (2007) Delayed winter spur-pruning in New Zealand can alter yield components of Merlot grapes. *Australian Journal of Grape and Wine Research* 13(3), 157–164. DOI 10.1111/j.1755-0238.2007.tb00246.x.

Friend, A.P., Creasy, G.L., Trought, M.C.T. and Lang, A. (2003) Use of tagging to trace capfall and development of individual *Vitis vinifera* L. cv. Pinot noir flowers. *American Journal of Enology and Viticulture* 54(4), 313–317.

Fuller, K.B., Alston, J.M. and Sambucci, O.S. (2014) The value of powdery mildew resistance in grapes: Evidence from California. *Wine Economics and Policy* 3, 90–107. DOI 10.1016/j.wep.2014.09.001.

Gabel, B. and Renczés, V. (1985) Factors affecting the monitoring of flight activity of *Lobesia botrana* and *Eupoecilia ambiguella* (Lepidoptera: Tortricidae) by pheromone traps. *Acta Entomologica Bohemoslovaca* 82, 269–277.

Gaby, J.M., Bakke, A.J., Baker, A.N., Hopfer, H. and Hayes, J.E. (2020) Individual differences in thresholds and consumer preferences for rotundone added to red wine. *Nutrients* 12, 2522. https://doi.org/10.3390/nu12092522.

Gadoury, D.M., Pearson, R.C., Riegel, D.G., Seem, R.C., Becker, C.M. and Pscheidt, J.W. (1994) Reduction of powdery mildew and other diseases by over-the-trellis applications of lime sulfur to dormant grapevines. *Plant Disease* 78(1), 83–87. DOI 10.1094/PD-78-0083.

Gadoury, D.M., Sapkota, S., Cadle-Davidson, L., Underhill, A., McCann, T., Gold, K.M., Gambhir, N. and Combs, D.B. (2023) Effects of nighttime applications of germicidal ultraviolet light upon Powdery Mildew (*Erysiphe necator*), Downy Mildew (*Plasmopara viticola*), and Sour Rot of grapevine. *Plant Disease* 107, 1452–1462. https://doi.org/10.1094/PDIS-04-22-0984-RE.

Gadoury, D.M., Seem, R.C., Magarey, P.A., Emmett, R. and Magarey, R. (1997) Effects of environment and fungicides on epidemics of grape powdery mildew: considerations for practical model development and disease management. *Viticultural and Enological Sciences* 52, 225–229.

Galar-Martínez, M., Torres, N., Sebastián, B., Palacios, J., Arzoz, I., Juanena, N., Villa-Llop, A., Loidi, M., Dewasme, C., Roby, J. P., and Santesteban, L.G. (2024) Respectful pruning improves grapevine development: A case study in young vineyards. *Australian Journal of Grape and Wine Research, 2024*(1), 8448405. https://doi.org/10.1155/2024/8448405.

Galet, P. (1982) Phylloxera. *Les Maladies et le Parasites de la Vigne, Tome II. Les Parasites Animaux*. Paysan du Midi, Montpellier, France, pp. 1059–1313. ISBN 978-2902771004.

Galvan, T.L., Burkness, E.C., Vickers, Z., Stenberg, P., Mansfield, A.K. and Hutchison, W.D. (2007) Sensory-based action threshold for Multicolored Asian Lady Beetle-related taint in winegrapes. *American Journal of Enology and Viticulture* 58, 518–522.

Gamon, J.A. and Pearcy, R.W. (1989) Leaf movement, stress avoidance and photosynthesis in *Vitis californica. Oecologia* 79(4), 475–481. DOI 10.1007/BF00378664.

Gao, L., Girard, B., Mazza, G. and Reynolds, A.G. (1997) Changes in anthocyanins and color characteristics of Pinot noir wines during different vinification processes. *Journal of Agricultural and Food Chemistry* 45, 2003–2008. DOI 10.1021/jf960836e.

Gardea, A.A. (1987) *Freeze Damage of Pinot Noir (*Vitis vinifera *L.) as Affected by Bud Development, INA-Bacteria, and a Bacterial Inhibitor.* PhD thesis, Department of Horticulture, Oregon State University, Corvallis, Oregon, 86 pp. http://hdl.handle.net/1957/25621.

Garganese, F., Sanzani, S.M., Ligorio, A., Di Gennaro, D., Tarricone, L. and Ippolito, A. (2016) Effect of irrigation management on field and postharvest quality of organic table grapes. *Acta Horticulturae* 1144, 273–278. DOI 10.17660/ActaHortic.2016.1144.40.

Gawel, R. (1998) Red wine astringency: a review. *Australian Journal of Grape and Wine Research* 4, 74–95. DOI 10.1111/j.1755-0238.1998.tb00137.x.

Gawel, R., Oberholster, A. and Francis, I.L. (2000) A 'mouth-feel wheel': terminology for communicating the mouth-feel characteristics of red wine. *Australian Journal of Grape and Wine Research* 6(3), 203–207. DOI 10.1111/j.1755-0238.2000.tb00180.x.

Geffroy, O., Dufourcq, T., Carcenac, D., Siebert, T., Herderich, M. and Serrano, E. (2014) Effect of ripeness and viticultural techniques on the rotundone concentration in red wine made from *Vitis vinifera* L. cv. Duras. *Australian Journal of Grape and Wine Research* 20, 401–408. DOI 10.1111/ajgw.12084.

Genesio, L., Miglietta, F., Baronti, S., and Vaccari, F.P. (2015) Biochar increases vineyard productivity without affecting grape quality: Results from a four years field experiment in Tuscany. *Agriculture, Ecosystems & Environment*, 201, 20–25. https://doi.org/10.1016/j.agee.2014.11.021.

Gentry, J.P. and Nelson, K.E. (1964) Conduction cooling of table grapes. *American Journal of Enology and Viticulture* 15(1), 41–46.

Gessler, C., Pertot, I. and Perazzolli, M. (2011) *Plasmopara viticola*: a review of knowledge on downy mildew of grapevine and effective disease management. *Phytopathologia Mediterranea* 50, 3–44. DOI 10.14601/Phytopathol_Mediterr-9360.

Ghassempour, A., Heydari, R., Talebpour, Z., Fakhari, A.R., Rassouli, A., Davies, N. and Aboul-Enein, H.Y. (2008) Study of new extraction methods for separation of anthocyanins from red grape skins: analysis by HPLC and LC-MS/MS. *Journal of Liquid Chromatography & Related Technologies* 31, 2686–2703. DOI 10.1080/10826070802353247.

Ghiglieno, I., Simonetto, A., Facciano, L., Tonni, M., Donna, P., Valenti, L. and Gilioli, G. (2023) Comparing the carbon footprint of conventional and organic vineyards in northern Italy. *Sustainability*, 15(6), Article 6. https://doi.org/10.3390/su15065252.

Giampetruzzi, A., Roumi, V., Roberto, R., Malossini, U., Yoshikawa, N., La Notte, P., Terlizzi, F., Credi, R. and Sa ldarelli, P. (2012) A new grapevine virus discovered by deep sequencing of virus- and viroid-derived small RNAs in Cv Pinot gris. *Virus Research* 163, 262–268. DOI 10.1016/j.virusres.2011.10.010.

Giannakis, C., Buchell, C.S., Skene, K.G.M., Robinson, S.P. and Scott, N.S. (1998) Chitinase and beta-1,3-glucanase in grapevine leaves: a possible defence against powdery mildew infection. *Australian Journal of Grape and Wine Research* 4(1), 14–22. DOI 10.1111/j.1755-0238.1998.tb00130.x.

Giménez-Romero, À., Iturbide, M., Moralejo, E., Gutiérrez, J.M. and Matías, M.A. (2024) Global warming significantly increases the risk of Pierce's disease epidemics in European vineyards. *Scientific Reports* 14, 9648. https://doi.org/10.1038/s41598-024-59947-y.

Giovannoni, J. (2001) Molecular biology of fruit maturation and ripening. *Annual Review of Plant Physiology and Plant Molecular Biology* 52(1), 725–749. DOI 10.1146/annurev.arplant.52.1.725.

Giovos, R., Tassopoulos, D., Kalivas, D., Lougkos, N. and Priovolou, A. (2021) Remote sensing vegetation indices in viticulture: A critical review. *Agriculture* 11(5), Article 5. https://doi.org/10.3390/agriculture11050457.

Glad, C., Regnard, J.L., Querou, Y., Brun, O. and Morot-Gaudry, J.F. (1992) Flux and chemical composition of xylem exudates from Chardonnay grapevines: temporal evolution and effect of recut. *American Journal of Enology and Viticulture* 43(3), 275–282.

Gladstone, J. (1992) *Viticulture and Environment*. Winetitles, Underdale, South Australia, 320 pp. ISBN 9781875130122.

Goetz, G., Fkyerat, A., Métais, N., Kunz, M., Tabacchi, R., Pezet, R. and Pont, V. (1999) Resistance factors to grey mould in grape berries: identification of some phenolics inhibitors of *Botrytis cinerea* stilbene oxidase. *Phytochemistry* 52, 759–767. DOI 10.1016/S0031-9422(99)00351-9.

Goffinet, M.C. (2004) *Anatomy of Grapevine Winter Injury and Recovery*. New York State Agricultural Experiment Station, Geneva, New York, 21 pp.

Goffinet, M.C. and Martinson, T. (2007) Burying canes in the Finger Lakes: grower experience and impact on buds. *Proceedings of Viticulture 2007*, 7–9 February, Rochester, New York, pp. 54–55.

Goheen, A.C. and Cook, J.A. (1959) Leafroll (red-leaf or rougeau) and its effects on vine growth, fruit quality, and yields. *American Journal of Enology and Viticulture* 10(4), 173–181.

Goldhamer, D. and Fereres, E. (2001) Irrigation scheduling protocols using continuously recorded trunk diameter measurements. *Irrigation Science* 20(3), 115–125. DOI 10.1007/s002710000034.

Golino, D.A., Sim, S.T., Gill, R. and Rowhani, A. (2002) California mealybugs can spread grapevine leafroll disease. *California Agriculture* 56, 196–201. DOI 10.3733/ca.v056n06p196.

González-Chang, M., Boyer, S., Creasy, G.L., Lefort, M.-C. and Wratten, S.D. (2017) Mussel shell mulch can increase vineyard sustainability by changing scarab pest behaviour. *Agronomy for Sustainable Development* 37, 42. DOI 10.1007/s13593-017-0450-x.

González-Domínguez, E., Caffi, T., Ciliberti, N. and Rossi, V. (2015) A mechanistic model of *Botrytis cinerea* on grapevines that includes weather, vine growth stage, and the main infection pathways. *PLOS ONE* 10, e0140444. DOI 10.1371/journal.pone.0140444.

Gonzalez-Dominguez, E., Caffi, T., Paolini, A., Mugnai, L., Latinović, N., Latinović, J., Languasco, L. and Rossi, V. (2022) Development and validation of a mechanistic model that predicts infection by *Diaporthe ampelina*, the causal agent of Phomopsis

Cane and Leaf Spot of grapevines. *Frontiers in Plant Science* 13. https://doi.org/10.3389/fpls.2022.872333.

González-Neves, G., Gómez-Cordovés, C. and Barreiro, L. (2001) Anthocyanic composition of Tannat, Cabernet Sauvignon and Merlot young red wines from Uruguay. *Journal of Wine Research* 12(2), 125–133. DOI 10.1080/09571260120095030.

Goodess, C.M., Hanson, C., Hulme, M. and Osborn, T.J. (2003) Representing climate and extreme weather events in integrated assessment models: a review of existing methods and options for development. *Integrated Assessment* 4(3), 145–172. DOI 10.1076/iaij.4.3.145.23772.

Gossett, D.R., Egli, D.B. and Leggett, J.E. (1977) The influence of calcium deficiency on the translocation of photosynthetically fixed 14C in soybeans. *Plant and Soil* 48(1), 243–251. DOI 10.1007/BF00015172.

Goto-Yamamoto, N., Mouri, H., Azumi, M. and Edwards, K.J. (2006) Development of grape microsatellite markers and microsatellite analysis including oriental cultivars. *American Journal of Enology and Viticulture* 57(1), 105–108. DOI: 10.5344/ajev.2006.57.1.105.

Grando, M.S., Bellin, D., Edwards, K.J., Pozzi, C., Stefanini, M. and Velasco, R. (2003) Molecular linkage maps of *Vitis vinifera* L. and *Vitis riparia* Mchx. *Theoretical and Applied Genetics* 106(7), 1213–1224. DOI 10.1007/s00122-002-1170-3.

Granett, J., Timper, P. and Lider, L.A. (1985) Grape phylloxera (*Daktulosphaira vitifoliae*) (Homoptera: Phylloxeridae) biotypes in California. *Journal of Economic Entomology* 78(6), 1463–1467. DOI 10.1093/jee/78.6.1463.

Granett, J., Walker, M.A., Kocsis, L. and Omer, A.D. (2001) Biology and management of grape phylloxera. *Annual Review of Entomology* 46, 387–412. DOI 10.1146/annurev.ento.46.1.387.

Green, S.R. and Clothier, B.E. (1988) Water use of kiwifruit vines and apple trees by the heat-pulse technique. *Journal of Experimental Botany* 39(1), 115–123. DOI 10.1093/jxb/39.1.115.

Griffiths, M., Sistrom, W.R., Cohen-Bazire, G. and Stanier, R.Y. (1955) Function of carotenoids in photosynthesis. *Nature* 176, 1211–1214. DOI 10.1038/1761211a0.

Grimplet, J., Deluc, L.G., Tillett, R.L., Wheatley, M.D., Schlauch, K.A., Cramer, G.R. and Cushman, J.C. (2007) Tissue-specific mRNA expression profiling in grape berry tissues. *BMC Genomics* 8, 187. DOI 10.1186/1471-2164-8-187.

Grncarevic, M. (1963) Effect of various dipping treatments on the drying rate of grapes for raisins. *American Journal of Enology and Viticulture* 14(4), 230–234.

Grncarevic, M. and Radler, F. (1971) A review of the surface lipids of grapes and their importance in the drying process. *American Journal of Enology and Viticulture* 22(2), 80–86.

Groot Obbink, J. and Alexander, D.M. (1973) Response of six grapevine cultivars to a range of chloride concentrations. *American Journal of Enology and Viticulture* 24(2), 65–68.

Gu, S., Du, G., Zoldoske, D., Hakim, A., Cochran, R., Fugelsang, K. and Jorgensen, G. (2004) Effects of irrigation amount on water relations, vegetative growth, yield and fruit composition of Sauvignon blanc grapevines under partial rootzone drying and conventional irrigation in the San Joaquin Valley of California, USA. *Journal of Horticultural Science and Biotechnology* 79(1), 26–33. DOI 10.1080/14620316.2004.11511732.

Gualandri, V., Asquini, E., Bianchedi, P., Covelli, L., Brilli, M., Malossini, U., Bragagna, P., Saldarelli, P. and Si-Ammour, A. (2017) Identification of herbaceous hosts of the Grapevine Pinot gris virus (GPGV). *European Journal of Plant Pathology* 147, 21–25. DOI 10.1007/s10658-016-0989-4.

Guarga, R., Scaglione, G., Supino, E. and Mastrangelo, P. (2003) Evaluation of the SIS – a new frost protection method applied in a citrus orchard. *Proceedings of the 9th International Citrus Congress*, Orlando, Florida, pp. 583.

Gubler, W.D., Marois, J.J., Bledsoe, A.M. and Bettiga, L.J. (1987) Control of Botrytis bunch rot of grape with canopy management. *Plant Disease* 71(7), 599–601.

Gubler, W.D., Bettiga, L.J. and Heil, D. (1991) Comparisons of hand and machine leaf removal for the control of Botrytis bunch rot. *American Journal of Enology and Viticulture* 42, 233–236.

Gubler, W.D., Rademacher, M.R., Vasquez, S.J., and Thomas, C.S. (1999) Control of Powdery Mildew using the UC Davis Powdery Mildew Risk Index. *APSnet Features, 10.1094/APSnetFeature-1999–0199*.https://doi.org/10.1094/APSnetFeature-1999-0199.

Guelfat-Reich, S. and Safran, B. (1971) Indices of maturity for table grapes as determined by variety. *American Journal of Enology and Viticulture* 22(1), 13–18.

Guerrero, R.F., Cantos-Villar, E., Fernández-Marín, M.I., Puertas, B. and Serrano-Albarrán, M.J. (2015) Optimising UV-C preharvest light for stilbene synthesis stimulation in table grape: Applications. *Innovative Food Science & Emerging Technologies* 29, 222–229. DOI 10.1016/j.ifset.2015.02.010.

Guilford, T., Nicol, C., Rothschild, M. and Moore, B. (1987) The biological roles of pyrazines: evidence for a warning odour function. *Biological Journal of the Linnean Society* 31, 113–128. DOI 10.1111/j.1095-8312.1987.tb01984.x.

Gunkle, W.W. and Throop, J.A. (1993) Robotic pruning of grapevines: development of intelligent pruning and thinning machines. *Proceedings of the 2nd New Jersey Shaulis Grape Symposium on Pruning Mechanization and Crop Control*, Fredonia, New York. New York State Agriculture Experiment Station, Geneva, New York, pp. 51–54.

Habbadi, K., Aoujil, F., Yahyaoui, H., Benbouazza, A., Houssaini, salma E.I.E. and Achbani, E.H. (2023) Grapevine crown gall: current data and research perspectives. *Journal of Microbiology, Biotechnology and Food Sciences* 13, e10198–e10198. https://doi.org/10.55251/jmbfs.10198.

Haeseler, C.W., Smith, C.B., Kardos, L.T. and Fleming, H.K. (1980) Response of mature vines of *Vitis labrusca* L. cv Concord to applications of phosphorus and potassium over an eight-year span in Pennsylvania. *American Journal of Enology and Viticulture* 31(3), 237–244.

Hale, C.R. and Weaver, R.J. (1962) The effect of developmental stage on direction of translocation of photosynthate in *Vitis vinifera*. *Hilgardia* 33, 89–131. DOI 10.3733/hilg.v33n03p039.

Hale, C.R., Coombe, B.G. and Hawker, J.S. (1970) Effects of ethylene and 2-chloroethylphosphonic acid on the ripening of grapes. *Plant Physiology* 45(5), 620–623. DOI 10.1104/pp.45.5.620.

Halleen, F., Crous, R.W. and Petrin, O. (2003) Fungi associated with healthy grapevine cuttings in nurseries, with special reference to pathogens involved in the decline of young vines. *Australasian Plant Pathology* 32, 47–52. DOI 10.1071/AP02062.

Halleen, F., Fourie, P.H. and Crous, P.W. (2006) A review of black foot disease of grapevine. *Phytopathology Mediterranea* 45(4), S55–S67. DOI 10.14601/Phytopathol_Mediterr-1845.

Hamilton, R.P. and Coombe, B.G. (1992) Harvesting of winegrapes. In: Coombe, B.G. and Dry, P.R. (eds) *Viticulture Vol. 2. Practices*. Winetitles, Underdale, South Australia, pp. 302–327. ISBN 9781875130016.

Hamman Jr., R.A., Dami, I.E., Walsh, T.M. and Stushnoff, C. (1996) Seasonal carbohydrate changes and cold hardiness of Chardonnay and Riesling grapevines. *American Journal of Enology and Viticulture* 47, 31–36.

Hanna, R., Zalom, F.G., Wilson, L.T. and Leavitt, G.M. (1997) Sulfur can suppress mite predators in vineyards. *California Agriculture* 51, 19–21. DOI 10.3733/ca.v051n01p19.

Hannah, L., Roehrdanz, P.R., Ikegami, M., Shepard, A.V., Shaw, M.R., Tabor, G., Zhi, L., Marquet, P.A. and Hijmans, R.J. (2013) Climate change, wine, and conservation. *PNAS* 110, 6907–6912. DOI 10.1073/pnas.1210127110.

Hao, Z., Fayolle, L., van Tuinen, D., Chatagnier, O., Li, X., Gianinazzi, S. and Gianinazzi-Pearson, V. (2012) Local and systemic mycorrhiza-induced protection against the ectoparasitic nematode *Xiphinema index* involves priming of defence gene responses in grapevine. *Journal of Experimental Botany* 63, 3657–3672. https://doi.org/10.1093/jxb/ers046.

Hao, L., Johnson, K., Cursino, L., Mowery, P. and Burr, T.J. (2017) Characterization of the *Xylella fastidiosa* PD1311 gene mutant and its suppression of Pierce's disease on grapevines. *Molecular Plant Pathology* 18, 684–694. https://doi.org/10.1111/mpp.12428.

Harbertson, J.F., Kennedy, J.A. and Adams, D.O. (2002) Tannin in skins and seeds of Cabernet Sauvignon, Syrah, and Pinot noir berries during ripening. *American Journal of Enology and Viticulture* 53(1), 54–59.

Hardie, W.J. and Considine, J.A. (1976) Response of grapes to water-deficit stress in particular stages of development. *American Journal of Enology and Viticulture* 27(2), 55–61.

Häring, S., Folawiyo, S., Podguzova, M., krauß, S., and Stricker, D. (2024) Vid2Cuts: A framework for enabling AI-guided grapevine pruning. *IEEE Access*, 12, 5814–5836. https://doi.org/10.1109/ACCESS.2024.3350432.

Harrell, D.C. and Williams, L.E. (1987) The influence of girdling and gibberellic acid application at fruitset on Ruby Seedless and Thompson Seedless grapes. *American Journal of Enology and Viticulture* 38(2), 83–88.

Harris, J.M., Kriedemann, P.E. and Possingham, J.V. (1968) Anatomical aspects of grape berry development. *Vitis* 7, 106–119.

Hartwig, N.L. and Ammon, H.U. (2002) Cover crops and living mulches. *Weed Science* 50(6), 688–699. DOI 10.1614/0043-1745(2002)050[0688:AIACCA]2.0.CO;2.

Hashizume, K. and Samuta, T. (1999) Grape maturity and light exposure affect berry methoxypyrazine concentration. *American Journal of Enology and Viticulture* 50, 194–198.

Haslam, E. (1980) *In vino veritas*: oligomeric procyanidins and the ageing of red wines. *Phytochemistry* 19(12), 2577–2582. DOI 10.1016/S0031-9422(00)83922-9.

Hawkes, C.V., Waring, B.G., Rocca, J.D. and Kivlin, S.N. (2017) Historical climate controls soil respiration responses to current soil moisture. *PNAS* 114, 6322–6327. DOI 10.1073/pnas.1620811114.

Heien, D. and Venner, R. (1993) Grape juice concentrate emerging as a sweetener in juices, food products. *California Agriculture* 47, 28–31.

Helsen, L. and Van den Bulck, E. (2005). Review of disposal technologies for chromated copper arsenate (CCA) treated wood waste, with detailed analyses of thermochemical conversion processes. *Environmental Pollution* 134, 301–314. DOI 10.1016/j.envpol.2004.07.025.

Hendrickson, D.A., Lerno, L.A., Hjelmeland, A.K., Ebeler, S.E., Heymann, H., Hopfer, H., Block, K.L., Brenneman, C.A. and Oberholster, A. (2016) Impact of mechanical harvesting and optical berry sorting on grape and wine composition. *American Journal of Enology and Viticulture* 67, 385–397. DOI 10.5344/ajev.2016.14132.

Henick-Kling, T., Gerling, C., Martinson, T., Acree, T., Lakso, A. and Chiang, L. (2008) Studies on the origin and sensory aspects of atypical aging in white wines, in: *Proceedings of 15th International Enology Symposium*, April 14-16, Trier, Germany, 9 pp.

Herderich, M.J. and Smith, P.A. (2005) Analysis of grape and wine tannins: methods, applications and challenges. *Australian Journal of Grape and Wine Research* 11(2), 205–214. DOI 10.1111/j.1755-0238.2005.tb00288.x.

Hill, G.K., Stellwaag-Kittler, F., Huth, G. and Schlösser, E. (1981) Resistance of grapes in different development stages to *Botrytis cinerea*. *Phytopathology Zeitschrift* 102, 329–338. DOI 10.1111/j.1439-0434.1981.tb03394.x.

Himelrick, D.G. (1991) Growth and nutritional responses of nine grape cultivars to low soil pH. *HortScience* 26, 269–271.

Himelrick, D.G. (2001) Economic consequences of Phylloxera in cold climate wine grape production areas of eastern Washington. *Small Fruits Review* 1(4), 3–15. DOI 10.1300/J301v01n04_02.

Himelrick, D.G. (2003) Handling, storage and postharvest physiology of muscadine grapes: a review. *Small Fruits Review* 2(4), 45–62. DOI 10.1300/J301v02n04_06.

Hislop, E.C. (1988) Electrostatic ground-rig spraying: an overview. *Weed Technology* 2(1), 94–105.

Hjeljord, L.G., Stensvand, A. and Tronsmo, A. (2000) Effect of temperature and nutrient stress on the capacity of commercial *Trichoderma* products to control *Botrytis cinerea* and *Mucor piriformis* in greenhouse strawberries. *Biological Control* 19(2), 149–160. DOI 10.1006/bcon.2000.0859.

Hoag, J.C., Simonson, B., Cornforth, B. and St John, L. (2001) Waterjet Stinger: a tool for planting dormant nonrooted cuttings. *Native Plants Journal* 2, 84–89. DOI 10.3368/npj.2.2.84.

Hoddle, M.S. (2004) The potential adventive geographic range of glassy-winged sharpshooter, *Homalodisca coagulata* and the grape pathogen *Xylella fastidiosa*: implications for California and other grape growing regions of the world. *Crop Protection* 23(8), 691–699. DOI 10.1016/j.cropro.2003.11.017.

Hoffman, L.E. and Wilcox, W.F. (2003) Factors influencing the efficacy of myclobutanil and azoxystrobin for control of grape black rot. *Plant Disease* 87(3), 273–281. DOI 10.1094/PDIS.2003.87.3.273.

Hoffman, L.E., Wilcox, W.F., Gadoury, D.M. and Seem, R.C. (2002) Influence of grape berry age on susceptibility to *Guignardia bidwellii* and its incubation period length. *Phytopathology* 92(10), 1068–1076. DOI 10.1094/PHYTO.2002.92.10.1068.

Hoffman, L.E., Wilcox, W.F., Gadoury, D.M., Seem, R.C. and Riegel, D.G. (2004) Integrated control of grape black rot: influence of host phenology, inoculum availability, sanitation, and spray timing. *Phytopathology* 94(6), 641–650. DOI 10.1094/PHYTO.2004.94.6.641.

Hoffmann, S., Gaspero, G.D., Kovács, L., Howard, S., Kiss, E., Galbács, Z., Testolin, R. and Kozma, P. (2008) Resistance to *Erysiphe necator* in the grapevine 'Kishmish vatkana' is controlled by a single locus through restriction of hyphal growth. *Theoretical and Applied Genetics* 116, 427–438. DOI 10.1007/s00122-007-0680-4.

Hofstein, R., Daoust, R.A. and Aeschlimann, J.P. (1996) Constraints to the development of biofungicides: The example of "AQ10", a new product for controlling powdery mildews. *Entomophaga* 41, 455–460. DOI 10.1007/BF02765797.

Hollman, P.C.H. and Arts, I.C.W. (2000) Review. Flavonols, flavones and flavanols – nature, occurrence and dietary burden. *Journal of Science in Food and Agriculture* 80, 1081–1093. DOI 10.1002/(SICI)1097-0010(20000515)80:7<1081::AID-JSFA566>3.0.CO;2-G.

Holzapfel, B.P., Smith, J.P. Mandel, R.M. and Keller, M. (2006) Manipulating the postharvest period and its impact on vine productivity of Semillon grapevines. *American Journal of Enology and Viticulture* 57, 148–157.

Hoos, G. and Blaich, R. (1988) Metabolism of stilbene phytoalexins in grapevines: oxidation of resveratrol in single-cell cultures. *Vitis* 27, 1–12.

Hopkins, D.L. (2005) Biological control of Pierce's Disease in the vineyard with strains of *Xylella fastidiosa* benign to grapevine. *Plant Disease* 89, 1348–1352. DOI 10.1094/PD-89-1348.

Hoskins, N., Bonfiglioli, R. and Wright, C. (2003) *The Riversun Rootstock Project – Part 1*. Riversun Nurseries Ltd., Gisborne, New Zealand, 1 p.

Howard, K.L., Mike, J.H. and Riesen, R. (2005) Validation of a solid-phase micro-extraction method for headspace analysis of wine aroma components. *American Journal of Enology and Viticulture* 56(1), 37–45.

Howell, G.S. (2001) Sustainable grape productivity and the growth–yield relationship: a review. *American Journal of Enology and Viticulture* 52(3), 165–174.

Howell, G.S., Stergios, B.G. and Stackhouse, S.S. (1978) Interrelation of productivity and cold hardiness of Concord grapevines. *American Journal of Enology and Viticulture* 29, 187–191.

Hrazdina, G., Parsons, G.F. and Mattick, L.R. (1984) Physiological and biochemical events during development and maturation of grape berries. *American Journal of Enology and Viticulture* 35(4), 220–227.

Huang, H.T. (2000) Fermentations and food science. In: Needham, J. (ed.) *Science and Civilisation in China. Volume 6: Biology and Biological Technology*. Cambridge University Press, Cambridge, UK. ISBN 9780521652704.

Huang, P.-D., Cash, J.N. and Santerre, C.R. (1988) Influence of stems, petioles and leaves on the phenolic content of Concord and Aurora blanc juice and wine. *Journal of Food Science* 53(1), 173–175. DOI 10.1111/j.1365-2621.1988.tb10202.x.

Huang, S. and Gale, F. (2006) *China's Rising Fruit and Vegetable Exports Challenge U.S. Industries*. USDA, Washington, DC, p. 21.

Huggett, J.M. (2006) Geology and wine: a review. *Proceedings of the Geological Association* 117(2), 239–247. DOI 10.1016/S0016-7878(06)80012-X.

Hughes, K.A. and Wilde, R.H. (1989). The effect of poor drainage on the root distribution of kiwifruit vines. *New Zealand Journal of Crop and Horticultural Science* 17, 239–244. DOI 10.1080/01140671.1989.10428038.

Huisman, J.A., Hubbard, S.S., Redman, J.D. and Annan, A.P. (2003) Measuring soil water content with ground penetrating radar: a review. *Vadose Zone Journal* 2(4), 476–491. DOI 10.2136/vzj2003.4760.

Hunt, M.D., Neuenschwander, U.H., Delaney, T.P., Weymann, K.B., Friedrich, L.B., Lawton, K.A., Steiner, H.-Y. and Ryals, J.A. (1996) Recent advances in systemic acquired resistance research – a review. *Gene* 179, 89–95. DOI 10.1016/S0378-1119(96)00429-5.

Ibacache, A.E. (1990) *Girdling and Shading Effects on Inflorescence Necrosis and Rachis Composition of Pinot Noir Grapevine.* PhD thesis, Department of Horticulture, Oregon State University, Corvallis, Oregon. http://hdl.handle.net/1957/25621.

IFOAM (2025) Statement from the Organic Movement - Elevating Truly Regenerative Agriculture. https://www.ifoam.bio/sites/default/files/2025-01/20250113_StatmentfromtheorganicMovement_02.pdf (accessed 3 April 2025).

Ifoulis, A.A. and Savopoulou-Soultani, M. (2004) Biological control of *Lobesia botrana* (Lepidoptera: Tortricidae) larvae by using different formulations of *Bacillus thuringiensis* in 11 vine cultivars under field conditions. *Journal of Economic Entomology*, 340–343. DOI 10.1603/0022-0493-97.2.340.

Ilyas, T., Chowdhary, P., Chaurasia, D., Gnansounou, E., Pandey, A. and Chaturvedi, P. (2021) Sustainable green processing of grape pomace for the production of value-added products: An overview. *Environmental Technology & Innovation* 23, 101592. https://doi.org/10.1016/j.eti.2021.101592.

Ingalsbe, D.W., Neubert, A.M. and Carter, G.H. (1963) Concord grape pigments. *Journal of Agricultural and Food Chemistry* 11, 263–268. DOI 10.1021/jf60127a009.

Intrieri, C. and Filippetti, I. (2000) Innovations and outlook in grapevine training systems and mechanization in North-Central Italy. *Proceedings of the American Society for Enology and Viticulture 50th Anniversary Annual Meeting*, Seattle, Washington, pp. 170–184.

Intrigliolo, D. and Castel, J. (2007) Evaluation of grapevine water status from trunk diameter variations. *Irrigation Science* 26(1), 49–59. DOI 10.1007/s00271-007-0071-2.

Iriti, M., Vitalini, S., Di Tommaso, G., D'amico, S., Borgo, M. and Faoro, F. (2011) New chitosan formulation prevents grapevine powdery mildew infection and improves polyphenol content and free radical scavenging activity of grape and wine. *Australian Journal of Grape and Wine Research* 17, 263–269.

Ishii, A., Roudnitzky, N., Béno, N., Bensafi, M., Hummel, T., Rouby, C. and Thomas-Danguin, T. (2008) Synergy and masking in odor mixtures: an electrophysiological study of orthonasal vs. retronasal perception. *Chemical Senses* 33, 553–561. DOI 10.1093/chemse/bjn022.

Jackson, D.I. (1991) Environmental and hormonal effects on development of Early Bunch Stem Necrosis. *American Journal of Enology and Viticulture* 42(4), 290–294.

Jackson, D.I. (1997) *Monographs in Cool Climate Viticulture 1. Pruning and Training.* Lincoln University Press, Canterbury, New Zealand, 69 pp.

Jackson, D.I. and Cherry, N.J. (1988) Prediction of a district's grape-ripening capacity using a Latitude–Temperature Index (LTI). *American Journal of Enology and Viticulture* 39(1), 19–28.

Jackson, D.I. and Coombe, B.G. (1988) Early bunchstem necrosis – a cause of poor set. *Proceedings of the 2nd International Cool Climate Viticulture and Oenolology Symposium*, Auckland, New Zealand, pp. 72–75.

Jackson, D.I. and Lombard, P.B. (1993) Environmental and management practices affecting grape composition and wine quality – a review. *American Journal of Enology and Viticulture* 44(4), 409–430.

Jackson, R.B. and Caldwell, M.M. (1993) The scale of nutrient heterogeneity around individual plants and its quantification with geostatistics. *Ecology* 74(2), 612–614. DOI 10.2307/1939320.

Janick, J. (2002) Ancient Egyptian agriculture and the origins of horticulture. *Acta Horticulturae* 582, 23–39. DOI 10.17660/ActaHortic.2002.582.1.

Jansen, M.A.K., Gaba, V. and Greenberg, B.M. (1998) Higher plants and UV-B radiation: balancing damage, repair and acclimation. *Trends in Plant Science* 3(4), 131–135. DOI 10.1016/S1360-1385(98)01215-1.

Jaspers, M.V., Seyb, A.M., Trought, M.C.T. and Balasubramaniam, R. (2013) Overwintering grapevine debris as an important source of *Botrytis cinerea* inoculum. *Plant Pathology* 62, 130–138. DOI 10.1111/j.1365-3059.2012.02631.x.

Jenkins, P.E. and Isaacs, R. (2007) Reduced-risk insecticides for control of grape berry moth (Lepidoptera: Tortricidae) and conservation of natural enemies. *Journal of Economic Entomology* 100(3), 855–865. DOI 10.1603/0022-0493(2007)100[855:RIFCOG]2.0.CO;2.

Jensen, F., Swanson, F., Peacock, W. and Leavitt, G. (1975) The effect of width of cane and trunk girdles on berry weight and soluble solids in table 'Thompson Seedless' vineyards. *American Journal of Enology and Viticulture* 26(2), 90–91.

Jensen, F., Andris, H. and Beede, R. (1981) A comparison of normal girdles and knife-line girdles on Thompson Seedless and Cardinal grapes. *American Journal of Enology and Viticulture* 32(3), 206–207.

Jermini, M. and Gessler, C. (1996) Epidemiology and control of grape black rot in southern Switzerland. *Plant Disease* 80(3), 322–325. DOI 10.1094/PD-80-0322.

Jermini, M., Linder, C., Colombi, L. and Marazzi, C. (2007) Lutte obligatoire contre le vecteur de la flavescence dorée au Tessin. *Revue suisse Viticulture, Arboriculture, Horticulture* 39(2), 102–106.

Jersch, S., Scherer, C., Huth, G. and Schlösser, E. (1989) Proanthocyanidins as basis for quiescence of *Botrytis cinerea* in immature strawberry fruits. *Journal of Plant Disease Protection* 96(4), 365–378.

Jiao, B., Hao, X., Liu, Z., Liu, M., Wang, J., Liu, L., Liu, N., Song, R., Zhang, J., Fang, Y. and Xu, Y. (2022) Engineering CRISPR immune systems conferring GLRaV-3 resistance in grapevine. *Horticulture Research* 9, uhab023. https://doi.org/10.1093/hr/uhab023.

Jie, W., Lite, L. and Yang, D. (2003) The correlation between freezing point and soluble solids of fruits. *Journal of Food Engineering* 60(4), 481–484. DOI 10.1016/S0260-8774(03)00081-5.

John, S., Wicks, T.J., Hunt, J.S., Lorimer, M.F., Oakey, H. and Scott, E.S. (2005) Protection of grapevine pruning wounds from infection by *Eutypa lata* using *Trichoderma harzianum* and *Fusarium lateritium*. *Australasian Plant Pathology* 34, 569–575. DOI 10.1071/AP05075.

Johnson, D.T., Lewis, B.A. and Snow, J.W. (1991) Control of grape root borer (Lepidoptera: Sesiidae) by mating disruption with two synthetic sex pheromone compounds. *Environmental Entomology* 20(3), 930–934. DOI 10.1093/ee/20.3.930.

Johnson, W. and Grgich, M. (1975) Another look at machine harvesting/wine quality. *Wines and Vines* 57, 40.

Johnston, W.H. (2000) Calibration of gypsum blocks and data loggers and their evaluation for monitoring soil water status. *Australian Journal of Experimental Agriculture* 40(8), 1131–1136. DOI 10.1071/EA99167.

Jones, G.V., White, M.A., Cooper, O.R. and Storchmann, K. (2005) Climate change and global wine quality. *Climatic Change* 73, 1573–1480. DOI 10.1007/s10584-005-4704-2.

Jordan, D. (1993) Cool climate effects on fruit set in grapevines through inflorescence necrosis. *Die Wein Wissenschaft* 48, 3–6.

Jordan, T.D., Pool, R.M., Zabadal, T.J. and Tomkins, J.P. (1981) *Cultural Practices for Commercial Vineyards*. Miscellaneous Bulletin, New York State College of Agriculture and Life Sciences, Cornell University, Ithaca, New York, 69 pp.

Jourjon, F., Chou, H.-C., Gezart, A., Kadison, A. E., Martinat, L., Pomarici, E. and Vecchio, R. (2016) Wineries evaluation of costs and benefits of sustainability certification program: The case of Terra Vitis in France. *Recent Patents on Food, Nutrition & Agriculture* 8(2), 138–147. https://doi.org/10.2174/2212798408666160524141500.

Kaiser, B.N., Gridley, K.L., Brady, J.N., Phillips, T. and Tyerman, S.D. (2005) The role of molybdenum in agricultural plant production. *Australian Journal of Experimental Agriculture* 96(5), 745–754. DOI 10.1093/aob/mci226.

Kania, G., Kowalski, R. and Pietra, R. (2016) Defensive secretions in millipede species of the order Julida (Diplopoda). *Acta Societatis Zoologicae Bohemicae*, 80, 17–20.

Kanner, J., Frankel, E., Granit, R., German, B. and Kinsella, J.E. (1994) Natural antioxidants in grapes and wines. *Journal of Agricultural and Food Chemistry* 42, 64–69. DOI 10.1021/jf00037a010.

Karban, R., English-Loeb, G., Walker, M.A. and Thaler, J. (1995) Abundance of phytoseiid mites on *Vitis* species: effects of leaf hairs, domatia, prey abundance and plant phylogeny. *Experimental and Applied Acarology* 19(4), 189–197. DOI 10.1007/ BF00130822.

Kasimatis, A.N., Weaver, R.J., Pool, R.M. and Halsey, D.D. (1971) Response of 'Perlette' grape berries to gibberellic acid applied during bloom or at fruit set. *American Journal of Enology and Viticulture* 22(1), 19–23.

Kassemeyer, H.-H. and Staudt, G. (1981) Development of the embryo sac and fertilization in grapevine. *Vitis* 20, 202–210.

Kast, W.K. and Stark-Urnau, M. (1999) Survival of sporangia from *Plasmopara viticola*, the downy mildew of grapevine. *Vitis* 38(4), 185–186.

Kato, A.E., Fathallah, F.A., Miles, J.A., Meyers, J.M., Faucett, Janowitz, I. and Garcia, E.G. (2006) Ergonomic evaluation of winegrape trellis systems pruning operation. *Journal of Agricultural Safety and Health* 12, 17–28. DOI 10.13031/2013.20199.

Kauer, R., Gaubatz, B., Kornitzer, U., Kirchner, B., Wöhrle, M. and Schultz, H.R. (2001) *Organic Viticulture without Sulphur? Three Years of Experience with Sodium and Potassium Bicarbonate*. GESCO Compte Rendu, Montpellier, France, pp. 213–218.

Keller, M. and Torres-Martinez, N. (2004) Does UV radiation affect winegrape composition? *Acta Horticulturae* 640, 313–319. DOI 10.17660/ActaHortic.2004.640.36.

Keller, M., Arnink, K.J. and Hrazdina, G. (1998) Interaction of nitrogen availability during bloom and light intensity during veraison. I. Effects on grapevine growth, fruit development, and ripening. *American Journal of Enology and Viticulture* 49(3), 333–340.

Keller, M., Rogiers, S.Y. and Schultz, H.R. (2003) Nitrogen and ultraviolet radiation modify grapevines' susceptibility to powdery mildew. *Vitis* 42(2), 87–94.

Keller, M., Mills, L.J., Wample, R.L. and Spayd, S.E. (2005) Cluster thinning effects on three deficit-irrigated *Vitis vinifera* cultivars. *American Journal of Enology and Viticulture* 56(2), 91–103.

Khiari, R., Ounaissi, D., Lançon-Verdier, V., Zemni, H., Mihoubi, D., and Maury, C. (2025) Ability of hyperspectral imaging to assess physicochemical and phytochemical quality parameters of raisins. *Journal of Food Measurement and Characterization*, 19(2), 1234–1247. https://doi.org/10.1007/s11694-024-03036-1.

Kido, H., Flaherty, D.L., Bosch, D.F. and Valero, K.A. (1984) French prune trees as overwintering sites for the grape leafhopper egg parasite. *American Journal of Enology and Viticulture* 35(3), 156–160.

Kimura, P.H., Okamoto, G. and Hirano, K. (1998) The mode of pollination and stigma receptivity in *Vitis coignetiae* Pulliat. *American Journal of Enology and Viticulture* 49(1), 1–5.

King, P.D. and Buchanan, G.A. (1986) The dispersal of phylloxera crawlers and spread of phylloxera infestations in New Zealand and Australian vineyards. *American Journal of Enology and Viticulture* 37(1), 26–33.

Kirchmair, M., Neuhauser, S., Strasser, H., Voloshchuk, N., Hoffmann, M. and Huber, L. (2009) Biological control of grape phylloxera – a historical review and future prospects. *Acta Horticulturae* 816, 13–18. DOI 10.17660/ActaHortic.2009.816.1.

Kirk, A. (2016) *Evaluating cane pruning decision criteria and the identification of grapevine pruning styles*. M. Sc. thesis. Lincoln University. 108 pp. https://hdl.handle.net/10182/7059.

Klepper, B., Douglas, V. and Taylor, H.M. (1971) Stem diameter in relation to plant water status. *Plant Physiology* 48, 683–685. DOI 10.1104/pp.48.6.683.

Kliewer, W.M. (1968) Changes in the concentration of free amino acids in grape berries during maturation. *American Journal of Enology and Viticulture* 19(3), 166–174.

Kliewer, W.M. (1977) Influence of temperature, solar radiation and nitrogen on coloration and composition of Emperor grapes. *American Journal of Enology and Viticulture* 28(2), 96–103.

Kliewer, W.M. and Antcliff, A.J. (1970) Influence of defoliation, leaf darkening, and cluster shading on the growth and composition of Sultana grapes. *American Journal of Enology and Viticulture* 21(1), 26–36.

Kliewer, W.M. and Dokoozlian, N.K. (2005) Leaf area/crop weight ratios of grapevines: influence on fruit composition and wine quality. *American Journal of Enology and Viticulture* 56(2), 170–181. DOI: 10.5344/ajev.2005.56.2.170.

Kliewer, W.M. and Lider, L.A. (1968) Influence of cluster exposure to the sun on the composition of Thompson Seedless fruit. *American Journal of Enology and Viticulture* 19(3), 175–184.

Kliewer, W.M. and Torres, R.E. (1972) Effect of controlled day and night temperatures on grape coloration. *American Journal of Enology and Viticulture* 23(2), 71–77.

Kliewer, W.M., Bowen, P. and Benz, M. (1989) Influence of shoot orientation on growth and yield development in Cabernet Sauvignon. *American Journal of Enology and Viticulture* 40(4), 259–264.

Kluba, R.M., Mattick, L.R. and Hackler, L.R. (1978) Changes in the free and total amino acid composition of several *Vitis labruscana* grape varieties during maturation. *American Journal of Enology and Viticulture* 29(2), 102–111.

Kolattukudy, P.E. (1985) Enzymatic penetration of the plant cuticle by fungal pathogens. *Annual Review of Phytopathology* 23, 223–250. DOI 10.1146/annurev.py.23.090185.001255.

Kolb, C.A., Kopecký, J., Riederer, M. and Pfündel, E.E. (2003) UV screening by phenolics in berries of grapevines (*Vitis vinifera*). *Functional Plant Biology* 30, 1177–1186. DOI 10.1071/FP03076.

Köller, W. and Scheinpflug, H. (1987) Fungal resistance to sterol biosynthesis inhibitors: a new challenge. *Plant Disease* 71, 1066–1074.

Kombrink, E. and Schmelzer, E. (2001) The hypersensitive response and its role in local and systemic disease resistance. *European Journal of Plant Pathology* 107(1), 69– 78. DOI 10.1023/A:1008736629717.

Kondo, K. and Ting, K.C. (1998) Robotics for plant production. *Artificial Intelligence Review* 12, 227–243. DOI 10.1007/978-94-011-5048-4_12.

Konrad, H., Lindner, B., Bleser, E. and Rühl, E.H. (2003) Strategies in the genetic selection of clones and the preservation of genetic diversity within varieties. *Acta Horticulturae* 603, 105–110. DOI 10.17660/ActaHortic.2003.603.10.

Korosi, G. (2020) *Entomopathogenic fungi as potential biocontrol agents for grapevine phylloxera*. PhD thesis, Charles Sturt University, Australia.

Kostaropoulos, A.E. and Saravacos, G.D. (1995) Microwave pre-treatment for sun-dried raisins. *Journal of Food Science* 60, 344–347. DOI 10.1111/j.1365-2621.1995.tb05669.x.

Kovacs, L.G., Hanami, H., Fortenberry, M. and Kaps, M.L. (2001) Latent infection by leafroll agent GLRaV-3 is linked to lower fruit quality in French–American hybrid grapevines Vidal blanc and St. Vincent. *American Journal of Enology and Viticulture* 52(3), 254–259.

Krenz, B., Fuchs, M., Thompson, J.R. (2023) Grapevine red blotch disease: A comprehensive Q&A guide. *PLOS Pathogens* 19, e1011671. https://doi.org/10.1371/journal.ppat.1011671.

Krenz, B., Thompson, J.R., McLane, H.L., Fuchs, M. and Perry, K.L. (2014) Grapevine red blotch-associated virus is widespread in the United States. *Phytopathology* 104, 1232–1240. DOI 10.1094/PHYTO-02-14-0053-R.

Kross, S.M., Tylianakis, J.M. and Nelson, X.J. (2012) Effects of introducing threatened falcons into vineyards on abundance of passeriformes and bird damage to grapes. *Conservation Biology* 26, 142–149. DOI 10.1111/j.1523-1739.2011.01756.x.

Kubota, N. and Miyamuki, M. (1992) Breaking bud dormancy in grapevines with garlic paste. *Journal of the American Society for Horticultural Science* 117(6), 898–901.

Kubota, N., Matthews, M.A., Takahagi, T. and Kliewer, W.M. (2000) Budbreak with garlic preparations: effects of garlic preparations and of calcium and hydrogen cyanamides on budbreak of grapevines grown in greenhouses. *American Journal of Enology and Viticulture* 51(4), 409–414.

Kulakiotu, E.K., Thanassoulopoulos, C.C. and Sfakiotakis, E.M. (2004) Biological control of *Botrytis cinerea* by volatiles of 'Isabella' grapes. *Phytopathology* 94(9), 924–931. DOI 10.1094/PHYTO.2004.94.9.924.

Kurtural, S.K. and Fidelibus, M.W. (2021) Mechanization of pruning, canopy management, and harvest in winegrape vineyards. *American Journal of Enology and Viticulture* 5, 29–44. https://doi.org/10.5344/catalyst.2021.20011.

Kwasniewski, M.T., Sacks, G.L. and Wilcox, W.F. (2014) Persistence of elemental sulfur spray residue on grapes during ripening and vinification. *American Journal of Enology and Viticulture* 65, 453–462. DOI 10.5344/ajev.2014.14027.

Lafon, R. (1921). *Modifications à apporter à la taille de la vigne des Charentes, taille Guyot-Poussart mixte et double: L'apoplexie, traitement préventif (méthode Poussard), traitement curatif*. Roumégous et Déhan. https://archive.org/details/bnf-bpt6k6565326w/mode/2up (accessed 8 April 2025).

Lakso, A.N., Pratt, C., Pearson, R.C., Pool, R.M., Seem, R.C. and Welser, M.J. (1982) Photosynthesis, transpiration, and water use efficiency of mature grape leaves infected with *Uncinula necator* (powdery mildew). *Phytopathology* 72, 232–236. DOI 10.1094/Phyto-72-232.

Lakso, A.N., Santiago, M. and Stroock, A.D. (2022) Monitoring stem water potential with an embedded microtensiometer to inform irrigation scheduling in fruit crops. *Horticulturae* 8, 1207. https://doi.org/10.3390/horticulturae8121207.

Lamb, D.W. (2009) Electrically heated cables protect vines from frost damage at early flowering. *Australian Journal of Grape and Wine Research* 15, 79–84. DOI 10.1111/j.1755-0238.2008.00034.x.

Landis, D.A., Wratten, S.D. and Gurr, G.M. (2000) Habitat management to conserve natural enemies of arthropod pests in agriculture. *Annual Review of Entomology* 45(1), 175–201. DOI 10.1146/annurev.ento.45.1.175.

Lang, A. (1970) Gibberellins: structure and metabolism. *Annual Review of Plant Physiology* 21, 537–570. DOI 10.1146/annurev.pp.21.060170.002541.

Langcake, P. and Lovell, P.A. (1980) Light and electron microscopical studies of the infection of *Vitis* spp. by *Plasmopara viticola*, the downy mildew pathogen. *Vitis* 19, 321–337.

Langcake, P. and McCarthy, W.V. (1979) The relationship of resveratrol production to infection of grapevine leaves by *Botrytis cinerea*. *Vitis* 18, 244–253.

Langcake, P. and Pryce, R.J. (1976) The production of resveratrol by *Vitis vinifera* and other members of the Vitaceae as a response to infection or injury. *Physiological Plant Pathology* 9, 77–86. DOI 10.1016/0048-4059(76)90077-1.

Larrauri, J.A., Ruperez, P. and Calixto, F.S. (1996) Antioxidant activity of wine pomace. *American Journal of Enology and Viticulture* 47(4), 369–372.

Larrey, M., Tandonnet, J.-P., Cast, C.S., Cookson, S.J. and Vivin, P. (2025) Exploring how graft length shapes root system architecture and morphology in grapevine rootstocks. *OENO One* 59. https://doi.org/10.20870/oeno-one.2025.59.1.8226.

Lave, L.B. (1963) The value of better weather information to the raisin industry. *Econometrica* 31(1/2), 151–164.

Lavee, S. (1973) Dormancy and bud break in warm climates: considerations of growth regulators involvement. *Acta Horticulturae* 34, 225–232. DOI 10.17660/ActaHortic.1973.34.31.

Lavee, S. and Nir, G. (1986) Grape. In: Monselise, S.P. (ed.) *CRC Handbook of Fruit Set and Development*. CRC Press, Boca Raton, Florida, pp. 167–191. ISBN 9780849332609.

Lavee, S., Regev, U. and Samish, R.M. (1967) The determination of induction and differentiation in grapevines. *Vitis* 6, 1–13.

Lecomte, P., Cholet, C., Bruez, E., Martignon, T., Giudici, M., Simonit, M., Ugaglia, A.A., Forget, D., Miramon, J., Arroyo, M., Dubourdieu, D., Geny-Denis, L. and Rey, P. (2022). Recovery after curettage of grapevines with esca leaf symptoms. *Phytopathologia Mediterranea* 61(3), 473-488. https://doi.org/10.36253/phyto-13357

Lecomte, P., Diarra, B., Carbonneau, A., Rey, P., and Chevrier, C. (2018) Esca of grapevine and training practices in France: Results of a 10-year survey. *Phytopathologia Mediterranea* 57(3), Article 3. https://doi.org/10.14601/Phytopathol_Mediterr-22025.

Lessio, F., Lioy, S., Colombo, M. and Alma, A. (2025) Factors affecting the impact of *Popillia japonica* Newman, 1841 (Coleoptera: Scarabaeidae) on grapevine in Northwestern Italy. *Bulletin of Entomological Research* 115, 175–183. https://doi.org/10.1017/S0007485325000021.

Li, X. (2004) *Potassium Nutrition of Selfrooted Pinot Noir Grapevines (*Vitis vinifera *L.) Grown in Pots*. Agriculture and Life Science Division, Lincoln University, Canterbury, New Zealand, 145 pp. http://hdl.handle.net/10182/1802.

Lider, L.A. (1958) Phylloxera-resistant grape rootstocks for the coastal valleys of California. *Hilgardia* 27, 287–318. DOI 10.3733/hilg.v27n11p287.

Lieth, J.H., Meyer, M.M., Yeo, K.-H. and Kirkpatrick, B.C. (2011) Modeling cold curing of Pierce's Disease in *Vitis vinifera* 'Pinot Noir' and 'Cabernet Sauvignon' grapevines in California. *Phytopathology* 101, 1492–1500. DOI 10.1094/PHYTO-08-10-0207.

Lin, C.H. and Wang, T.Y. (1985) Enhancement of bud sprouting in grape single bud cuttings by cyanamide. *American Journal of Enology and Viticulture* 36(1), 15–17.

Lindow, S., Koutsoukis, R., Meyer, K. and Baccari, C. (2024) Control of Pierce's Disease of grape with *Paraburkholderia phytofirmans* PsJN in the field. *Phytopathology* 114, 503–511. https://doi.org/10.1094/PHYTO-06-23-0219-R.

Lipe, W.N., Baumhardt, L., Wendt, C.W. and Rayburn, D.J. (1992) Differential thermal analysis of deacclimating Chardonnay and Cabernet Sauvignon grape buds as affected by evaporative cooling. *American Journal of Enology and Viticulture* 43(4), 355–361.

Liu, H.-F., Wu, B.-H., Fan, P.-G., Xu, H.-Y. and Li, S.-H. (2007) Inheritance of sugars and acids in berries of grape (*Vitis vinifera* L.). *Euphytica* 153(1), 99–107. DOI 10.1007/ s10681-006-9246-9.

Liu, H.M., Guo, J.H., Cheng, Y.J., Luo, L., Liu, P., Wang, B.Q., Deng, B.X. and Long, C.A. (2010) Control of gray mold of grape by *Hanseniaspora uvarum* and its effects on postharvest quality parameters. *Annals of Microbiology* 60, 31–35. DOI 10.1007/ s13213-010-0018-3.

Lizamore, D.K. (2013) *A study of endogenous transposon activity in grapevine (*Vitis vinifera *L.)*. PhD thesis, Lincoln University, Christchurch, New Zealand, 255 pp. https:// hdl.handle.net/10182/5969.

Logan, G. (2015) *Rotundone in New Zealand* Vitis vinifera *L. Syrah: Fruit, Fermentation and Functional Food Chemistry*. PhD thesis, University of Auckland, New Zealand. http://hdl.handle.net/2292/25774.

Loinger, C., Cohen, S., Dror, N. and Berlinger, M.J. (1977) Effect of grape cluster rot on wine quality. *American Journal of Enology and Viticulture* 28(4), 196–199.

Lomkatsi, S.I., Ordzhonikidze, A.A., Mgeladze, T.A., Gurabanidze, L.R., Dzhimshitashvili, L.G., Pipiya, A.D., Chubinishvili, T.-M. and Mazanashvili, T.G. (1983) Protection of vineyards from hail damage. *Vestnik-Sel'skokhozyaistvennoi-Nauki* 2, 66–68.

Long, Z. (1997) Developing wine flavor in the vineyard. *Practical Winery & Vineyard* 18, 6–9.

Loughrin, J.H., Potter, D.A., Hamilton-Kemp, T.R. and Byers, M.E. (1997) Diurnal emission of volatile compounds by Japanese Beetle-damaged grape leaves. *Phytochemistry* 45, 919–923. DOI 10.1016/S0031-9422(97)00076-9.

Lovisolo, C. and Schubert, A. (2000) Downward shoot positioning affects water transport in field-grown grapevines. *Vitis* 39(2), 49–53.

Lowe, T., Moghadam, P., Edwards, E. and Williams, J. (2021) Canopy density estimation in perennial horticulture crops using 3D spinning lidar SLAM. *Journal of Field Robotics* 38, 598–618. https://doi.org/10.1002/rob.22006.

Lydakis, D., Fysarakis, I., Papadimitriou, M. and Kolioradakis, G. (2003) Optimization study of sulfur dioxide application in processing of Sultana raisins. *International Journal of Food Properties* 6(3), 393–403. DOI 10.1081/JFP-120020117.

Magarey, P.A., Wicks, T.J. and Wachtel, M.F. (1990) Phosphonic (phosphorous) acid controls *Plasmopara viticola* the cause of downy mildew of grapevines. *Australasian Plant Pathology* 19, 126–127. DOI 10.1071/APP9900126.

Magarey, P.A., Wachtel, M.F., Weir, P.C. and Seem, R.C. (1991) A computer-based simulator for rational management of grapevine downy mildew (*Plasmopara viticola*). *Plant Protection Quarterly* 6, 29–33.

Magarey, P., Wicks, T., Hitch, C. and Emmett, B. (2003) Effects of temperature and application rates on the phytotoxicity of sulphur on grapevines. In: Emmett, R.W. (ed.) *Strategic Use of Sulphur in Integrated Pest and Disease Management (IPM) Programs for Grapevines*. Project Number: DAV 98/1, Victoria Dept. of Primary Industries, Mildura, Victoria, Australia, pp. 110–116.

Mahfoudhi, N., Digiaro, M. and Dhouibi, M.H. (2009) Transmission of grapevine leafroll viruses by *Planococcus ficus* (Hemiptera: Pseudococcidae) and *Ceroplastes rusci* (Hemiptera: Coccidae). *Plant Disease* 93, 999–1002. DOI 10.1094/PDIS-93-10-0999.

Mai, W., Xue, X., Feng, G., Yang, R. and Tian, C. (2019) Arbuscular mycorrhizal fungi – 15-Fold enlargement of the soil volume of cotton roots for phosphorus uptake in intensive planting conditions. *European Journal of Soil Biology* 90, 31–35. doi.org/10.1016/j.ejsobi.2018.12.002.

Mailhac, N. and Chervin, C. (2006) Ethylene and grape berry ripening. *Stewart Postharvest Review* (2), 5 pp.

Malagnini, V., Lillo, E. de, Saldarelli, P., Beber, R., Duso, C., Raiola, A., Zanotelli, L., Valenzano, D., Giampetruzzi, A., Morelli, M. *et al.* (2016) Transmission of grapevine Pinot gris virus by *Colomerus vitis* (Acari: Eriophyidae) to grapevine. *Archives of Virology* 161, 2595–2599. DOI 10.1007/s00705-016-2935-3.

Malan, F.S. and Verryn, S.D. (1996) Effect of genotype-by-environment interaction on the wood properties and qualities of four-year-old *Eucalyptus grandis* and *E. grandis* hybrids. *South African Forestry Journal* 176, 47–53. DOI 10.1080/00382167.1996.9629709.

Malossini, U., Moscon, R., Ferrazza, M., Bianchedi, P., Varner, M. and Credi, R. (2012) Caratteristiche vegeto-produttive di viti Pinot grigio e Traminer aromatico affette da una nuova virosi segnalata in Trentino. In: *IV Convegno Nazionale di Viticoltura (CONAVI)*: Asti, 10-12 July 2012: 37 (032) https://hdl.handle.net/10449/21490.

Manuelian Fusté, C.L., Albanell Trullás, E., Rovai, M., Vidal Giralt, L.L., Salama, L., Salama, A.A.K., Domingo Gustems, C. and Caja López, G. (2014) Creation and persistence of conditioned sheep aversion to grapevine leaves. Application for grazing vineyards during spring. *Pastos* 44, 59–66.

Marais, J. (1994) Sauvignon blanc cultivar aroma – a review. *South African Journal of Enology and Viticulture* 15, 41–45. DOI 10.21548/15-2-2283.

Marais, J., Wyk, C.J.v. and Rapp, A. (1992) Effect of sunlight and shade on norisoprenoid levels in maturing Weisser Riesling and Chenin blanc grapes and Weisser Riesling wines. *South African Journal of Enology and Viticulture* 13, 23–32.

Marschner, H. (1986) *Mineral Nutrition of Higher Plants*. Academic Press, Harcourt Brace Jovanovich, London. ISBN 9780124735408.

Marshall, J.K. (1967) The effect of shelter on the productivity of grasslands and field crops. *Field Crops Abstracts* 20, 1–14.

Martínez-Casasnovas, J.A. and Concepción Ramos, M. (2009) Soil alteration due to erosion, ploughing and levelling of vineyards in northeast Spain. *Soil Use and Management* 25, 183–192. https://doi.org/10.1111/j.1475-2743.2009.00215.x.

Martínez-Lüscher, J., Torres, N., Hilbert, G., Richard, T., Sánchez-Díaz, M., Delrot, S., Aguirreolea, J., Pascual, I. and Gomès, E. (2014) Ultraviolet-B radiation modifies the quantitative and qualitative profile of flavonoids and amino acids in grape berries. *Phytochemistry* 102, 106–114. DOI 10.1016/j.phytochem.2014.03.014.

Martini, A., Ciani, M. and Scorzetti, G. (1996) Direct enumeration and isolation of wine yeasts from grape surfaces. *American Journal of Enology and Viticulture* 47(4), 435–440.

Martinson, T. (1995) Management of insect pests in organic vineyards. *Organic Grape and Wine Production Symposium*, Geneva, New York. Communications Services, New York State Agricultural Experiment Station, Cornell University, Geneva, New York.

Matthews, G.A. (2001) The application of chemicals for plant disease control. In: Waller, J.M., Lenné, J.M. and Waller, S.J. (eds) *Plant Pathologist's Pocketbook*. CAB International, Wallingford, UK, pp. 345–353. ISBN 9780851994598.

Matthews, M.A., Cheng, G. and Weinbaum, S.A. (1987) Changes in water potential and dermal extensibility during grape berry development. *Journal of the American Society of Horticultural Science* 112(2), 314–319.

Mattick, L.R., Shaulis, N.J. and Moyer, J.C. (1972) The effect of potassium fertilization on the acid content of 'Concord' grape juice. *American Journal of Enology and Viticulture* 23(1), 26–30.

Mattivi, F., Caputi, L., Carlin, S., Lanza, T., Minozzi, M., Nanni, D., Valenti, L. and Vrhovsek, U. (2011) Effective analysis of rotundone at below-threshold levels in red and white wines using solid-phase microextraction gas chromatography/tandem mass spectrometry. *Rapid Communications in Mass Spectrometry* 25, 483–488. DOI 10.1002/rcm.4881.

May, P. (1965) Reducing inflorescence formation by shading individual Sultana buds. *Australian Journal of Biological Science* 18, 463–473. DOI 10.1071/BI9650463.

May, P. (1987) The grapevine as a perennial, plastic and productive plant. *Proceedings of the 6th Australian Wine Industry Technical Conference*, Adelaide. Australian Industrial Publishers, Adelaide, South Australia, Australia, pp. 40–49.

May, P. (1994) *Using grapevine rootstocks: the Australian perspective*. Winetitles, Adelaide, Australia.

May, P. (2000) From bud to berry, with special reference to inflorescence and bunch morphology in *Vitis vinifera* L. *Australian Journal of Grape and Wine Research* 6, 82–98. DOI 10.1111/j.1755-0238.2000.tb00166.x.

May, P. and Kerridge, G.H. (1967) Harvest pruning of Sultana vines. *Vitis* 6, 390–393.

May, P., Scholefield, P.B., Clingeleffer, P.R. and Smith, L. (1974) Experiments on the mechanical harvesting of sultanas for drying. *Journal of Science and Food Agriculture* 25(5), 541–552. DOI 10.1002/jsfa.2740250514.

Maynard, D.N., Barker, A.V. and Lachman, W.H. (1968) Influence of potassium on the utilization of ammonium by tomato plants. *Proceedings of the American Society for Horticultural Science* 92, 331–337.

Mazzetto, F., Gallo, R., Calcante, A., Landonio, S. and Lazzari, M. (2013) Integrated full line of mechanization for vineyard preparation. *Acta Horticulturae* 978, 399–407. DOI 10.17660/ActaHortic.2013.978.46.

McBride, R.L. and Johnson, R.L. (1987) Perception of sugar–acid mixtures in lemon juice drink. *International Journal of Food Science and Technology* 22(4), 399–408. DOI 10.1111/j.1365-2621.1987.tb00503.x.

McCarthy, M.G., Cirami, R.M. and Furkaliev, D.G. (1997) Rootstock response of Shiraz (*Vitis vinifera*) grapevines to dry and drip-irrigated conditions. *Australian Journal of Grape and Wine Research* 3(2), 95–98. DOI 10.1111/j.1755-0238.1997.tb00121.x.

McGovern, P.E., Zhang, J., Tang, J., Zhang, Z., Hall, G.R., Moreau, R.A., Nunez, A., Butrym, E.D., Richards, M.P., Wang, C. *et al.* (2004) Fermented beverages of pre- and proto-historic China. *Proceedings of the National Academy of Sciences* 101(51), 17593–17598. DOI 10.1073/pnas.0407921102.

McGovern, P., Jalabadze, M., Batiuk, S., Callahan, M.P., Smith, K.E., Hall, G.R., Kvavadze, E., Maghradze, D., Rusishvili, N., Bouby, L. *et al.* (2017) Early Neolithic wine of Georgia in the South Caucasus. *Proceedings of the National Academy of Sciences* 201714728. DOI 10.1073/pnas.1714728114.

McKenry, M.V. (1999) *The Replant Problem and its Management*. Catalina Circle, Fresno, California, 125 pp.

Melander, B., Heisel, T. and Jørgensen, M.H. (2002) Band-steaming for intra-row weed control. *Proceedings of the 5th EWRS Workshop on Physical Weed Control*, Pisa, Italy, 11–13 March, pp. 216–219.

Melanson, D.L., Rawnsley, B. and Scheper, R.W.A. (2002) Molecular detection of Phomopsis taxa 1 and 2 in grapevine canes and buds. *Australasian Plant Pathology* 31, 67–73. DOI 10.1071/AP01075.

Mengel, K., Breininger, T. and Bubl, W. (1984) Bicarbonate, the most important factor inducing iron chlorisis in vine grapes on calcareous soil. *Plant and Soil* 81, 333–344. DOI 10.1007/BF02323048.

Merot, A., and Wery, J. (2017) Converting to organic viticulture increases cropping system structure and management complexity. *Agronomy for Sustainable Development*, 37(3), 19. https://doi.org/10.1007/s13593-017-0427-9.

Messmer, N., Bohnert, P., Askani, L., Schumacher, S., Voegele, R.T. and Fuchs, R. (2024) Grapevine Pinot gris virus spreads in infected vineyards: latent infections have no direct impact on grape production. *Virology Journal* 21, 178. https://doi.org/10.1186/s12985-024-02453-4.

Meyers, J.M. and Vanden Heuvel, J.E. (2008) Enhancing the precision and spatial acuity of point quadrat analyses via calibrated exposure mapping. *American Journal of Enology and Viticulture* 59, 425–431.

Mikram, M., Moujahdi, C., and Rhanoui, M. (2025) Deep learning and machine learning approaches for data-driven risk management and decision support in precision agriculture. *International Journal of Sustainable Agricultural Management and Informatics*, 11(2), 226–247. https://doi.org/10.1504/IJSAMI.2025.145317.

Milne, G., Lang, S. and Kingston, C. (2003) Can pollen source influence grape fruit set or berry characteristics? *Australian and New Zealand Grapegrower and Winemaker* 471, 32–34.

Mirabel, M., Saucier, C., Guerra, C. and Glories, Y. (1999) Copigmentation in model wine solutions: occurrence and relation to wine aging. *American Journal of Enology and Viticulture* 50(2), 211–218.

Mitchell, F.G. (1992) Cooling horticultural commodities. In: Kader, A.A. (ed.) *Postharvest Technology of Horticultural Crops*. University of California, Division of Agriculture and Natural Resources, Oakland, California, pp. 53–63.

Mitchell, F.G. and Crisosto, C.H. (1995) The use of cooling and cold storage to stabilize and preserve fresh stone fruits. *Seminario sobre Calidad Post-Cosecha y Productos Derivados en Frutos de Hueso*, Lleida, Spain.

Moio, L. and Etievant, P.X. (1995) Ethyl anthranilate, ethyl cinnamate, 2,3-dihydrocinnamate, and methyl anthranilate: four important odorants identified in Pinot noir wines of Burgundy. *American Journal of Enology and Viticulture* 46(3), 392–398.

Molitor, D., Hoffmann, L. and Beyer, M. (2015) Flower debris removal delays grape bunch rot epidemic. *American Journal of Enology and Viticulture* 66, 548–553. DOI 10.5344/ajev.2015.15019.

Mollah, M. (1997) *Practical Aspects of Grapevine Trellising*. Winetitles, Underdale, South Australia. ISBN 9781875130245.

Mollenhauer, H.H. and Hopkins, D.L. (1974) Ultrastructural study of Pierce's Disease bacterium in grape xylem tissue. *Journal of Bacteriology* 119, 612–618.

Moncur, M.W., Rattigan, K., Mackenzie, D.H. and McIntyre, G.N. (1989) Base temperature for budbreak and leaf appearance of grapevines. *American Journal of Enology and Viticulture* 40(1), 21–26.

Mondello, V., Songy, A., Battiston, E., Pinto, C., Coppin, C., Trotel-Aziz, P., Clément, C., Mugnai, L. and Fontaine, F. (2018) Grapevine trunk diseases: A review of fifteen years of trials for their control with chemicals and biocontrol agents. *Plant Disease* 102, 1189–1217. https://doi.org/10.1094/PDIS-08-17-1181-FE.

Monta, M., Kondo, N. and Shibano, Y. (1995) Agricultural robot in grape production system. *Proceedings of the IEEE International Conference on Robotics and Automation*, Nagoya, Japan, IEEE, Los Alamitos, California, pp. 2504–2509.

Moore, M.O. and Giannasi, D.E. (1994) Foliar flavonoids of eastern North American *Vitis* (Vitaceae) north of Mexico. *Plant Systematics and Evolution* 193(1–4), 21–36. DOI 10.1007/BF00983538.

Moran, W. (2006) Crafting terroir: people in cool climates, soils and markets. *Proceedings of the 6th International Symposium on Cool Climate Viticulture and Oenology*, Christchurch, New Zealand. New Zealand Society for Viticulture and Oenology, Auckland, New Zealand. ISBN 9780864761686.

Moriyama, K., Kono, A., Matsuzaki, R., Azuma, A., Onoue, N., Sekozawa, Y., Sato, A. and Sugaya, S. (2024) Diversity of flavour characteristics of table grapes and their contributing volatile compounds analysed by the solvent-assisted flavour evaporation method. *Horticulture Research* 11. https://doi.org/10.1093/hr/uhae048.

Morin, F., Fortin, J.A., Hamel, C., Granger, R.L. and Smith, D.L. (1994) Apple rootstock response to vesicular–arbuscular mycorrhizal fungi in a high phosphorus soil. *Journal of the American Society for Horticultural Science* 119, 578–583.

Morris, J.R. (1989) Producing quality grape juice. *110th Annual Meeting of the Arkansas State Horticultural Society* (pdfs.semanticscholar.org/1683/257d871febf5e60473 0cbbacc467e2274f36.pdf).

Morris, J.R. (2000) Past, present, and future of vineyard mechanization. *Proceedings of the American Society for Enology and Viticulture* 50th *Annual Meeting*, Seattle, Washington, pp. 155–167.

Morris, J.R. (2007) Development and commercialization of a complete vineyard mechanization system. *HortTechnology* 17, 411–420.

Morris, J.R., Fleming, J.W., Benedict, R.H. and McCaskill, D.R. (1972) Effects of sulfur dioxide on postharvest quality of mechanically harvested grapes. *Arkansas Farm Research* 21(2), 5.

Morris, J.R., Cawthon, D.L. and Fleming, J.W. (1975) Effect of mechanical pruning on yield and quality of 'Concord' grapes. *Arkansas Farm Research* 24(3), 12.

Morris, J.R., Sistrunk, W.A., Junek, J. and Sims, C.A. (1986) Effects of fruit maturity, juice storage, and juice extraction temperature on quality of 'Concord' grape juice. *Journal of the American Society for Horticultural Science* 111(5), 742–746.

Mullins, M.G., Bouquet, A. and Williams, L.E. (1992) *Biology of the Grapevine*. Cambridge University Press, Cambridge, UK, 251 pp. ISBN 9780521305075.

Mundy, D.C., Alspach, P.A. and Dufay, J. (2005) Grass grub damage and mycorrhizal colonisation of grapevine rootstocks. *New Zealand Plant Protection* 58, 234–238.

Mundy, D.C., Trought, M.C.T., McLachlan, A.R.G., Neal, S.M. and Pecchenino, D. (2021) Effects of mechanical thinning on botrytis bunch rot on Sauvignon blanc wine grapes. *New Zealand Plant Protection* 74, 30–36. https://doi.org/10.30843/nzpp.2021.74.11729.

Munkvold, G.P. and Marois, J.J. (1995) Factors associated with variation in susceptibility of grapevine pruning wounds to infection by *Eutypa lata*. *Phytopathology* 85, 249–256. DOI 10.1094/Phyto-85-249.

Nagarkatti, S., Tobin, P.C., Muza, A.J. and Saunders, M.C. (2002) Carbaryl resistance in populations of grape berry moth (Lepidoptera: Tortricidae) in New York and Pennsylvania. *Journal of Economic Entomology* 95, 1027–1032. DOI 10.1603/0022-0493-95.5.1027.

Nagarkatti, S., Tobin, P.C., Saunders, M.C. and Muza, A.J. (2003) Release of native *Trichogramma minutum* to control grape berry moth. *Canadian Entomologist* 135(4), 589–598. DOI 10.4039/n02-099.

Nair, N.G. and Allen, R.N. (1993) Infection of grape flowers and berries by *Botrytis cinerea* as a function of time and temperature. *Mycological Research* 97, 1012–1014. DOI 10.1016/S0953-7562(09)80871-X.

Nair, N.G., Guilbaud-Oulton, S., Barchia, I. and Emmett, R. (1995) Significance of carry over inoculum, flower infection and latency on the incidence of *Botrytis cinerea* in berries of grapevines at harvest in New South Wales. *Australian Journal of Experimental Agriculture* 35(8), 1177–1180. DOI 10.1071/EA9951177.

Nakagawa, S. and Nanjo, Y. (1965) A morphological study of Delaware grape berries. *Journal of the Japanese Society for Horticultural Science* 34, 85–95. DOI 10.2503/jjshs.34.85.

NASS (2017) *Noncitrus Fruits and Nuts 2016 Summary*, United States Department of Agriculture National Agricultural Statistics Service, 120 pp.

Naylor, A.P., Creasy, G.L. and Vanhanen, L. (2003) Effects of row orientation and cluster exposure on light interception and Sauvignon blanc fruit composition. *Australian and New Zealand Grapegrower & Winemaker* 474, 97–100.

Neel, D. (2000) Windbreaks mitigate climate. *Practical Winery & Vineyard* XXI(1) (May/June), 20–22.

Neeson, R., Glasson, A., Morgan, A., Macalpine, S. and Darnley-Naylor, M. (1995) On-farm options. *Murrumbidgee Irrigation Areas and Districts Land and Water Management Plan*, NSW Agriculture, Australia.

Nelson, K.E. (1983) Effects of in-package sulfur dioxide generators, package liners, and temperature on decay and desiccation of table grapes. *American Journal of Enology and Viticulture* 34(1), 10–16.

Nelson, K.E. (1985) Harvesting and handling California table grapes for market. *University of California Agricultural and Experimental Station Research Bulletin 1913*, Davis, California, p. 72.

Nelson, K.E. and Ahmedullah, M. (1970) Effect on 'Cardinal' grapes of position of sulfur dioxide generators and retention of gas and water vapor in unvented containers. *American Journal of Enology and Viticulture* 21(2), 70–77.

Nelson, K.E., Baker, G.A., Winkler, A.J., Amerine, M.A., Richardson, H.B. and Jones, F.R. (1963) Chemical and sensory variability in table grapes. *Hilgardia* 34, 1–42. DOI 10.3733/hilg.v34n01p001.

Nelson, R.R., Acree, T.E., Lee, C.Y. and Butts, R.M. (1977) Methyl anthranilate as an aroma constituent of American wine. *Journal of Food Science* 42, 57–59. DOI 10.1111/j.1365-2621.1977.tb01217.x.

Németh, M.Z., Seress, D. and Nonomura, T. (2023) Fungi parasitizing Powdery Mildew fungi: Ampelomyces strains as biocontrol agents against Powdery Mildews. *Agronomy* 13, 1991. https://doi.org/10.3390/agronomy13081991.

Newman, J.L. (1986) Vines, wines, and regional identity in the Finger Lakes Region. *Geographical Review* 76(3), 301–316.

New Zealand Winegrowers (2025) *Sustainability Report 2025*. https://www.nzwine.com/en/sustainability/sustainability-report/ (accessed 16 April 2025).

Nguyen, A.V. (2021) *Evaluation of grapevine rootstocks for use against Grapevine Fanleaf Virus.* PhD thesis, UC Davis, Davis, CA USA. https://escholarship.org/uc/item/9548b4k6 (accessed 29 March 2025).

Nicholas, P. (2004) *Soil, Irrigation and Nutrition.* Winetitles, Underdale, South Australia. ISBN 9781875130405.

Nicholas, P.R., Chapman, A.P. and Cirami, R.M. (1992) Grapevine propagation. In: Coombe, B.G. and Dry, P.R. (eds) *Viticulture Volume 2, Practices.* Winetitles, Underdale, South Australia, pp. 1–22. ISBN 9781875130016.

Nicholas, P., Magarey, P.A. and Wachtel, M. (1994) *Diseases and Pests.* Grape Production Series 1, Hyde Park Press, Adelaide, South Australia, 106 pp. ISBN 1875130152.

Nicholls, C.I., Parrella, M.P. and Altieri, M.A. (2000) Reducing the abundance of leafhoppers and thrips in a northern California organic vineyard through maintenance of full season floral diversity with summer cover crops. *Agricultural and Forest Entomology* 2(2), 107–113. DOI 10.1046/j.1461-9563.2000.00054.x.

Nicholson, T. (2015) Shaken not dropped. *New Zealand Winegrower* 92(Jun/Jul), 12–15.

Nicol, J.M., Stirling, G.R., Rose, B.J., May, P. and Heeswijck, R.V. (1999) Impact of nematodes on grapevine growth and productivity: current knowledge and future directions, with special reference to Australian viticulture. *Australian Journal of Grape and Wine Research* 5(3), 109–127. DOI 0.1111/j.1755-0238.1999.tb00295.x.

Nisbet, M.A., Martinson, T.E. and Mansfield, A.K. (2013) Preharvest prediction of yeast assimilable nitrogen in Finger Lakes Riesling using linear and multivariate modeling. *American Journal of Enology and Viticulture* 64, 485–494. DOI 10.5344/ajev.2013.13030.

Nisbet, M.A., Martinson, T.E. and Mansfield, A.K. (2014) Accumulation and prediction of yeast assimilable nitrogen in New York winegrape cultivars. *American Journal of Enology and Viticulture* 65, 325–332. DOI 10.5344/ajev.2014.13130.

Nita, M., Ellis, M.A., Wilson, L.L. and Madden, L.V. (2007) Evaluation of the curative and protectant activity of fungicides and fungicide–adjuvant mixtures on Phomopsis cane and leaf spot of grape: a controlled environment study. *Crop Protection* 26(9), 1377–1384. DOI 10.1016/j.cropro.2006.11.010.

Noble, A.C., Ough, C.S. and Kasimatis, A.N. (1975) Effect of leaf content and mechanical harvest on wine 'quality'. *American Journal of Enology and Viticulture* 26(3), 158–163.

Noble, A.C., Arnold, R.A., Masuda, B.M., Pecore, S.D., Schmidt, J.O. and Stern, P.M. (1984) Progress towards a standardized system of wine aroma terminology. *American Journal of Enology and Viticulture* 35(2), 107–109.

Noble, A.C., Arnold, R.A., Buechsenstein, J., Leach, E.J., Schmidt, J.O. and Stern, P.M. (1987) Modification of a standardized system of wine aroma terminology. *American Journal of Enology and Viticulture* 38(2), 143–146.

Northover, J. (1987) Infection sites and fungicidal prevention of *Botrytis cinerea* bunch rot of grapes in Ontario. *Canadian Journal of Plant Pathology* 9, 129–136. DOI 10.1080/07060668709501892.

Nougaret, R.L. and Lapham, M.H. (1928) *A Study of Phylloxera Infestation in California as Related to Types of Soils.* Technical Bulletin No. 2, US Department of Agriculture, Washington, DC.

Nuttall, S.L., Kendall, M.J., Bombardelli, E. and Morazzoni, P. (1998) An evaluation of the antioxidant activity of a standardized grape seed extract, Leucoselect®. *Journal of Clinical Pharmacological Therapy* 23(5), 385–389. DOI 10.1046/j.1365-2710.1998.00180.x.

Ochs, E.S., Throop, J.A. and Gunkel, W.W. (1992) *Vision Control and Rough Terrain Compensation for a Robotic Grape Pruner.* ASAE Paper No. 92–7043, American Society of Agricultural Engineers, St Joseph, Michigan.

O'Connor, M. (2023) *The Value of Inclusive Collaborative Design: Lessons from a Virtual Reality Vine-Pruning Training Tool* [PhD, University of Otago]. https://hdl.handle.net/10523/16143.

O'Gorman, P.A. (2015) Precipitation Extremes Under Climate Change. *Current Climate Change Reports* 1(2), 49—59. DOI 10.1007/s40641-015-0009-3.

OIV (2000) *International list of vine varieties and their synonyms.* L'Organisation Internationale de la Vigne et du Vin, Paris, 187 pp. ISBN: 979-10-91799-26-3.

OIV (2003) *Situation Report for the World Vitivinicultural Sector in 2003.* L'Organisation Internationale de la Vigne et du Vin, Paris, 59 pp.

OIV (2009) Liste des descripteurs OIV pour les variétés et espèces de Vitis. 2nd Edition. Office International de la Vigne et du Vin, Dedon, Paris, 232 pp.

OIV (2013) *Description of World Vine Varieties.* L'Organisation Internationale de la Vigne et du Vin, Paris, 542 pp.

OIV (2017) *OIV Focus 2017. Vine Varieties Distribution in the World.* L'Organisation Internationale de la Vigne et du Vin, Paris, 4 pp.

OIV (2023) *Annual Assessment of the World Vine and Wine Sector in 2022.* L'Organisation Internationale de la Vigne et du Vin, Paris, 30 pp.

OIV (2024) *Annual Assessment of the World Vine and Wine Sector in 2023.* L'Organisation Internationale de la Vigne et du Vin, Paris, 30 pp.

OIV (2025a) *State of the World Vine and Wine Sector in 2024.* L'Organisation Internationale de la Vigne et du Vin, Paris, 21 pp.

OIV (2025b) *Database, Dried Grapes Production 1995-2023.* https://www.oiv.int/what-we-do/data-discovery-report?oiv (accessed 20 March 2025).

Okamoto, G. (1994) Grape. In: Konishi, K., Iwahori, S., Kitagawa, H. and Yakuwa, T. (ed.) *Horticulture in Japan.* Asakura Publishing Co., Tokyo, pp. 27–33.

Olien, W.C. (1990) The muscadine grape: botany, viticulture, history, and current industry. *HortScience* 25(7), 732–739.

Ollat, N. (1992) Nouaison chez *Vitis vinifera* L. cv. 'Merlot Noir': rôle de l'intensité lumineuse et de la photosynthèse à la floraison. *Proceedings of the IVth International Symposium on Grapevine Physiology,* Istituto Agrario San Michele all'Adige, University of Turin, Italy, pp. 113–116.

Olmo, H.P. (1946) Correlations between seed and berry development in some seeded varieties of *Vitis vinifera. Proceedings of the American Society for Horticultural Science* 48, 291–297.

Olmo, D., Nieto, A., Adrover, F., Urbano, A., Beidas, O., Juan, A., Marco-Noales, E., López, M.M., Navarro, I., Monterde, A., Montes-Borrego, M., Navas-Cortés, J.A. and Landa, B.B. (2017) First detection of *Xylella fastidiosa* infecting cherry (*Prunus avium*) and *Polygala myrtifolia* plants, in Mallorca Island, Spain. *Plant Disease* 101, 1820–1820. https://doi.org/10.1094/PDIS-04-17-0590-PDN.

Olmstead, M.A. and Tarara, J.M. (2001) Physical principles of row covers and grow tubes with application to small fruit crops. *Small Fruits Review* 1(3), 29–46. DOI 10.1300/J301v01n03_04.

Omer, A.D., Granett, J., De Benedictis, J.A. and Walker, M.A. (1995) Effects of fungal root infections on the vigor of grapevines infested by root-feeding grape phylloxera. *Vitis* 34(3), 165–170.

Ore, O.T., Adeola, A.O., Bayode, A.A., Adedipe, D.T. and Nomngongo, P.N. (2023) Organophosphate pesticide residues in environmental and biological matrices: Occurrence, distribution and potential remedial approaches. *Environmental Chemistry and Ecotoxicology* 5, 9–23. https://doi.org/10.1016/j.enceco.2022.10.004.

Orlandini, S., Gozzini, B., Rosa, M., Egger, E., Storchi, P., Maracchi, G. and Miglietta, F. (1993) PLASMO: a simulation model for control of *Plasmopara viticola* on grapevine. *EPPO Bulletin* 23, 619–626. DOI 10.1111/j.1365-2338.1993.tb00559.x.

Orlando, F., Movedi, E., Coduto, D., Parisi, S., Brancadoro, L., Pagani, V., Guarneri, T., and Confalonieri, R. (2016) Estimating Leaf Area Index (LAI) in vineyards using the PocketLAI smart-app. *Sensors* 16, 2004. DOI 10.3390/s16122004.

Oroian, C., Chiciudean, G., Poruţiu, A. and Bordeanu, B. (2017) Consumers' perception towards organic farming. *ProEnvironment*, 10(29), 41–45.

Ortega-Heras, M., González-SanJosé, M.L. and Beltrán, S. (2002) Aroma composition of wine studied by different extraction methods. *Analytica Chimica Acta* 458, 85–93. DOI 10.1016/S0003-2670(01)01526-4.

Ough, C.S. and Berg, H.W. (1971) Simulated mechanical harvest and gondola transport. 11. Effect of temperature, atmosphere, and skin contact on chemical and sensory qualities of white wines. *American Journal of Enology and Viticulture* 22(4), 194–198.

Ozkana, H.E., Mirallesb, A., Sinfortb, C., Zhuc, H. and Fox, R.D. (1997) Shields to reduce spray drift. *Journal of Agricultural Engineering Research* 67(4), 311–322. DOI 10.1006/jaer.1997.0174.

Paggi, M.S. and Yamazaki, F. (2005) *Grape Juice Concentrate Trade Profile 2004*. California Association of Winegrape Growers, p. 9.

Pala, Ç.U. and Toklucu, A.K. (2013) Effects of UV-C light processing on some quality characteristics of grape juices. *Food and Bioprocess Technology* 6, 719–725. DOI 10.1007/s11947-012-0808-7.

Palleja, T. and Landers, A.J. (2017) Real time canopy density validation using ultrasonic envelope signals and point quadrat analysis. *Computers and Electronics in Agriculture* 134, 43–50. https://doi.org/10.1016/j.compag.2017.01.012.

Palliotti, A., Cartechini, A. and Ferranti, F. (2000) Morpho-anatomical and physiological characteristics of primary and lateral shoot leaves of Cabernet Franc and Trebbiano Toscano grapevines under two irradiance regimes. *American Journal of Enology and Viticulture* 51(2), 122–130.

Palma, B.A. and Jackson, D.I. (1989) Inflorescence initiation in grapes – response to plant growth regulators. *Vitis* 28, 1–12.

Pankhurst, C., Hawke, B., McDonald, H., Kirkby, C., Buckerfield, J., Michelsen, P., O'Brien, K., Gupta, V. and Doube, B. (1995) Evaluation of soil biological properties as potential bioindicators of soil health. *Australian Journal of Experimental Agriculture* 35(7), 1015–1028. DOI 10.1071/EA9951015.

Panneton, B., Thériault, R. and Lacasse, B. (2001) Efficacy evaluation of a new spray-recovery sprayer for orchards. *Transactions of the American Society of Agricultural Engineering* 44(3), 473–479. DOI 10.13031/2013.6102.

Parat, C., Chaussod, R., Lévêque, J., Dousset, S. and Andreux, F. (2002) The relationship between copper accumulated in vineyard calcareous soils and soil organic matter and iron. *European Journal of Soil Science* 53, 663–670. DOI 10.1046/j.1365-2389.2002.00478.x.

Parish, M.E. and Carroll, D.E. (1985) Indigenous yeasts associated with muscadine (*Vitis rotundifolia*) grapes and musts. *American Journal of Enology and Viticulture* 36(2), 165–169.

Parker, A.K., Hofmann, R.W., van Leeuwen, C., McLachlan, A.R.G. and Trought, M.C.T. (2015) Manipulating the leaf area to fruit mass ratio alters the synchrony of total soluble solids accumulation and titratable acidity of grape berries. *Australian Journal of Grape and Wine Research* 21, 266–276. DOI 10.1111/ajgw.12132.

Parr, W.V., Green, J.A., White, K.G. and Sherlock, R.R. (2007) The distinctive flavour of New Zealand Sauvignon blanc: sensory characterisation by wine professionals. *Food Quality Preference* 18(6), 849–861. DOI 10.1016/j.foodqual.2007.02.001.

Patton, D. (2006) *China to Start Making Ice Wine.* APFI, Singapore. https://www.beveragedaily.com/Article/2006/09/12/china-to-start-making-ice-wine/ (Accessed 20 April 2025).

Paull, J. (2011) Attending the first organic agriculture course: Rudolf Steiner's agriculture course at Koberwitz, 1924. *European Journal of Social Sciences*, 21(1), 64–70.

Peacock, W.L. and Swanson, C.A. (2005) The future of California raisins is drying on the vine. *California Agriculture* 59(2), 70–74. DOI 10.3733/ca.v059n02p70.

Peacock, W.L., Jensen, F., Else, J. and Leavitt, G. (1977a) The effects of girdling and ethephon treatments on fruit characteristics of Red Malaga. *American Journal of Enology and Viticulture* 28(4), 228–230.

Peacock, W.L., Rolston, D.E., Aljibury, F.K. and Rauschkolb, R.S. (1977b) Evaluating drip, flood, and sprinkler irrigation of wine grapes. *American Journal of Enology and Viticulture* 28(4), 193–195.

Peacock, W.L., Christensen, L.P. and Broadbent, F.E. (1989) Uptake, storage, and utilization of soil-applied nitrogen by Thompson Seedless as affected by time of application. *American Journal of Enology and Viticulture* 40, 16–20.

Pederson, C.S. (1971) Grape juice. In: Tressler, D.K.J. (ed.) *Fruit and Vegetable Juice: Processing Technology*. Avi Publishing Co., Westport, Connecticut, pp. 234–271. ISBN 9780870550980.

Pedrelli, A., Carli, M., Panattoni, A., Pellegrini, E., Rizzo, D., Nali, C., and Cotrozzi, L. (2024) Investigating a new alarming outbreak of flavescence dorée in Tuscany (Central Italy): Molecular characterization and map gene typing elucidate the complex phytoplasma ecology in the vineyard agroecosystem. *Frontiers in Plant Science*, 15. https://doi.org/10.3389/fpls.2024.1489790.

Penfold, C. (2004) Mulching and its impact on weed control, vine nutrition, yield and quality. *Australian and New Zealand Grapegrower and Winemaker* 485a, 32–38.

Perold, A.I. (1927) *A Treatise on Viticulture*. Macmillan and Co. Ltd., London. ISBN 9780956152336.

Perring, T.M., Gruenhagen, N.M. and Farrar, C.A. (1999) Management of plant viral diseases through chemical control of insect vectors. *Annual Review of Entomology* 44(1), 457–481. DOI 10.1146/annurev.ento.44.1.457.

Perry, K.L., McLane, H., Hyder, M.Z., Dangl, G.S., Thompson, J.R. and Fuchs, M.F. (2016) Grapevine red blotch-associated virus is present in free-living *Vitis* spp. proximal to cultivated grapevines. *Phytopathology* 106, 663–670. https://doi.org/10.1094/PHYTO-01-16-0035-R.

Pertot, I., El Bilali, H., Simeone, V., Vecchione, A. and Zulini, L. (2006) Efficacy evaluation and phytotoxicity assessment of copper peptidate on seven grapevine varieties and identification of the potential factors that induced copper damages on leaves. *Integrated Protection in Viticulture, IOBC/WPRS Bulletin* Vol. 29 (11), Boario Terme, Italy.

Petrie, P.R. and Clingeleffer, P.R. (2005) Effects of temperature and light (before and after budburst) on inflorescence morphology and flower number of Chardonnay grapevines (*Vitis vinifera* L.). *Australian Journal of Grape and Wine Research* 11, 59–65. DOI 10.1111/j.1755-0238.2005.tb00279.x.

Petrie, P.R. and Clingeleffer, P.R. (2006) Crop thinning (hand versus mechanical), grape maturity and anthocyanin concentration: outcomes from irrigated Cabernet Sauvignon (*Vitis vinifera* L.) in a warm climate. *Australian Journal of Grape and Wine Research* 12(1), 21–29. DOI 10.1111/j.1755-0238.2006.tb00040.x.

Petrie, P.R. and Sadras, V.O. (2008) Advancement of grapevine maturity in Australia between 1993 and 2006: putative causes, magnitude of trends and viticultural consequences. *Australian Journal of Grape and Wine Research* 14, 33–45.

Petrie, P.R., Trought, M.C.T. and Howell, G.S. (2000a) Influence of leaf ageing, leaf area and crop load on photosynthesis, stomatal conductance and senescence of grapevine (*Vitis vinifera* L. cv. Pinot noir) leaves. *Vitis* 39, 31–36.

Petrie, P.R., Trought, M.C.T. and Howell, G.S. (2000b) Growth and dry matter partitioning of Pinot Noir (*Vitis vinifera* L.) in relation to leaf area and crop load. *Australian Journal of Grape and Wine Research* 6, 40–45. DOI 10.1111/j.1755-0238.2000. tb00160.x.

Petrucci, V.E. and Siegfried, R. (1976) The extraneous matter in mechanically harvested wine grapes. *American Journal of Enology and Viticulture* 27(1), 40–41.

Phelps, G.W. (1999) *The influence of site, crop load, and cluster light exposure on Pinot noir fruit composition in Canterbury*. M. Appl. Sci. thesis, Lincoln University, Christchurch, New Zealand, 100 pp. https://hdl.handle.net/10182/2797.

Pickering, G. (2006) Icewine – the frozen truth. *Proceedings of the 6th International Cool Climate Symposium on Viticulture and Oenology*, Christchurch, New Zealand. New Zealand Society for Viticulture and Oenology, Auckland, New Zealand. ISBN 9780864761686.

Pickering, G.J., and Botezatu, A. (2021) A review of ladybug taint in wine: Origins, prevention, and remediation. *Molecules*, 26(14), Article 14. https://doi.org/10.3390/molecules26144341.

Pierquet, P. and Stushnoff, C. (1980) Relationship of low temperature exotherms to cold injury in *Vitis riparia* Michx. *American Journal of Enology and Viticulture* 31(1), 1–6.

Pierquet, P., Stushnoff, C. and Burke, M.J. (1977) Low temperature exotherms in stem and bud tissues of *Vitis riparia* Michx. *Journal of the American Society for Horticultural Science* 102, 54–55.

Pierro, R., Moussa, A., Mori, N., Marcone, C., Quaglino, F., and Romanazzi, G. (2024) Bois noir management in vineyard: A review on effective and promising control strategies. *Frontiers in Plant Science*, 15. https://doi.org/10.3389/fpls.2024.1364241.

Pietrzak, U. and McPhail, D.C. (2004) Copper accumulation, distribution and fractionation in vineyard soils of Victoria, Australia. *Geoderma* 122(2–4), 151–166. DOI 10.1016/j.geoderma.2004.01.005.

Pillitteri, L.J., Lovatt, C.J. and Walling, L.L. (2004) Isolation and characterization of a terminal flower homolog and its correlation with juvenility in citrus. *Plant Physiology* 135(3), 1540–1551. DOI 10.1104/pp.103.036178.

Pinel, M.P.C., Bond, W., White, J.G. and de Courcy Williams, M. (1999) *Field Vegetables: Assessment of the Potential for Mobile Soil Steaming Machinery to Control Diseases, Weeds and Mites of Field Salad and Related Crops*. Final Report on HDC Project FV229, UK Horticultural Development Council, East Malling, UK, 55 pp.

Pinney, T. (1989) *A History of Wine in America: from the Beginnings to Prohibition*. University of California Press, Berkeley, California. ISBN 9780520062245. http://ark.cdlib.org/ark:/13030/ft967nb63q/ (accessed 15 March 2025).

Plank, C.M., Hellman, E.W. and Montague, T. (2019) Light and temperature independently influence methoxypyrazine content of *Vitis vinifera* (cv. Cabernet Sauvignon) berries. *HortScience* 54, 282–288. https://doi.org/10.21273/HORTSCI13634-18.

Plauborg, F., Iversen, B.V. and Laerke, P.E. (2005) *In situ* comparison of three dielectric soil moisture sensors in drip irrigated sandy soils. *Vadose Zone Journal* 4(4), 1037–1047. DOI 10.2136/vzj2004.0138.

Pocock, K.F., Hayasaka, Y., Peng, Z., Williams, P.J. and Waters, E.J. (1998) The effect of mechanical harvesting and long-distance transport on the concentration of haze-forming proteins in grape juice. *Australian Journal of Grape and Wine Research* 4(1), 23–29. DOI 10.1111/j.1755-0238.1998.tb00131.x.

Poling, E.B. (2007) Spring frost control. In: Poling, E.B. (ed.) *North Carolina Winegrape Grower's Guide*. North Carolina Cooperative Extension Service, Raleigh, North Carolina, 196 pp.

Pollock, D. (2017) Positive attributes aplenty for new Sunpreme raisin grape. www.westernfarmpress.com/grapes/positive-attributes-aplenty-new-sunpreme- raisin-grape (accessed 20 November, 2017).

Pollock, J.G., Shepardson, E.S., Shaulis, N.J. and Crowe, D.E. (1977) Mechanical pruning of American hybrid grapevines. *Transactions of the American Society of Agricultural Engineers* 20, 817–821. DOI 10.13031/2013.35656.

Poni, S., Bernizzoni, F., Briola, G. and Cenni, A. (2005) Effects of early leaf removal on cluster morphology, shoot efficiency and grape quality in two *Vitis vinifera* cultivars. *Acta Horticulturae* 689, 217–226. DOI 10.17660/ActaHortic.2005.689.24.

Poni, S., Casalini, L., Bernizzoni, F., Civardi, S. and Intrieri, C. (2006) Effects of early defoliation on shoot photosynthesis, yield components, and grape composition. *American Journal of Enology and Viticulture* 57, 397–407.

Pool, R.M. (1987) Thin grapes mechanically. *Western Fruit Grower* 107, 17–19.

Pool, R.M., Dunst, R.M., Kamas, J.S., Gunkel, W.W., Lakso, A.N. and Goffinet, M.C. (1990) *Shoot Positioning Native American (Concord-type) Grapevines*. New York's Food and Life Sciences Bulletin No. 131, Cornell University Agricultural Experiment Station, Ithaca, New York, 5 pp.

Pool, R.M., Dunst, D.C., Crowe, H., Hubbard, G.E., Howard, G.E. and DeGolier, G. (1993) Predicting and controlling crop on machine or minimal pruned grapevines. *Proceedings of the 2nd New Jersey Shaulis Grape Symposium on Pruning Mechanization and Crop Control*, Fredonia, New York. New York State Agriculture Experiment Station, Geneva, New York, pp. 31–45.

Possingham, J.V. and Groot Obbink, J. (1971) Endotrophic mycorrhiza and the nutrition of grape vines. *Vitis* 10, 120–130.

Possingham, J.V., Chambers, T.C., Radler, F. and Grncarevic, M. (1967) Cuticular transpiration and wax structure and composition of leaves and fruit of *Vitis vinifera*. *Australian Journal of Biological Science* 20, 1149–1153. DOI 10.1071/ BI9671149.

Pouget, R. and Delas, J. (1989) Le choix des porte-greffes de la vigne pour une production de qualité. *Connaisance Vigne et du Vin Hors Série* 23, 27–31.

Pratt, C. (1971) Reproductive anatomy in cultivated grapes – a review. *American Journal of Enology and Viticulture* 22(2), 92–109.

Pratt, C. (1974) Vegetative anatomy of cultivated grapes – a review. *American Journal of Enology and Viticulture* 25, 131–150. DOI: 10.5344/ajev.1974.25.3.131.

Pratt, C. (1979) Shoot and bud development during the prebloom period of *Vitis*. *Vitis* 18(1), 1–5.

Pratt, C. and Coombe, B.G. (1978) Shoot growth and anthesis in *Vitis*. *Vitis* 17, 125–133.

Prazaru, S. C., dal Mas, G., Padoin, M., Rizzardo, D., Meggio, F., Pitacco, A., Pozzebon, A., and Duso, C. (2023) Effect of leaf removal and insecticide applications on population densities of leafhoppers and mites associated with grapevines. *Insects*, 14(10), Article 10. https://doi.org/10.3390/insects14100791.

Preto, C.R., Sudarshana, M.R. and Zalom, F.G. (2018) Feeding and reproductive hosts of *Spissistilus festinus* (Say) (Hemiptera: Membracidae) found in Californian vineyards. *Journal of Economic Entomology* 111, 2531–2535. https://doi.org/10.1093/jee/toy236.

Prial, F.J. (1987) Wine; stealing from thieves. *The New York Times* March 15, 1987.

Price, S.F., Breen, P.J., Valladao, M. and Watson, B.T. (1995) Cluster sun exposure and quercetin in Pinot noir grapes and wine. *American Journal of Enology and Viticulture* 46, 187–194.

Price, S.F., Watson, B.T. and Valladao, M. (1996) Vineyard and winery effects on wine phenolics – flavonols in Oregon Pinot noir. *Proceedings of the 9th Australian Wine Industry Technical Conference*, Adelaide, South Australia. Winetitles, Underdale, South Australia, pp. 93–97.

Prins, T.W., Tudzynski, P., von Tiedemann, A., Tudzynski, B., ten Have, A., Hansen, M.E., Tenberge, K. and van Kan, J.A.L. (2000) Infection strategies of *Botrytis cinerea* and related necrotrophic pathogens. In: Kronstad, J. (ed.) *Fungal Pathology*. Kluwer Academic Publishers, Dordrecht, Netherlands, pp. 32–64. ISBN 9780792363705.

Pscheidt, J.W. and Pearson, R.C. (1989a) Effect of grapevine training systems and pruning practices on occurrence of Phomopsis cane and leaf spot. *Plant Disease* 73(10), 825–828. DOI 10.1094/PD-73-0825.

Pscheidt, J.W. and Pearson, R.C. (1989b) Time of infection and control of Phomopsis fruit rot of grape. *Plant Disease* 73, 829–833. DOI 10.1094/PD-73-0829.

Pulighe, G., Lupia, F. and Manente, V. (2025) Climate-driven invasion risks of Japanese Beetle (*Popillia japonica* Newman) in Europe predicted through species distribution modelling. *Agriculture* 15, 684. https://doi.org/10.3390/agriculture15070684.

Purcell, A. H. (1980) Environmental therapy for Pierce's disease of grapevines. *Plant Disease* 64, 388-390.

Purcell, A.H. and Hopkins, D.L. (1996) Fastidious xylem-limited bacterial plant pathogens. *Annual Review of Phytopathology* 34(1), 131–151. DOI 10.1146/annurev.phyto.34.1.131.

Quaglino, F., Maghradze, D., Casati, P., Chkhaidze, N., Lobjanidze, M., Ravasio, A., Passera, A., Venturini, G., Failla, O., and Bianco, P.A. (2016) Identification and characterization of new '*Candidatus* Phytoplasma solani' strains associated with Bois Noir disease in *Vitis vinifera* L. cultivars showing a range of symptom severity in Georgia, the Caucasus region. *Plant Disease*, 100(5), 904–915. https://doi.org/10.1094/PDIS-09-15-0978-RE.

Quinlan, J.D. and Weaver, R.J. (1970) Modification of pattern of photosynthate movement within and between shoots of *Vitis vinifera* L. *Plant Physiology* 46(4), 527–530. DOI 10.1104/pp.46.4.527.

Raihen, M.N. and Akter, S. (2023) Prediction modeling using deep learning for the classification of grape-type dried fruits. *International Journal of Mathematics and Computer in Engineering* 2, 1–12. https://doi.org/10.2478/ijmce-2024-0001.

Rakimov, A., Hoffmann, A.A. and Malipatil, M.B. (2015) Natural enemies of soft scale insects (Hemiptera: Coccoidea: Coccidae) in Australian vineyards. *Australian Journal of Grape and Wine Research* 21, 302–310. https://doi.org/10.1111/ajgw.12134.

Raju, B.C., Nome, S.F., Docampo, D.M., Goheen, A.C. and Nyland, G. (1980) Alternative hosts of Pierce's disease of grapevines that occur adjacent to grape growing areas in California. *American Journal of Enology and Viticulture* 31(2), 144–148.

Raski, D.J., Hewitt, W.B., Goheen, A.C., Taylor, C.E. and Taylor, R.H. (1965) Survival of *Xiphinema Index* and reservoirs of fanleaf virus in fallowed vineyard soil. *Nematologica* 11(3), 349–352. DOI 10.1163/187529265X00267.

Raynal, G., Ravisé, A. and Bompeix, G. (1980) Action du tris-O-éthylphosphonate d'aluminium (phoséthyl d'aluminium) sur la pathogénie de *Plasmopara viticola* et sur la stimulation des réactions de défense de la vigne. *Annales de Phytopathologie* 12, 163–175.

Reedy, R.C. and Scanlon, B.R. (2003) Soil water content monitoring using electromagnetic induction. *Journal of Geotechnical and Geoenvironmental Engineering* 129(11), 1028–1039. DOI 10.1061/(ASCE)1090-0241(2003)129:11(1028).

Reichard, D.L., Zhu, H., Fox, R.D. and Brazee, R.D. (1992) Computer simulation of variables that influence spray drift. *Transactions of the American Society of Agricultural Engineers* 35(5), 1401–1407. DOI 10.13031/2013.28747.

Reis, E.M., Sônego, O.R. and de Mendes, C.S. (2013) Application and validation of a warning system for grapevine downy mildew control using fungicides. *Summa Phytopathologica* 39, 10–15. DOI 10.1590/S0100-54052013000100002.

Reisenzein, H., Pfeffer, M., Aust, G. and Baumgarten, A. (2007) The influence of soil properties on the development of grape phylloxera populations in Austrian viticulture. *Acta Horticulturae* 733, 13–23. DOI 10.17660/ActaHortic.2007.733.1.

ReportLinker Research (2025) *Global Export of Grape Juice or Must by Country 2023*. https://www.reportlinker.com/dataset/b2fffe2388bfd8d811a405405d58be276d93ed09 (accessed 5 April 2025).

Renault, A.S., Deloire, A. and Bierne, J. (1996) Pathogenesis-related proteins in grapevines induced by salicylic acid and *Botrytis cinerea*. *Vitis* 35, 49–52.

Reuveni, M. and Reuveni, R. (1995) Efficacy of foliar sprays of phosphates in controlling powdery mildews in field-grown nectarine, mango trees and grapevines. *Crop Protection* 14(4), 311–314. DOI 10.1016/0261-2194(94)00009-W.

Reuveni, M. and Reuveni, R. (2002) Mono-potassium phosphate fertilizer (PeaK) – a component in integrated control of powdery mildews in fruit trees and grapevines. *Acta Horticulturae* 594, 619–625. DOI 10.17660/ActaHortic.2002.594.83.

Reuveni, M., Zahavi, T. and Cohen, Y. (2001) Controlling downy mildew (*Plasmopara viticola*) in field-grown grapevine with b-aminobutyric acid (BABA). *Phytoparasitica* 29, 125–133. DOI 10.1007/BF02983956.

Reynolds, A.G. (1988) Response of Okanagan Riesling vines to training system and simulated mechanical pruning. *American Journal of Enology and Viticulture* 39(3), 205–212.

Reynolds, L.B., Potter, J.W. and Ball-Coelho, B.R. (2000) Crop rotation with *Tagetes* sp. is an alternative to chemical fumigation for control of root-lesion nematodes. *Agronomy Journal* 92, 957–966. DOI 10.2134/agronj2000.925957x.

Ribereau-Gayon, J. and Ribereau-Gayon, P. (1958) The anthocyans and leucoanthocyans of grapes and wines. *American Journal of Enology and Viticulture* 9(1), 1–9.

Richardson, E.A., Seeley, S.D., Walker, D.R., Anderson, J.L. and Ashcroft, G.L. (1975) Pheno-climatography of spring peach bud development. *HortScience* 10, 236–237.

Rieger, M. (1989) Freeze protection for horticultural crops. *Horticultural Reviews* 11, 45–109. DOI 0.1002/9781118060841.ch3.

Ristic, R. and Iland, P.G. (2005) Relationships between seed and berry development of *Vitis vinifera* L. cv Shiraz: developmental changes in seed morphology and phenolic composition. *Australian Journal of Grape Wine Research* 11(1), 43–58. DOI 10.1111/ j.1755-0238.2005.tb00278.x.

Ritchie, G.A. and Hinckley, T.M. (1975) The pressure chamber as an instrument for ecological research. *Advances in Ecological Research* 9, 165–253. DOI 10.1016/ S0065-2504(08)60290-1.

Robinson, B., Greven, M., Green, S., Sivakumaran, S., Davidson, P. and Clothier, B. (2006) Leaching of copper, chromium and arsenic from treated vineyard posts in Marlborough, New Zealand. *Science of The Total Environment* 364, 113–123. DOI 10.1016/j.scitotenv.2005.07.012.

Robinson, J., Harding, J. and Vouillamoz, J. (2012) *Wine Grapes: A Complete Guide to 1,368 Vine Varieties, including their Origins and Flavours* Allen Lane, London, UK. ISBN 978-1846144462.

Robinson, J.B. (1992) Grapevine nutrition. In: Coombe, B.G. and Dry, P.R. (ed.) *Viticulture Volume 2. Practices*. Winetitles, Underdale, South Australia, pp. 178–208. ISBN 9781875130016.

Robinson, S.P., Jacobs, A.K. and Dry, I.B. (1997) A class IV chitinase is highly expressed in grape berries during ripening. *Plant Physiology* 114, 771–778. DOI 10.1104/ pp.114.3.771.

Roby, G. and Matthews, M.A. (2004) Relative proportions of seed, skin and flesh, in ripe berries from Cabernet Sauvignon grapevines grown in a vineyard either well irrigated or under water deficit. *Australian Journal of Grape and Wine Research* 10, 74–82. DOI 10.1111/j.1755-0238.2004.tb00009.x.

Rodríguez-Rajo, F.J., Jato, V., Fernández-González, M. and Aira, M.J. (2010) The use of aerobiological methods for forecasting Botrytis spore concentrations in a vineyard. *Grana* 49, 56–65. DOI 10.1080/00173130903472393.

Romanazzi, G., Murolo, S., and Feliziani, E. (2013) Effects of an innovative strategy to contain grapevine Bois Noir: Field treatment with resistance inducers. *Phytopathology* 103(8), 785–791. https://doi.org/10.1094/PHYTO-01-13-0031-R.

Roper, M. and Gupta, V. (1995) Management practices and soil biota. *Australian Journal of Soil Research* 33(2), 321–339. DOI 10.1071/SR9950321.

Rosslenbroich, H.-J. and Stuebler, D. (2000) *Botrytis cinerea* – history of chemical control and novel fungicides for its management. *Crop Protection* 19(8–10), 557–561. DOI 10.1016/S0261-2194(00)00072-7.

Rumbolz, J., Wirtz, S., Kassemeyer, H.-H., Guggenheim, R., Schäfer, E. and Büche, C. (2002) Sporulation of *Plasmopara viticola*: differentiation and light regulation. *Plant Biology* 4, 413–422. DOI 10.1055/s-2002-32342.

Ryle, G.J.A. and Hesketh, J.D. (1969) Carbon dioxide uptake in nitrogen-deficient plants. *Crop Science* 9(4), 451–454. DOI 10.2135/cropsci1969.0011183X000900040019x.

Saayman, D. and Van Huyssteen, L. (1983) Preliminary studies on the effect of a permanent cover crop and root pruning on an irrigated Colombar vineyard. *South African Journal of Enology and Viticulture* 4, 7–12. DOI 10.21548/4-1-2375.

Sabir, F.K. and Sabir, A. (2013) Quality response of table grapes (*Vitis vinifera* L.) during cold storage to postharvest cap stem excision and hot water treatments. *International Journal of Food Science & Technology* 48, 999–1006. DOI 10.1111/ijfs.12052.

Sabir, F.K., Sabir, A., Unal, S., Taytak, M., Kucukbasmaci, A. and Bilgin, O.F. (2019) Postharvest quality extension of minimally processed table grapes by chitosan coating. *International Journal of Fruit Science* 19, 347–358. https://doi.org/10.1080/15538362.2018.1506961.

Saldarelli, P., Giampetruzzi, A., Morelli, M., Malossini, U., Pirolo, C., Bianchedi, P. and Gualandri, V. (2014) Genetic variability of grapevine Pinot gris virus and its association with grapevine leaf mottling and deformation. *Phytopathology* 105, 555–563. DOI 10.1094/PHYTO-09-14-0241-R.

Salo, W., Considine, J.A. and Considine, M.J. (2024) Influence of mixed and single infection of grapevine leafroll-associated viruses and viral load on berry quality. *Tree Physiology* 44, tpae035. https://doi.org/10.1093/treephys/tpae035.

Salvagnin, U., Giacomelli, L., Lagreze Perez, J., Guche, M.D., Rojas San Martin, B., Micheli, S., Weil, T.F., Pirrello, C., Pilati, S. and Malacarne, G. (2024) Gene functional studies and gene editing at work to improve the sustainability of viticulture. In: *LXVII SIGA Annual Congress*, Bologna, 10-13 September 2024. ISBN: 9788890457074. handle: https://hdl.handle.net/10449/86555.

Sampson, B., Noffsinger, S., Gupton, C. and Magee, J. (2001) Pollination biology of the muscadine grape. *HortScience* 36, 120–124.

Sams, C.E. and Deyton, D.E. (2002) Botanical and fish oils: history, chemistry, refining, formulation and current uses. In: Beattie, G.A.C., Watson, D.M., Stevens, M.L., Rae, D.J. and Spooner-Hart, R.N. (eds) *Spray Oils Beyond 2000: Sustainable Pest and Disease Management*. University of Western Sydney, Sydney, Australia, pp. 19–28.

Santiago, J.L., Boso, S., Martinez, M.d.-C., Pinto-Carnide, O. and Ortiz, J.M. (2005) Ampelographic comparison of grape cultivars (*Vitis vinifera* L.) grown in Northwestern Spain and Northern Portugal. *American Journal of Enology and Viticulture* 56(3), 287–290. DOI: 10.5344/ajev.2005.56.3.287.

Sartorato, I., Zanin, G., Baldoin, C. and Zanche, C. (2006) Observations on the potential of microwaves for weed control. *Weed Research* 46(1), 1–9. DOI 10.1111/j.1365-3180.2006.00484.x.

Sato, A. and Yamada, M. (2003) Berry texture of table, wine, and dual-purpose grape cultivars quantified. *HortScience* 38(4), 578–581.

Sawant, I.S., Wadkar, P.N., Ghule, S.B., Rajguru, Y.R., Salunkhe, V.P. and Sawant, S.D. (2017) Enhanced biological control of powdery mildew in vineyards by integrating a strain of *Trichoderma afroharzianum* with sulphur. *Biological Control* 114, 133–143. https://doi.org/10.1016/j.biocontrol.2017.08.011.

Saxton, V.P., Creasy, G.L., Paterson, A.M. and Trought, M.C.T. (2004a) Experimental method to investigate and monitor bird behavior and damage in vineyards. *American Journal of Enology and Viticulture* 55(3), 288–291.

Saxton, V.P., Hickling, G.J., Trought, M.C.T. and Creasy, G.L. (2004b) Comparative behavior of free-ranging blackbirds (*Turdus merula*) and silvereyes (*Zosterops lateralis*) with hexose sugars in artificial grapes. *Applied Animal Behaviour Science* 85, 157–166. DOI 10.1016/j.applanim.2003.09.010.

Scarratt, S.L., Wratten, S.D. and Shishehbor, P. (2008) Measuring parasitoid movement from floral resources in a vineyard. *Biological Control* 46, 107–113. DOI 10.1016/j.biocontrol.2008.03.016.

Schenk, H.J. and Jackson, R.B. (2002) Rooting depths, lateral root spreads and below-ground/above-ground allometries of plants in water-limited ecosystems. *Journal of Ecology* 90, 480–494. DOI 10.1046/j.1365-2745.2002.00682.x.

Scheck, H.J., Vasquez, S.J., Gubler, W.D. and Foggle, D. (1998) Grape growers report losses to black-foot and young vine decline. *California Agriculture* 54, 19–23. DOI 10.3733/ca.v052n04p19.

Schilder, A.M.C., Smokevitch, S.M., Catal, M. and Mann, W.K. (2005) First report of anthracnose caused by *Elsinoë ampelina* on grapes in Michigan. *Plant Disease* 89(9), 1011. DOI 10.1094/PD-89-1011A.

Schmidt, H.-P., Kammann, C., Niggli, C., Evangelou, M. W. H., Mackie, K. A. and Abiven, S. (2014) Biochar and biochar-compost as soil amendments to a vineyard soil: Influences on plant growth, nutrient uptake, plant health and grape quality. *Agriculture, Ecosystems & Environment*, 191, 117–123. https://doi.org/10.1016/j.agee.2014.04.001

Schmidt, K., Teulon, D.A.J. and Jaspers, M.V. (2006) Phenology of the New Zealand flower thrips (*Thrips obscuratus*) in two vineyards. *New Zealand Plant Protection* 59, 323–329.

Schnabel, B.J. and Wample, R.L. (1987) Dormancy and cold hardiness in *Vitis vinifera* L. cv. White Riesling as influenced by photoperiod and temperature. *American Journal of Enology and Viticulture* 38(4), 265–272.

Scholander, P.F., Bradstreet, E.D., Hemmingsen, E.A. and Hammel, H.T. (1965) Sap pressure in vascular plants: negative hydrostatic pressure can be measured in plants. *Science* 148(3668), 339–346. DOI 10.1126/science.148.3668.339.

Schreiner, R.P. (2002) Seasonal dynamics of mineral uptake and allocation in whole Pinot noir vines in a Red Hill soil. *Proceedings of the Oregon Wine Industry Symposium*, Oregon State University, Corvallis, Oregon.

Schreiner, R.P. and Scagel, C.F. (2017) Leaf blade versus petiole nutrient tests as predictors of nitrogen, phosphorus, and potassium status of 'Pinot noir' grapevines. *HortScience* 52, 174–184. https://doi.org/10.21273/HORTSCI11405-16.

Schreiner, R.P., Scagel, C.F. and Baham, J. (2006) Nutrient uptake and distribution in a mature 'Pinot noir' vineyard. *HortScience* 41, 336–345.

Schubert, A., Lovisolo, C. and Peterlunger, E. (1999) Shoot orientation affects vessel size, shoot hydraulic conductivity and shoot growth rate in *Vitis vinifera* L. *Plant, Cell & Environment* 22(2), 197–204. DOI 10.1046/j.1365-3040.1999.00384.x.

Schulte, F. (1989) The association between *Rhabditis necromena* Sudhaus & Schulte, 1989 (Nematoda: Rhabditidae) and native and introduced millipedes in South Australia. *Nematologica* 35(1), 82–89.

Schultz, H.R. (2000) Climate change and viticulture: a European perspective on climatology, carbon dioxide and UV-B effects. *Australian Journal of Grape and Wine Research* 6(1), 2–12. DOI 10.1111/j.1755-0238.2000.tb00156.x.

Schultz, H.R. (2003) Differences in hydraulic architecture account for near-isohydric and anisohydric behaviour of two field-grown *Vitis vinifera* L. cultivars during drought. *Plant, Cell and Environment* 26(8), 1393–1405. DOI 10.1046/j.1365-3040.2003.01064.x.

Schwarz, M., Picazo-Bacete, J.J., Winterhalter, P. and Hermosin-Gutierrez, I. (2005) Effect of copigments and grape cultivar on the color of red wines fermented after the addition of copigments. *Journal of Agricultural and Food Chemistry* 53(21), 8372–8381. DOI 10.1021/jf051005o.

Scienza, A., Miravalle, R., Visai, C. and Fregoni, M. (1978) Relationships between seed number, gibberellin and abscisic acid levels and ripening in Cabernet Sauvignon grape berries. *Vitis* 17, 361–368.

Scortichini, M., Loreti, S., Pucci, N., Scala, V., Tatulli, G., Verweire, D., Oehl, M., Widmer, U., Codina, J.M., Hertl, P., Cesari, G., De Caroli, M., Angilè, F., Migoni, D., Del Coco, L., Girelli, C.R., Dalessandro, G. and Fanizzi, F.P. (2021) Progress towards sustainable control of *Xylella fastidiosa* subsp. pauca in olive groves of Salento (Apulia, Italy). *Pathogens* 10, 668. https://doi.org/10.3390/pathogens10060668.

Scott, K.D., Ablett, E.M., Lee, L.S. and Henry, R.J. (2000) AFLP markers distinguishing an early mutant of Flame Seedless grape. *Euphytica* 113(3), 243–247. DOI 10.1023/A:1003977408214.

Sekhon, K., Sun, Q., Sekhon, K. and Sun, Q. (2024) Opportunities and challenges for low-alcohol wine. In: *Global Warming and the Wine Industry - Challenges, Innovations and Future Prospects*. IntechOpen, p. 148.

Settle, W.H., Wilson, L.T., Flaherty, D.L. and English-Loeb, G.M. (1986) The variegated leafhopper, an increasing pest of grapes. *California Agriculture* 40, 30–32.

Seufert, V., Ramankutty, N. and Mayerhofer, T. (2017). What is this thing called organic? – How organic farming is codified in regulations. *Food Policy*, 68, 10–20. https://doi.org/10.1016/j.foodpol.2016.12.009.

Severin, H.H.P. (1949) Transmission of the virus of Pierce's Disease of grapevines by leafhoppers. *Hilgardia* 19, 190–202. DOI 10.3733/hilg.v19n06p190.

Seyb, A.M. (2004) Botrytis Cinerea *Inoculum Sources in the Vineyard System*. PhD thesis, Soil Plant and Ecological Sciences Division, Lincoln University, Christchurch, New Zealand, 227 pp. https://hdl.handle.net/10182/1762.

Sharley, D.J., Hoffmann, A.A. and Thomson, L.J. (2008) The effects of soil tillage on beneficial invertebrates within the vineyard. *Agricultural and Forest Entomology*, 10(3), 233–243. https://doi.org/10.1111/j.1461-9563.2008.00376.x.

Sharp-Heward, S., Almond, P. and Robinson, B. (2014). Soil disturbance and salinisation on a vineyard affected by landscape recontouring in Marlborough, New Zealand. *CATENA* 122, 170–179. DOI 10.1016/j.catena.2014.06.022.

Shaulis, N.J. (1961) Associations between symptoms of potassium deficiency, plant analysis, growth and yield of Concord grapes. *American Institute of Biological Science* 8, 44–57.

Shaulis, N., Kimball, K. and Tomkins, J.P. (1953) The effect of trellis height and training systems on the growth and yield of Concord grapes under a controlled pruning severity. *Proceedings of the American Society for Horticultural Science* 62, 221–227.

Shaulis, N.J., Shepardson, E.S. and Jordan, T.D. (1967) *The Geneva Double Curtain for Vigorous Grapevines: Vine Training and Trellis Construction*. New York State Agricultural Experiment Station, Geneva, New York, 12 pp.

Shaulis, N.J., Pollock, J., Crowe, D. and Shepardson, E.S. (1973) Mechanical pruning of grapevines: progress 1968–1972. *Proceedings of the New York State Horticultural Science* 118, 61–69.

Shaw, A.B. (2001) Pelee Island and Lake Erie north shore, Ontario: a climatic analysis of Canada's warmest wine region. *Journal of Wine Research* 12(1), 19–37. DOI 10.1080/09571260120069732.

Shepardson, E.S., Studer, H.E., Shaulis, N.J. and Moyer, J.C. (1962) Mechanical grape harvesting research at Cornell III. *Journal of the American Society of Agricultural Engineers* 43(2), 66–71.

Shepardson, E.S., Markwardt, E.D., Millier, W.F. and Rehkugler, G.E. (1970) *Mechanical Harvesting of Fruits and Vegetables*. New York's Food and Life Sciences Bulletin No. 5, Cornell University Agricultural Experiment Station, Ithaca, New York, 12 pp.

Shimizu-Marin, V.D., Nishiyama-Hortense, Y.P., Olivati, C., Gonçales, A.C., Garcia-Santos, M.de S.L., Janzantti, N.S., Taboga, S.R. and Lago-Vanzela, E.S. (2024) Natural pre-treatment in raisins production: Effects on quality during storage. *Journal of Stored Products Research* 107, 102342. https://doi.org/10.1016/j.jspr.2024.102342.

Shinde, K.G. and Patil, B.G. (2017) Sorting of raisins using computer vision approach. *International Research Journal of Engineering and Technology* 4, 643.

Shorrocks, V.M. (1997) The occurrence and correction of boron deficiency. *Plant and Soil* 193(1), 121–148. DOI 10.1023/A:1004216126069.

Shrestha, A., Kurtural, S.K., Fidelibus, M.W., Dervishian, G. and Konduru, S. (2013) Efficacy and cost of cultivators, steam, or an organic herbicide for weed control in organic vineyards in the San Joaquin Valley of California. *HortTechnology* 23, 99–108.

Shulman, Y., Nir, G., Fanberstein, L. and Lavee, S. (1983) The effect of cyanamide on the release from dormancy of grapevine buds. *Scientia Horticulturae* 19, 97–104. DOI 10.1016/0304-4238(83)90049-3.

Singleton, V.L. and Draper, D.E. (1964) The transfer of polyphenolic compounds from grape seeds into wines. *American Journal of Enology and Viticulture* 15(1), 34–40.

Singleton, V.L. and Esau, P. (1969) Phenolic substances in grapes and wines, and their significance. *Advances in Food Research, Supplement* 1, 1–261. DOI 10.1086/406539.

Singleton, V.L. and Trousdale, E.K. (1992) Anthocyanin–tannin interactions explaining differences in polymeric phenols between white and red wines. *American Journal of Enology and Viticulture* 43, 63–70.

Singleton, V.L., Sieberhagen, H.A., De Wet, P. and Van Wyk, C.J. (1975) Composition and sensory qualities of wines prepared from white grapes by fermentation with and without grape solids. *American Journal of Enology and Viticulture* 26, 62–69.

Skene, K.G.M. (1968) Increase in the levels of cytokinins in bleeding sap of *Vitis vinifera* L. after CCC treatment. *Science* 159, 1477–1478. DOI 10.1126/science.159.3822.1477.

Skene, K.G.M. and Kerridge, G.H. (1967) Effect of root temperature on cytokinin activity in root exudate of *Vitis vinifera* L. *Plant Physiology* 42, 1131–1139. DOI 10.1104/pp.42.8.1131.

Slaughter, A.R., Hamiduzzaman, M.M., Gindro, K., Neuhaus, J.-M. and Mauch-Mani, B. (2008) Beta-aminobutyric acid-induced resistance in grapevine against downy mildew: involvement of pterostilbene. *European Journal of Plant Pathology* 122, 185–195.

Smart, R.E. (1982) Vine manipulation to improve wine grape quality. *Proceedings of the UC Davis Grape and Wine Centennial Symposium*, Davis, California, pp. 362–375.

Smart, R.E. (1988) Shoot spacing and canopy light microclimate. *American Journal of Enology and Viticulture* 39(4), 325–333.

Smart, R.E. (1996) The rain in Spain falls ... Who cares anymore, now we can irrigate! *Australian & New Zealand Wine Industry Journal* 11(3), 238–241.

Smart, R.E. (1999) Trees in vineyards. *Practical Winery & Vineyard* July/August, 82–84.

Smart, R.E. (2003) Portugese homoclimes in Australia. *Australian & New Zealand Wine Industry Journal* 18(1), 48–50.

Smart, R.E. and Vouillamoz, J. (2023) Clone. In: Harding, J. and Robinson, J. (eds.) *The Oxford Companion to Wine, 5th Edition*. Oxford University Press, Oxford, UK. ISBN 9780198871316.

Smart, R.E. and Robinson, M. (1991) *Sunlight into Wine: Handbook for Wine Grape Canopy Management*. Winetitles, Underdale, South Australia, 96 pp. ISBN 978-1875130108.

Smart, R.E. and Sinclair, T.R. (1976) Solar heating of grape berries and other spherical fruits. *Agricultural Meteorology* 17, 241–259. DOI 10.1016/0002-1571(76)90029-7.

Smart, R.E., Smith, S.M. and Winchester, R.V. (1988) Light quality and quantity effects on fruit ripening for Cabernet Sauvignon. *American Journal of Enology and Viticulture* 39, 250–258.

Smith, B.R. and Cheng, L. (2007) Iron assimilation and carbon metabolism in 'Concord' grapevines grown at different pHs. *Journal of the American Society for Horticultural Science* 132, 473–483.

Smith, B.P., Morales, N.B., Thomas, M.R., Smith, H.M., and Clingeleffer, P.R. (2017) Grapevine rootstocks resistant to the root-knot nematode *Meloidogyne javanica*. *Australian Journal of Grape and Wine Research* 23, 125–131. DOI 10.1111/ajgw.12242.

Smith, C.B., Fleming, H.K., Kardos, L.T. and Haeseler, C.W. (1972) Responses of Concord grapevines to lime and potassium. *Pennsylvania State University Agriculture Experimental Station Bulletin* 785, University Park, Pennsylvania.

Smith, L. (2002) Site selection for establishment and management of vineyards. *14th Annual Colloquium of the Spatial Information Research Centre*, University of Otago, Dunedin, New Zealand.

Sniegowski, P.D., Dombrowski, P.G. and Fingerman, E. (2002) *Saccharomyces cerevisiae* and *Saccharomyces paradoxus* coexist in a natural woodland site in North America and display different levels of reproductive isolation from European conspecifics. *FEMS Yeast Research* 1(4), 299–306. DOI 10.1111/j.1567-1364.2002.tb00048.x.

Snyder, R.L., Tau Paw, K. and Thompson, J.F. (1992) *Passive Frost Protection of Trees and Vines*. Cooperative Extension University of California, Division of Agriculture and Natural Resources, Berkeley, California, 9 pp.

Sofia, J., Sofia, R. and Vila-Maior, J. (2024) Evaluating the effect of *Trichoderma atroviride* (i-1237) on grapevine phomopsis cane and leaf spot: a promising and reproducible trial. *Ciência e Técnica Vitivinícola* 39, 64–73. https://doi.org/10.1051/ctv/ctv2024390264.

Somers, C. (1998) *The Wine Spectrum*. Winetitles, Underdale, South Australia, 136 pp. ISBN 9781875130276.

Somers, T. (1971) The polymeric nature of wine pigments. *Phytochemistry* 10, 2175–2186. DOI 10.1016/S0031-9422(00)97215-7.

Somers, T.C., Evans, M.E. and Cellier, K.M. (1983) Red wine quality and style: diversities of composition and adverse influences from free SO_2. *Vitis* 22, 348–356.

Sommer, K.J., Islam, M.T. and Clingeleffer, P.R. (2000) Light and temperature effects on shoot fruitfulness in *Vitis vinifera* L. cv. Sultana: influence of trellis type and grafting. *Australian Journal of Grape and Wine Research* 6, 99–108. DOI 10.1111/j.1755-0238.2000.tb00167.x.

Song, J., Smart, R., Wang, H., Dambergs, B., Sparrow, A., and Qian, M.C. (2015) Effect of grape bunch sunlight exposure and UV radiation on phenolics and volatile composition of *Vitis vinifera* L. cv. Pinot noir wine. *Food Chemistry* 173, 424–431. DOI 10.1016/j.foodchem.2014.09.150.

Sosnowski, M.R., Creaser, M.L., Wicks, T.J., Lardner, R. and Scott, E.S. (2008) Protection of grapevine pruning wounds from infection by *Eutypa lata*. *Australian Journal of Grape and Wine Research* 14, 134–142. DOI 10.1111/j.1755-0238.2008.00015.x.

Sosnowski, M.R. and Mundy, D.C. (2019) Pruning wound protection strategies for simultaneous control of Eutypa and Botryosphaeria Dieback in New Zealand. *Plant Disease* 103, 519–525. https://doi.org/10.1094/PDIS-05-18-0728-RE.

Souquet, J.M., Cheynier, V., Brossaud, F. and Moutounet, M. (1996) Polymeric proanthocyanidins from grape skins. *Phytochemistry* 43, 509–512. DOI 10.1016/0031-9422(96)00301-9.

Southey, J.M. (1992) Grapevine rootstock performance under diverse conditions in South Africa. *Rootstock Seminar: a Worldwide Perspective*, Reno, Nevada. American Society for Enology and Viticulture, Davis, California.

Southey, J.M. and Jooste, J.H. (1991) The effect of grapevine rootstock on the performance of *Vitis vinifera* L. (cv. Colombard) on a relatively saline soil. *South African Journal of Enology and Viticulture* 12, 32–41.

Spayd, S.E., Wample, R.L., Stevens, R.G., Evans, R.G. and Kawakami, A.K. (1993) Nitrogen fertilization of White Riesling in Washington: effects on petiole nutrient concentration, yield, yield components, and vegetative growth. *American Journal of Enology and Viticulture* 44(4), 378–386.

Spayd, S.E., Tarara, J.M., Mee, D.L. and Ferguson, J.C. (2002) Separation of sunlight and temperature effects on the composition of *Vitis vinifera* cv. Merlot berries. *American Journal of Enology and Viticulture* 53(3), 171–182.

Sperry, J.S., Holbrook, N.M., Zimmermann, M.H. and Tyree, M.T. (1987) Spring filling of xylem vessels in wild grapevine. *Plant Physiology* 83(2), 414–417. DOI 10.1104/pp.83.2.414.

Sridhar, K. and Charles, A.L. (2021) Fortification using grape extract polyphenols – a review on functional food regulations. *International Journal of Food Science and Technology* 56, 3742–3751. https://doi.org/10.1111/ijfs.15001.

Srinivasan, C. and Mullins, M.G. (1980) Effects of temperature and growth regulators on formation of anlagen, tendrils and inflorescences in *Vitis vinifera* L. *Australian Journal of Experimental Agriculture* 45, 439–446. DOI 10.1093/oxfordjournals.aob.a085842.

Srinivasan, C. and Mullins, M.G. (1981) Physiology of flowering in the grapevine – a review. *American Journal of Enology and Viticulture* 32, 47–63.

Stafne, E.T. and Carroll, B. (2006) *Rootstocks for Grape Production*, Oklahoma State University, Stillwater, Oklahoma, 4 pp.

Stapleton, J.J., Barnett, W.W., Marois, J.J. and Gubler, W.D. (1990) Leaf removal for pest management in wine grapes. *California Agriculture* 44, 15–17.

Staub, T. (1991) Fungicide resistance: practical experience with antiresistance strategies and the role of integrated use. *Annual Review of Phytopathology* 29(1), 421–442. DOI 10.1146/annurev.py.29.090191.002225.

Staudt, G. (1982) Pollen germination and pollen tube growth *in vivo* and the dependence on temperature. *Vitis* 21, 205–216.

Staudt, G. (1999) Opening time of flowers and time of anthesis in grapevines, *Vitis vinifera* L. *Vitis* 38(1), 15–20.

Steiner, R. (1924) *Agriculture: Spiritual Foundations for the Renewal of Agriculture* Translated by George Adams. https://archive.org/details/agriculture-cw-327/mode/2up (accessed 9 April 2025).

Stern, V., Smith, R., van Den Bosch, R. and Hagen, K. (1959) The integration of chemical and biological control of the spotted alfalfa aphid: The integrated control concept. *Scilit*, 29(2), 81–101. https://doi.org/doi.org/10.3733/hilg.v29n02p081.

Steudle, E. and Meshcheryakov, A.B. (1996) Hydraulic and osmotic properties of oak roots. *Journal of Experimental Botany* 47(3), 387–401. DOI 10.1093/jxb/47.3.387.

Stoll, M., Loveys, B. and Dry, P. (2000) Hormonal changes induced by partial rootzone drying of irrigated grapevine. *Journal of Experimental Botany* 51(350), 1627–1634. DOI 10.1093/jexbot/51.350.1627.

Striegler, K. and Howell, G.S. (1991) The influence of rootstock on the cold hardiness of Seyval grapevines. I. Primary and secondary effects on growth, canopy development, yield, fruit quality and cold hardiness. *Vitis* 30, 1–10.

Studer, H.E. (2000). Raisin Harvest Mechanization: A Bit of History. *Raisin Production Manual*. University of California, Agricultural and Natural Resources, Oakland, California, pp. 245–251. ISBN 9781879906440.

Studer, H.E. and Olmo, H.P. (1971) The severed cane technique and its application to mechanical harvesting of raisin grapes. *Transactions of the American Society of Agricultural Engineers* 14, 38–43. DOI 10.13031/2013.38220.

Suckling, D.M., Twidle, A.M., Gibb, A.R., Manning, L.M., Mitchell, V.J., Sullivan, T.E.S., Wee, S.L. and El-Sayed, A.M. (2012) Volatiles from apple trees infested with Light Brown Apple Moth larvae attract the parasitoid *Dolichogenidia tasmanica*. *Journal of Agricultural and Food Chemistry* 60, 9562–9566. DOI 10.1021/jf302874g.

Sudarshana, M.R., Perry, K.L. and Fuchs, M.F. (2015) Grapevine Red Blotch-Associated Virus, an emerging threat to the grapevine industry. *Phytopathology* 105, 1026–1032. DOI 10.1094/PHYTO-12-14-0369-FI.

Šuklje, K., Antalick, G., Coetzee, Z., Schmidtke, L.M., Baša Česnik, H., Brandt, J., du Toit, W. j., Lisjak, K. and Deloire, A. (2014) Effect of leaf removal and ultraviolet radiation on the composition and sensory perception of *Vitis vinifera* L. cv. Sauvignon Blanc wine. *Australian Journal of Grape and Wine Research* 20, 223–233. DOI 10.1111/ajgw.12083.

Süle, S. and Burr, T.J. (1998) The effect of resistance of rootstocks to crown gall (*Agrobacterium* spp.) on the susceptibility of scions in grape vine cultivars. *Plant Pathology* 47, 84–88. https://doi.org/10.1046/j.1365-3059.1998.00205.x.

Sullivan, C.L. (2003) *Zinfandel: A History of a Grape and Its Wine*. University of California Press, Berkeley, California. ISBN 9780520930520.

Sun, B.S., Pinto, T., Leandro, M.C., Ricardo-da-Silva, J.M. and Springer, M.I. (1999) Transfer of catechins and proanthocyanidins from solids parts of the grape cluster into wine. *American Journal of Enology and Viticulture* 50, 179–184.

Svyantek, A.W., Coneva, E.D., Kessler, J.R., Spiers, J.D., Iii, E.L.V., Walker, M.A. and Pitts, J.A. (2020) Assessment of Pierce's Disease resistant 87.5% *Vitis vinifera* L. selections in Central Alabama. *American Journal of Enology and Viticulture* 4, 74–81. https://doi.org/10.5344/catalyst.2020.19008.

Swain, T. and Bate-Smith, E.C. (1962) Flavonoid compounds. I. In: Florkin, F.M. and Mason, H.S. (eds) *Comparative Biochemistry*. Academic Press, London, pp. 755–809. ISBN 9780323142151.

Swanepoel, J.J. and Archer, E. (1988) The ontogeny and development of *Vitis vinifera* L. cv. Chenin blanc inflorescence in relation to phenological stages. *Vitis* 27, 133–141.

Swartwout, H.G. (1925) Fruiting habit of the grape. *Proceedings of the American Society for Horticultural Science* 22, 70–74.

Szolnoki, G. (2013) A cross-national comparison of sustainability in the wine industry. *Journal of Cleaner Production*, 53, 243–251. https://doi.org/10.1016/j.jclepro.2013.03.045.

Szulmayer, W. (1971) From sun-drying to solar dehydration II. *Food Technology in Australia* 29(September), 494–501.

Tagliavini, M. and Rombolà, A.D. (2001) Iron deficiency and chlorosis in orchard and vineyard ecosystems. *European Journal of Agronomy* 15, 71–92. DOI 10.1016/S1161-0301(01)00125-3.

Tanton, T.W. and Crowdy, S.H. (1972) Water pathways in higher plants: II. Water pathways in roots. *Journal of Experimental Botany* 23(3), 600–618. DOI 10.1093/jxb/23.3.600.

Taylor, L.P. and Hepler, P.K. (1997) Pollen germination and tube growth. *Annual Review of Plant Physiology and Plant Molecular Biolology* 48, 461–491. DOI 10.1146/annurev.arplant.48.1.461.

Teissedre, P.L., Frankel, E.N., Waterhouse, A.L., Peleg, H. and German, J.B. (1996) Inhibition *of in vitro* human LDL oxidation by phenolic antioxidants from grapes and wine. *Journal of the Science of Food and Agriculture* 70, 55–61. DOI 10.1002/(SICI)1097-0010(199601)70:1<55::AID-JSFA471>3.0.CO;2-X.

Thellier, M., Duval, Y. and Demarty, M. (1979) Borate exchanges of *Lemna minor* L. as studied with the help of the enriched stable isotopes and of a (n,a) nuclear reaction. *Plant Physiology* 63(2), 283–288. DOI 10.1104/pp.63.2.283.

Thomas, B. and Schapel, A. (2003) Soil analysis in established vineyards – the importance of collecting the right samples. *Australian Grapegrower and Winemaker* 478, 41–43.

Thomas, C.S., Marois, J.J. and English, J.T. (1988) The effects of wind speed, temperature, and relative humidity on development of aerial mycelium and conidia of *Botrytis cinerea* on grape. *Phytopathology* 78(3), 260–265. DOI 10.1094/Phyto-78-260.

Thomas, C.S., Gubler, W.D., Silacci, M.W. and Miller, R. (1993) Changes in elemental sulfur residues on Pinot noir and Cabernet Sauvignon grape berries during the growing season. *American Journal of Enology and Viticulture* 44(2), 205–210.

Thomas, C.S., Gubler, W.D. and Leavitt, G. (1994) Field testing of a powdery mildew disease forecast model on grapes in California. *Phytopathology* 84(10), 1070.

Thompson, J., Cantwell, M., Arpaia, M.L., Kader, A., Crisosto, C. and Smilanick, J. (2001) Effect of cooling delays on fruit and vegetable quality. *Perishables Handling Quarterly*, Department of Plant Science, University of California, Davis, California, pp. 1–4.

Thorne, E.T., Stevenson, J.F., Rost, T.L., Labavitch, J.M. and Matthews, M.A. (2006) Pierce's Disease symptoms: comparison with symptoms of water deficit and the impact of water deficits. *American Journal of Enology and Viticulture* 57, 1–11.

Tipton, S., Morris, J., Main, G., Sharp, C. and McNew, R. (1999) Grape juice as an extender and sweetener for blueberry drinks. *Journal of Food Quality* 22(3), 275–285. DOI 10.1111/j.1745-4557.1999.tb00557.x.

Tobin, P.C., Nagarkatti, S. and Saunders, M.C. (2001) Modeling development in Grape Berry Moth (Lepidoptera: Tortricidae). *Environmental Entomology* 30, 692–699. DOI 10.1603/0046-225X-30.4.692.

Tollner, E.W. and Moss, R.B. (eds) (1988) Neutron probe vs. tensiometers vs. gypsum blocks for monitoring soil moisture status. In: *Sensors and Techniques for Irrigation Management*. Center for Irrigation Technology, California State University, Fresno, California, pp. 95–112.

Tomasino, E. (2011) *Characterization of regional examples of New Zealand Pinot noir by means of sensory and chemical analysis*. PhD thesis, Lincoln University, Christchurch, New Zealand, 150 pp. http://hdl.handle.net/10182/4334.

Tominaga, T., Baltenweck-Guyot, R., Gachons, C.P.D. and Dubourdieu, D. (2000) Contribution of volatile thiols to the aromas of white wines made from several *Vitis vinifera* grape varieties. *American Journal of Enology and Viticulture* 51(2), 178–181.

Torres, E., Santibanez, C., Rubio, F., Godoy, F., Cadavid-Labrada, A., Bruno, C., Medina, C. and Arce-Johnson, P. (2014) Gene silencing as a strategy to induce grapevine fanleaf virus (GFLV) resistance in grapevine rootstocks. *Acta Horticulturae* 1046, 187–193. DOI 10.17660/ActaHortic.2014.1046.24.

Torres, E., Recasens, I., Àvila, G., Lordan, J. and Alegre, S. (2017) Early stage fruit analysis to detect a high risk of bitter pit in 'Golden Smoothee.' *Scientia Horticulturae* 219, 98–106. DOI 10.1016/j.scienta.2017.03.003.

Trimmer, C., Keller, A., Murphy, N.R., Snyder, L.L., Willer, J.R., Nagai, M., Katsanis, N., Vosshall, L.B., Matsunami, H. and Mainland, J.D. (2017) Genetic variation across the human olfactory receptor repertoire alters odor perception. *BioRxiv* 212431. DOI 10.1101/212431.

Trjapitzin, S.V. and Trjapitzin, V.A. (1999) Parasites of mealybugs (Homoptera: Pseudococcidae) on cultivated grapes in Argentina, with description of a new species of the genus *Aenasius* Walker (Hymenoptera: Encyrtidae) *Entomological Review* 79(4), 386–390.

Trought, M.C.T. (1997) The New Zealand terroir: sources of variation in fruit composition in New Zealand vineyards. *Proceedings of the Fourth International Symposium for Cool Climate Viticulture and Enology*, Rochester, New York. New York State Agricultural and Experimental Station, Geneva, New York, pp. I, 23–27.

Trought, M.C.T. and Tannock, S.J.C. (1996) Berry size and soluble solids variation within a bunch of grapes. *Proceedings of the Fourth International Symposium for Cool Climate Viticulture and Enology*, New York State Agricultural and Experimental Station, Geneva, New York, pp. V, 70–73.

Trought, M.C.T., Howell, G.S. and Cherry, N. (1999) *Practical Considerations for Reducing Frost Damage in Vineyards*. Report to New Zealand Winegrowers, Lincoln University, Christchurch, New Zealand, 44 pp.

Tubajika, K.M., Civerolo, E.L., Puterka, G.J., Hashim, J.M. and Luvisi, D.A. (2007) The effects of kaolin, harpin, and imidacloprid on development of Pierce's disease in grape. *Crop Protection* 26, 92–99. DOI 10.1016/j.cropro.2006.04.006.

Tukey, R.B. and Clore, W.J. (1972) Grapes – their characteristics and suitability for production in Washington. Cooperative Extension Service, College of Agriculture, Washington State University, Pullman, Washington, 12 pp.

Tulasidas, T.N., Raghavan, G.S.V. and Norris, E.R. (1993) Microwave and convective drying of grapes. *Transactions of the ASAE* 36, 1861–1865. DOI 10.13031/2013.28534.

Turinek, M., Grobelnik-Mlakar, S., Bavec, M. and Bavec, F. (2009) Biodynamic agriculture research progress and priorities. *Renewable Agriculture and Food Systems*, 24(2), 146–154. https://doi.org/10.1017/S174217050900252X.

Urra, C., Sanhueza, D., Pavez, C., Tapia, P., Núñez-Lillo, G., Minio, A., Miossec, M., Blanco-Herrera, F., Gainza, F., Castro, A., Cantu, D. and Meneses, C. (2023) Identification of grapevine clones via high-throughput amplicon sequencing: A proof-of-concept study. *G3 Genes|Genomes|Genetics* 13 (9), jkad145. https://doi.org/10.1093/g3journal/jkad145.

USDA (1989) *Alternative Farming Systems Information Center Oral History Interview Series.* Robert Rodale Question 4. https://youtu.be/OGW8Rc2_zAs (accessed 9 April 2025).

USDA (2024) *National Agricultural Statistics Service Report Press Release 8* May, 2024. https://www.nass.usda.gov/Statistics_by_State/Idaho/Publications/Census_Press_Releases/2024/FRUIT.pdf (accessed 16 Mar 2025).

USDA (2025a) *FoodData Central*. United States Department of Agriculture Agricultural Research Service, Available at: https://fdc.nal.usda.gov/ (Accessed 18 March 2025).

USDA (2025b) *Foreign Agricultural Service Production, Supply and Distribution Report on Grapes, Fresh Table: Production, Supply and Distribution.* https://apps.fas.usda.gov/psdonline/app/index.html#/app/downloads (accessed 16 March 2025).

USDA (2025c) *Foreign Agricultural Service Production, Supply and Distribution Report on Raisin Production, Supply and Distribution.* https://apps.fas.usda.gov/psdonline/app/index.html#/app/downloads (accessed 17 March 2025).

Uva, R.H., Neal, J.C. and DiTomaso, J.M. (1997) *Weeds of the Northeast*. Cornell University Press, Ithaca, New York. ISBN 9780801483349.

Vaia, R. and McDaniel, M. (1996) Sensory effects of quercetin in a model wine and a Chardonnay. *Food Quality Preference* 7(3–4), 339–340. DOI 10.1016/S0950-3293(96)90233-5.

van der Vlugt-Bergmans, C.J.B., Brandwagt, B.F., van't Klooster, J.M., Wagemakers, C.A.M. and van Kan, J.A.L. (1993) Genetic variation and segregation of DNA polymorphisms in *Botrytis cinerea*. *Mycological Research* 97(10), 1193–1200. DOI 10.1016/S0953-7562(09)81284-7.

van Leeuwen, C. and Seguin, G. (2006) The concept of terroir in viticulture. *Journal of Wine Research* 17(1), 1–10. DOI 10.1080/09571260600633135.

van Leeuwen, C., Sgubin, G., Bois, B., Ollat, N., Swingedouw, D., Zito, S. and Gambetta, G.A. (2024) Climate change impacts and adaptations of wine production. *Nature Reviews Earth & Environment* 5(4), 258–275. https://doi.org/10.1038/s43017-024-00521-5.

Varela, L., Smith, R. and Phillips, P. (2001) *Pierce's Disease*. University of California, Agriculture and Natural Resources, Oakland, California, 20 pp.

Vargas-Arispuro, I., Corrales-Maldonado, C. and Martínez-Téllez, M.Á. (2008) Compounds derived from garlic as bud induction agents in organic farming of table grape. *Chilean Journal of Agricultural Research* 68, 94–101. https://doi.org/10.4067/S0718-58392008000100009.

Vecchio, R., Pomarici, E., Giampietri, E. and Borrello, M. (2022) Consumer acceptance of fungus-resistant grape wines: Evidence from Italy, the UK, and the USA. *PLOS ONE* 17, e0267198. https://doi.org/10.1371/journal.pone.0267198.

Velasco, R., Zharkikh, A., Troggio, M., Cartwright, D.A., Cestaro, A., Pruss, D., Pindo, M., FitzGerald, L.M., Vezzulli, S., Reid, J. *et al.* (2007) A high quality draft consensus sequence of the genome of a heterozygous grapevine variety. *PLoS ONE* 2(12), e1326. DOI 10.1371/journal.pone.0001326.

Velicheti, R.K., Lamison, C., Brill, L.M. and Sinclair, J.B. (1993) Immunodetection of Phomopsis species in asymptomatic soybean plants. *Plant Disease* 77, 70–73. DOI 10.1094/PD-77-0070.

VerCauteren, K.C., Lavelle, M.J. and Hygnstrom, S. (2006) Fences and deer-damage management: a review of designs and efficacy. *Wildlife Society Bulletin* 34(1), 191–200. DOI 10.2193/0091-7648(2006)34[191:FADMAR]2.0.CO;2.

Versari, A., Parpinello, G.P., Tornielli, G.B., Ferrarini, R. and Giulivo, C. (2001) Stilbene compounds and stilbene synthase expression during ripening, wilting, and UV treatment in grape cv. Corvina. *Journal of Agricultural and Food Chemistry* 49(11), 5531–5536. DOI 10.1021/jf010672o.

Viala, P. and Vermorel, V. (1909) Traité général de viticulture. In: *Ampelographie*. Masson, Paris.

Vidal, J., Kikkert, J., Malnoy, M., Wallace, P., Barnard, J. and Reisch, B. (2006) Evaluation of transgenic 'Chardonnay' (*Vitis vinifera*) containing magainin genes for resistance to crown gall and powdery mildew. *Transgenic Research* 15(1), 69–82. DOI 10.1007/s11248-005-4423-5.

Vigneault, C., Benoit, D.L. and McLaughlin, N.B. (1990) Energy aspects of weed electrocution. *Reviews of Weed Science* 5, 15–26.

Vrhovsek, U., Mattivi, F. and Waterhouse, A.L. (2001) Analysis of red wine phenolics: comparison of HPLC and spectrophotometric methods. *Vitis* 40, 87–91.

Wade, J., Holzapfel, B., DeGaris, K., Williams, D. and Keller, M. (2004) Nitrogen and water management strategies for wine-grape quality. *Acta Horticulturae* 640, 61–67. DOI 10.17660/ActaHortic.2004.640.6.

Wainwright, S.J. and Woolhouse, H.W. (1977) Some physiological aspects of copper and zinc tolerance in *Agrostis tenuis* Sibth.: cell elongation and membrane damage. *Journal of Experimental Botany* 28(4), 1029–1036. DOI 10.1093/jxb/28.4.1029.

Walker, A.R., Lee, E., Bogs, J., McDavid, D.A.J., Thomas, M.R. and Robinson, S.P. (2007) White grapes arose through the mutation of two similar and adjacent regulatory genes. *Plant Journal* 49(5), 772–785. DOI 10.1111/j.1365-313X.2006.02997.x.

Wallace, H. (2019, March 4). *Tracing the Origin Story of Biodynamic Wine*. SevenFifty Daily. https://daily.sevenfifty.com/tracing-the-origin-story-of-biodynamic-wine/ (accessed 9 April 2025).

Walton, V.M. and Pringle, K.L. (2004) Vine mealybug, *Planococcus ficus* (Signoret) (Hemiptera: Pseudococcidae), a key pest in South African vineyards. A review. *South African Journal of Enology and Viticulture* 25(2), 54–62. DOI 10.21548/25-2-2140.

Wample, R.L., Shoemake, J. and Mills, L. (2000) Micrometeorology characteristics of 'grow-tubes' and their relationship to the growth and development of Cabernet Sauvignon and Chardonnay vines. *Proceedings of the Fifth International Symposium for Cool Climate Viticulture and Oenology*, Melbourne, Australia.

Wang, S., Okamoto, G., Hirano, K., Lu, J. and Zhang, C. (2001) Effects of restricted rooting volume on vine growth and berry development of Kyoho grapevines. *American Journal of Enology and Viticulture* 52(3), 248–253.

Waring, R.H. and Cleary, B.D. (1967) Plant moisture stress: evaluation by pressure bomb. *Science* 155(3767), 1248–1254. DOI 10.1126/science.155.3767.1248.

Weaver, R.J. and McCune, S.B. (1959) Response of certain varieties of *Vitis vinifera* to gibberellin. *Hilgardia* 28, 297–350. DOI 10.3733/hilg.v28n13p297.

Weaver, R.J. and Montgomery, R. (1974) Effect of ethephon on coloration and maturation of wine grapes. *American Journal of Enology and Viticulture* 25(1), 39–41.

Weber, F., Greve, K., Durner, D., Fischer, U. and Winterhalter, P. (2013) Sensory and chemical characterization of phenolic polymers from red wine obtained by gel permeation chromatography. *American Journal of Enology and Viticulture* 64, 15–25. DOI 10.5344/ajev.2012.12074.

Westphal, A., Browne, G.T. and Schneider, S. (2002) Evidence for biological nature of the grape replant problem in California. *Plant and Soil* 242(2), 197–203. DOI 10.1023/A:1016297603427.

Westwood, M.N. (1993) *Temperate Zone Pomology. Physiology and Culture*. Timber Press, Portland, Oregon. ISBN 9781604690705.

Whalley, W.R., Cope, R.E., Nicholl, C.J. and Whitmore, A.P. (2004) In-field calibration of a dielectric soil moisture meter designed for use in an access tube. *Soil Use Management* 20, 203–206. DOI 10.1111/j.1475-2743.2004.tb00358.x.

Whiteman, S.A., Stewart, A., Ridgway, H.J. and Jaspers, M.V. (2007) Infection of rootstock mother-vines by *Phaeomoniella chlamydospora* results in infected young grapevines. *Australasian Plant Pathology* 36, 198–203. DOI 10.1071/AP07012.

Whiting, J.R. (1992) Harvesting and drying of grapes. In: Coombe, B.G. and Dry, P.R. (eds) *Viticulture Volume 2. Practices*. Winetitles, Underdale, South Australia, pp. 328–358. ISBN 9781875130016.

Whiting, J.R. (2004) Grapevine Rootstocks. In: Dry, P.R. and Coombe, B.G. (eds) *Viticulture Volume 1. Resources.* 2nd ed. Winetitles, Underdale, South Australia, pp. 167–188. ISBN 0975685007.

Whiting, J.R. and Buchanan, G.A. (1992) Evaluation of rootstocks for phylloxera infested vineyards in Australia. *Rootstock Seminar: A Worldwide Perspective,* Reno, Nevada, American Society for Enology and Viticulture, Davis, California, pp. 15–26.

Whiting, J.R., Buchanan, G.A. and Edwards, M.E. (1987) Assessment of rootstocks for wine grape production. *Proceedings of the 6th Australian Wine Industry Technical Conference,* Adelaide, South Australia. Australian Industrial Publishers, Adelaide, South Australia, pp. 184–190.

Wicks, T., Hitch, C. and Emmett, B. (2003) Effects of temperature and application rates on the efficacy of sulphur for powdery mildew control. In: Emmett, R.W. (ed.) *Strategic Use of Sulphur in Integrated Pest and Disease Management (IPM) Programs for Grapevines.* Victoria Dept. of Primary Industries, Mildura, Victoria, Australia, pp. 86–109.

Wilcox, W.F., Gubler, W.D. and Uyemoto, J.K. (2015) *Compendium of Grape Diseases, Disorders, and Pests,* Second Edition (St. Paul, Minnesota: American Phytopathological Society). ISBN 9780890544792.

Wildenradt, H.L., Christensen, E.N., Stackler, B., Caputi, A., Jr., Slinkard, K. and Scutt, K. (1975) Volatile constituents of grape leaves. I. *Vitis vinifera* var. 'Chenin Blanc'. *American Journal of Enology and Viticulture* 26(3), 148–153.

Wilkins, M.B. (1966) Geotropism. *Annual Review of Plant Physiology* 17(1), 379–408. DOI 10.1146/annurev.pp.17.060166.002115.

Willer, H., Trávníček, J. and Schlatter, B. (Eds.). (2024). *The World of Organic Agriculture. Statistics and Emerging Trends 2024* (pp. 1–352). Research Institute of Organic Agriculture FiBL, Frick and IFOAM - Organics International. https://orgprints.org/52272 (accessed 9 April 2025).

Williams, C.F. (1923) Hybridization of *Vitis rotundifolia.* Inheritance of anatomical stem characters. *North Carolina Agricultural Station Technical Bulletin* 23.

Williams, C.M.J., Maier, N.A. and Bartlett, L. (2004) Effect of molybdenum foliar sprays on yield, berry size, seed formation, and petiolar nutrient composition of 'Merlot' grapevines. *Journal of Plant Nutrition* 27(11), 1891–1916. DOI 10.1081/PLN-200030023.

Williams, D.W., Andris, H.L., Beede, R.H., Luvisi, D.A., Norton, M.V.K. and Williams, L.E. (1985) Validation of a model for the growth and development of the Thompson Seedless grapevine. II. Phenology. *American Journal of Enology and Viticulture* 36(4), 283–289.

Williams, H., Smith, D., Shahabi, J., Gee, T., Nejati, M., McGuinness, B., Black, K., Tobias, J., Jangali, R., Lim, H., Duke, M., Bachelor, O., McCulloch, J., Green, R., O'Connor, M., Gounder, S., Ndaka, A., Burch, K., Fourie, J., Hsiao, J., Werner, A., Agnew, R., Oliver, R. and MacDonald, B.A. (2023) Modelling wine grapevines for autonomous robotic cane pruning. *Biosystems Engineering* 235, 31–49. https://doi.org/10.1016/j.biosystemseng.2023.09.006.

Williams, L.E. (1996) 36. Grape. In: *Photoassimilate Distribution in Plants and Crops Source–Sink Relationships.* Marcel Dekker, Inc., New York, pp. 851–882. ISBN 9780824794408.

Williams, L.E. and Araujo, F.J. (2002) Correlations among predawn leaf, mid-day leaf and mid-day stem water potential and their correlations with other measures of soil and plant water status in *Vitis vinifera. Journal of the American Society for Horticultural Science* 127(3), 448–454.

Williams, L.E. and Trout, T.J. (2005) Relationships among vine- and soil-based measures of water status in a Thompson Seedless vineyard in response to high-frequency drip irrigation. *American Journal of Enology and Viticulture* 56(4), 357–366.

Williams III, L. and Martinson, T.E. (2000) Colonization of New York vineyards by *Anagrus* spp. (Hymenoptera: Mymaridae): overwintering biology, within-vineyard distribution of wasps, and parasitism of grape leafhopper, *Erythroneura* spp. (Homoptera: Cicadellidae), eggs. *BioControl* 18(2), 136–146. DOI 10.1006/ bcon.2000.0817.

Williams, P.J., Strauss, C.R. and Wilson, B. (1980) New linalool derivatives in Muscat of Alexandria grapes and wines. *Phytochemistry* 19, 1137–1139. DOI 10.1016/0031-9422(80)83070-6.

Williamson, P.M., Sivasithamparam, K. and Cowling, W.A. (1991) Formation of subcuticular hyphae by *Phomopsis leptostromiformis* upon latent infection of narrow-leafed lupins. *Plant Disease* 75, 1023–1025. DOI 10.1094/PD-75-1023.

Willocquet, L., Colombet, D., Rougier, M., Fargues, J. and Clerjeau, M. (1996) Effects of radiation, especially ultraviolet B, on conidial germination and mycelial growth of grape powdery mildew. *European Journal of Plant Pathology* 102(5), 441–449. DOI 10.1007/BF01877138.

Willstätter, R. and Zollinger, E.H. (1915) Untersuchungen über die anthocyane: VI. Über die farbstoffe der weintraube und der heidelberre. *Liebigs Annalen* 408, 83–109.

Wilmink, J., Breuer, M. and Forneck, A. (2022) Effects of grape phylloxera leaf infestation on grapevine growth and yield parameters in commercial vineyards: a pilot study. *OENO One* 56, 197–208. https://doi.org/10.20870/oeno-one.2022.56.1.4803.

Wilson, B., Strauss, C.R. and Williams, P.J. (1986) The distribution of free and glycosidically bound monoterpenes among skin, juice, and pulp fractions of some white grape varieties. *American Journal of Enology and Viticulture* 37(2), 107–111.

Winkler, A.J. (1926) Some responses of *Vitis vinifera* to pruning. *Hilgardia* 1, 525–543. DOI 10.3733/hilg.v01n20p525.

Winkler, A.J. (1932) Maturity tests for table grapes. *California Agricultural Experiment Station Bulletin* 529, 1–35.

Winkler, A.J. and Williams, W.O. (1935) Effect of seed development on the growth of grapes. *Proceedings of the American Society for Horticultural Science* 33, 430–434.

Winkler, A.J., Lamouria, L.H. and Abernathy, G.H. (1957) Mechanical grape harvest – problems and progress. *American Journal of Enology and Viticulture* 8(4), 182–187.

Winkler, A.J., Cook, J.A., Kliewer, W.M. and Lider, L.A. (1974) *General Viticulture*. University of California Press, Berkeley, California. ISBN 9780520025912.

Winter, S., Bauer, T., Strauss, P., Kratschmer, S., Paredes, D., Popescu, D., Landa, B., Guzmán, G., Gómez, J. A., Guernion, M., Zaller, J. G. and Batáry, P. (2018) Effects of vegetation management intensity on biodiversity and ecosystem services in vineyards: A meta-analysis. *Journal of Applied Ecology*, 55(5), 2484–2495. https://doi.org/10.1111/1365-2664.13124.

Witzgall, P., Bengtsson, M. and Trimble, R.M. (2000) Sex pheromone of grape berry moth (Lepidoptera: Tortricidae). *Environmental Entomology* 29(3), 433–436. DOI 10.1603/0046-225X-29.3.433.

Wolf, T.K. and Cook, M.K. (1992) Seasonal deacclimation patterns of three grape cultivars at constant, warm temperature. *American Journal of Enology and Viticulture* 43(2), 171–179.

Wolf, T.K. and Pool, R.M. (1987) Factors affecting exotherm detection in differential thermal analysis of grapevine dormant buds. *Journal of the American Society for Horticultural Science* 112, 520–525.

Wolf, T.K. and Warren, M.K. (1995) Shoot growth-rate and density affect bud necrosis of Riesling grapevines. *Journal of the American Society for Horticultural Science* 120(6), 989–996.

Wolf, T.K., Haeseler, C.W. and Bergman, E.L. (1983) Growth and foliar elemental composition of Seyval Blanc grapevines as affected by four nutrient solution concentrations of nitrogen, potassium and magnesium. *American Journal of Enology and Viticulture* 34(4), 271–277.

Wolf, T.K., Baudoin, A.B.A.M. and Martinez-Ochoa, N. (1997) Effect of floral debris removal from fruit clusters on botrytis bunch rot of Chardonnay grapes. *Vitis* 36, 27–33.

Wolpert, J.A. (1992) Rootstock use in California: history and future prospects. *Grape Rootstock Meeting Proceedings*. Oregon State University, Corvallis, Oregon, pp. 65–72.

Wood, C., Siebert, T.E., Parker, M., Capone, D.L., Elsey, G.M., Pollnitz, A.P., Eggers, M., Meier, M., Vössing, T., Widder, S. *et al.* (2008) From wine to pepper: rotundone, an obscure sesquiterpene, is a potent spicy aroma compound. *Journal of Agricultural and Food Chemistry* 56, 3738–3744. DOI 10.1021/jf800183k.

Woodham, R.C. and Alexander, D.M. (1966) The effect of root temperature on development of small fruiting Sultana vines. *Vitis* 5, 345–350.

Wooldridge, J., Louw, P.J.E. and Conradie, W.J. (2010) Effects of rootstock on grapevine performance, petiole and must composition, and overall wine score of *Vitis vinifera* cv. Chardonnay and Pinot noir. *South African Journal of Enology and Viticulture* 31, 45–48.

Xu, G., Magen, H., Tarchitzky, J. and Kafkafi, U. (1999) Advances in chloride nutrition of plants. In: Sparks, D.L. (ed.) *Advances in Agronomy*. Academic Press, New York, pp. 97–150. ISBN 9780080524337.

Yamane, T., Jeong, S.T., Goto-Yamamoto, N., Koshita, Y. and Kobayashi, S. (2006) Effects of temperature on anthocyanin biosynthesis in grape berry skins. *American Journal of Enology and Viticulture* 57(1), 54–59.

Yepes, L.M., Cieniewicz, E., Krenz, B., McLane, H., Thompson, J.R., Perry, K.L. and Fuchs, M. (2018) Causative role of grapevine Red Blotch Virus in Red Blotch Disease. *Phytopathology* 108, 902–909. https://doi.org/10.1094/PHYTO-12-17-0419-R.

Yilmaz, Y. and Toledo, R.T. (2004) Major flavonoids in grape seeds and skins: antioxidant capacity of catechin, epicatechin, and gallic acid. *Journal of Agricultural and Food Chemistry* 52(2), 255–260. DOI 10.1021/jf030117h.

Yin, L., Burkness, E.C., Hutchison, W.D. and Clark, M.D. (2021) Effects of foliar phylloxera (Hemiptera: Phylloxeridae) infestations on wine grape photosynthesis, yield, and fruit quality. *Journal of Entomological Science* 56, 504–518. https://doi.org/10.18474/JES20-77.

Yiridoe, E. K., Bonti-Ankomah, S. and Martin, R. C. (2005) Comparison of consumer perceptions and preference toward organic versus conventionally produced foods: A review and update of the literature. *Renewable Agriculture and Food Systems*, 20(4), 193–205. https://doi.org/10.1079/RAF2005113.

Yiridoe, E. K. and Marett, G. E. (2004) Mitigating the high cost of ISO 14001 EMS standard certification: Lessons from agribusiness case research. *International Food and Agribusiness Management Review*, 7(2), 37–62. https://doi.org/10.22004/ag.econ.8122.

Yokessa, M. and Marette, S. (2019) A review of eco-labels and their economic impact. *International Review of Environmental and Resource Economics* 13(1–2), 119–163. https://doi.org/10.1561/101.00000107.

Yokoyama, V.Y. (1979) Effect of thrips scars on table grape quality. *Journal of the American Society for Horticultural Science* 104, 243–245.

Yuan, C., Zhang, T., Tian, J., Deng, T., Xin, H., Liu, Y., Zhang, Y. and Xue, W. (2024) A potential antifungal agent: Insight into the antifungal mechanism against *Phomopsis sp. Arabian Journal of Chemistry* 17, 105480. https://doi.org/10.1016/j.arabjc.2023.105480.

Yun, H.K., Louime, C. and Lu, J. (2007) First report of anthracnose caused by *Elsinoe ampelina* on muscadine grapes (*Vitis rotundifolia*) in northern Florida. *Plant Disease* 91(7), 905. DOI 10.1094/PDIS-91-7-0905B.

Yunusa, I.A.M., Walker, R.R. and Guy, J.R. (1997) Partitioning of seasonal evapotranspiration from a commercial furrow-irrigated Sultana vineyard. *Irrigation Science* 18(1), 45–54. DOI 10.1007/s002710050043.

Zabadal, T.J. and Dittmer, T.W. (2003) The maintenance of fruiting potential through the winter for 'Merlot' grapevines grown in southwestern Michigan. *Small Fruits Review* 2(4), 37–44. DOI 10.1300/J301v02n04_05.

Zabadal, T.J., Vanee, G.R., Dittmer, T.W. and Ledebuhr, R.L. (2002) Evaluation of strategies for pruning and crop control of Concord grapevines in southwest Michigan. *American Journal of Enology and Viticulture* 53(3), 204–209.

Zamolo, F., Heinrichs, J. P. and Wüst, M. (2023) Sites of biosynthesis, distribution and phloem transport of 3-isobutyl-2-Methoxypyrazine in *Capsicum annuum* (bell pepper) plants. *Phytochemistry* 205, 113488. https://doi.org/10.1016/j.phytochem.2022.113488.

Zazueta, F.S. and Xin, J. (1994) *Soil Moisture Sensors*. Florida Cooperative Extension Service, University of Florida, Gainesville, Florida, 11 pp.

Zhang, J., Zhang, Y., Yu, H. and Wang, Y. (2014) Evaluation of powdery mildew-resistance of grape germplasm and rapid amplified polymorphic DNA markers associated with the resistant trait in Chinese wild *Vitis*. *Genetics and Molecular Research* 13, 3599–3614. DOI 10.4238/2014.May.9.3.

Zhang, X., Zhou, Y., Li, Y., Fu, X. and Wang, Q. (2017) Screening and characterization of endophytic *Bacillus* for biocontrol of grapevine downy mildew. *Crop Protection* 96, 173–179. https://doi.org/10.1016/j.cropro.2017.02.018.

Zhang, P., Barlow, S., Krstic, M., Herderich, M., Fuentes, S. and Howell, K. (2015) Within-vineyard, within-vine, and within-bunch variability of the rotundone concentration in berries of *Vitis vinifera* L. cv. Shiraz. *Journal of Agricultural and Food Chemistry* 63, 4276–4283. DOI 10.1021/acs.jafc.5b00590.

Zoecklein, B.W. (2002) *A Review of Methode Champenoise Production*. Virginia Cooperative Extension, Arrington, Virginia, 30 pp.

Zouari, N., Romette, J.-L. and Thomas, D. (1988) A continuous-flow method for the rapid determination of sanitary quality of grape must at industrial scales. *Journal of Chemical Technology and Biotechnology* 41(3), 243–248. DOI 10.1002/jctb.280410309.

Index

Note: Page numbers in **bold** type refer to figures
Page numbers in *italic* type refer to tables

Printed and bound by CPI Group (UK) Ltd, Croydon, CR0 4YY

13/09/2026

14961108-0001